2021 International Conference on Electronic Communication, Computer Science and Technology

Journal of Physics: Conference Series Volume 2209

Online
7 – 9 January 2022

ISBN: 978-1-7138-5172-1
ISSN: 1742-6588

Printed from e-media with permission by:

Curran Associates, Inc.
57 Morehouse Lane
Red Hook, NY 12571

Some format issues inherent in the e-media version may also appear in this print version.

Institute of Physics
Dirac House, Temple Back
Bristol BS1 6BE UK

Phone: 44 1 17 929 7481
Fax: 44 1 17 920 0979

techtracking@iop.org

Additional copies of this publication are available from:

Curran Associates, Inc.
57 Morehouse Lane
Red Hook, NY 12571 USA
Phone: 845-758-0400
Fax: 845-758-2633
Email: curran@proceedings.com
Web: www.proceedings.com

TABLE OF CONTENTS

Journal of Physics: Conference Series **2209** (2022) 011001 doi:10.1088/1742-6596/2209/1/011001

Preface

It is our great pleasure to welcome you to the 2021 International Conference on Electronic Communication, Computer Science and Technology (ECCST 2021) on January 07-09, 2022, which is planned to held in Nanchang, China. However, due to the spread of COVID-19, many countries have strict restrictions to travel aboard. Meanwhile, the safety of the participants is always the priority. We made a hard decision to hold the conference online.

2021 International Conference on Electronic Communication, Computer Science and Technology (ECCST 2021) was co-hosted by Jiangxi Science & Technology Normal University. It mainly focused on Electronics and communications, networking, artificial intelligence and computers and other research fields to discuss. This conference provided a valuable opportunity for researchers and industry experts to exchange opinions face-to-face.

ECCST 2021 was a leading conference for about 90 researchers from different countries and territories to present their research results about Electronic Communication, Computer Science and Technology. During the conference, the conference model was divided into three sessions, including oral presentations, keynote speeches, and online Q&A discussion. In the first part, some scholars, whose submissions were selected as the excellent papers, were given about 5-10 minutes to perform their oral presentations one by one. Then in the second part, keynote speakers were each allocated 30-45 minutes to hold their speeches.

In the second part, we invited six professors as our keynote speakers. Prof. Qiegen Liu, our first keynote speakers, from Nanchang university, China. He is engaged in the application of sparse and deep learning representations in medical imaging and image processing. And then we had Prof. Zhen Yang, Jiangxi Science & Technology Normal University, China. His main research interests include target detection and tracking, saliency calculation, fine-grained object recognition and image semantic segmentation. Prof. Xinguo Yu, Central China Normal University, China. For a long time, he has been engaged in the research of video analysis, computer vision, human-computer interaction, multimedia technology, etc. A.Prof. Muhammad Shafiq, Guangzhou University, China. His current research areas of interest include IoT Security, IoT anomaly and intrusion traffic classification, IoT management, Network Traffic Classification, Cloud Computing and Network Security. A. Prof. Pavel Loskot, Zhejiang University-University of lllinois at Urbana-Champaign Institute, China. His current research interests are in the application of statistical signal processing methods to problems in telecommunications engineering and computational molecular biology. Lastly, we had Prof Associate. Ata Jahangir Moshayedi, Information Engineering, Jiangxi University of Science and Technology, China. His research direction is machine sense OF smell. Their insightful speeches had triggered heated discussion in the third session of the conference. Every participant praised this conference for disseminating useful and insightful knowledge.

The proceedings are a compilation of the accepted papers and represent an interesting outcome of the conference. Topics include but are not limited to the following areas: Circuits and Systems, Physical Electronics and Optoelectronics, Information and Communication System, Information Technology, Computer Science and more related topics. All the papers have been through rigorous review and process to meet the requirements of International publication standard.

We would like to acknowledge all of those who supported ECCST 2021. The help and contribution of each individual and institution was instrumental in the success of the conference. In particular, we would like to thank the organizing committee for its valuable inputs in shaping the conference program and reviewing the submitted papers.

The Committee of ECCST 2021

Committee

Conference Chairman
Professor Heping Zuo, Jiangxi Science & Technology Normal University, China

Academic committees Chairman
Professor Yuming Fang, Jiangxi University of Finance and Economics, China

Organizing committees Chairman
Professor Zhijian Yin, Jiangxi Science & Technology Normal University, China

Organizing committees
Prof. Zhijian Yin, Jiangxi Science & Technology Normal University, China
Prof. Fan Yang, Jiangxi Science & Technology Normal University, China
Prof. Zhen Yang, Jiangxi Science & Technology Normal University, China
Lecturer. Hualin Zhan, Jiangxi Science & Technology Normal University, China
Lecturer. Qiang Luo, Jiangxi Science & Technology Normal University, China
Lecturer. Xin Zhou, Jiangxi Science & Technology Normal University, China

Academic committees
Prof. Qiegen Liu， Nanchang university, China
Prof. Zhijian Yin, Jiangxi Science & Technology Normal University, China
Prof. Yuming Fang， Jiangxi University of Finance and Economics, China
Prof. Fan Yang, Jiangxi Science & Technology Normal University, China
Prof. Zhen Yang, Jiangxi Science & Technology Normal University, China
Prof. Jiangxiong Fang, East China University of Technology, China

International Technical Committees
A. Prof. Ziyan Zhang, Hainan Tropical Ocean University, China
A. Prof. Sahil Verma, Lovely Professional University, Phagwara
Dr. Ariffin Nor Hapiza, Information System Department, Universiti Teknologimara, Selangor, Malaysia
Dr. Wan Nor Shuhadan Wan Nik, IEEE Member, Universiti Sulatan Zainal Abidin, Terengganu, Mala ysia

Peer Review Statement

All papers published in this volume have been reviewed through processes administered by the Editors. Reviews were conducted by expert referees to the professional and scientific standards expected of a proceedings journal published by IOP Publishing.

- **Type of peer review:** Single Anonymous
- **Conference submission management system:** Morressier
- **Number of submissions received:** 54
- **Number of submissions sent for review:** 39
- **Number of submissions accepted:** 33
- **Acceptance Rate (Submissions Accepted / Submissions Received × 100):** 61.1
- **Average number of reviews per paper:** 2
- **Total number of reviewers involved:** 8
- **Contact person for queries:**
  Name: Xuexia Ye
  Email: xx.ye@keoaeic.org
  Affiliation: AEIC Academic Exchange Information Centre

Robust Visual Place Recognition Method for Robot Facing Drastic Illumination Changes

Zhen Li[1]**, Lin Zhou**[2] **and Zeqin Lin**[3*]

[1]School of Cosmetics and Artistic Designing, Guangdong Food and Drug Vocational College, GuangDong, 510520, China.

[2]Department of Mechatronics Engineering, Foshan University, GuanDong,528000, China.

[3]Schoolof Mechatronic Engineering and Automation, Foshan University, GuangDong, 528000, China

[*]Corresponding author' e-mail: martinkeel@foxmail.com

Abstract. The robustness of visual place recognition determines the accuracy of the SLAM to construct the environmental map. However, when the robot moves in the outdoor environment for a long time, it must face the challenge of drastic illumination changes (time shift, season or rain and fog weather factors), which leads to the robot's ability to identify places is greatly restricted. This paper proposes a method for visual place recognition that is more robust to severe illumination changes. First, a generative adversarial network is introduced into visual SLAM to enhance the quality of candidate keyframes, and consistency of geographic locations corresponding to images before and after quality enhancement is evaluated, and then the image descriptor with illumination invariance is established for the robot's new observation image and keyframes of the map. Finally, the performance of the method in this paper is tested on the public dataset. The experimental results show that this method is conducive to improving the quality of environmental map nodes, and enables the robot to show a highly robust ability for visual place recognition in the face of severe illumination changes, which provides a powerful tool for robot to build an accurate environmental map.

1. Introduction

Visual simultaneous location and mapping (VSALM) is the key technology for mobile robots to implement autonomous navigation. It mainly consists of four parts: estimation of front-end, optimization of back-end, loop detection and construction of a map [1]. Visual SLAM uses the inter-frame estimation of the visual odometry and the existing map to achieve the robot's own location. Based on the results of location, it achieves an incremental map construction. However, the motion estimation errors of front-end based on adjacent frames will continue to accumulate and gradually increase the trajectory drift, which will lead to the construction of a wrong environment map. In addition, the robot may repeatedly visit the place that it has been to. Therefore, loop detection is required to perform online place recognition to determine whether a closed-loop situation occurs currently. The final estimation of front-end and results of loop detection will be used as input information for optimization of back-end. In the end, the output result is an environment map consistent with the real scenario. In the entire visual SLAM process, it can be seen that visual place recognition has a crucial impact on building an accurate environment map.

When a mobile robot moves in an outdoor environment, it will inevitably undergo the influence of changes in ambient illumination caused by time shifts, seasonal and weather conditions, which has an impact on the robot's visual place recognition. As shown in Fig. 1, when the illumination at the same place changes dramatically in different times, the scene shows great differences, which easily causes false positive and false negative problems. The purpose of this paper is to propose a highly robust method of visual place recognition to enhance the adaptability of visual SLAM to complex illumination changes.

Fig. 1. Illumination changes at the same place

Therefore, this paper proposes a method for visual place recognition with strong robustness in the face of severe illumination changes and presents three contributions as follows:

- A generative adversarial network is introduced into visual SLAM for the first time to enhance the image quality of candidate keyframe of severe illumination changes, which lays the foundation for generating a high-quality environment map.
- Proposing a consistency evaluation method of the corresponding geographic location before and after the enhancement of the candidate keyframe's quality, which provides a guarantee for the authenticity expression of the map nodes.
- An invariant image descriptor is constructed to ensure high robustness of image matching and visual place recognition.

The rest of this paper is organized as follows. Section II analyzes the current research status at home and abroad; Section III introduces the methods of this paper; Section IV explains the experimental scheme, results and conclusions of this paper; Section V summarizes the paper and discuss the future work.

2. Related Work

The existing representative visual SLAM systems (FAB-MAP [2], ORB-SLAM [3], SeqSLAM [4]) mainly generate image descriptors based on traditional hand-crafted features, and then calculate the similarity of image descriptors to accomplish the task of place recognition. Among them, the FAB-MAP system uses the SIFT feature to construct a bag-of-words model for online place recognition. It has achieved good results in a relatively stable outdoor environment. However, traditional hand-crafted features represented by SIFT is too weak to illumination changes, which lead to the FAB-MAP system can't adapt to the changing environment of illumination. In addition, the ORB-SLAM system used the bag-of-words model constructed by the FAST feature extractor and BRIEF feature descriptor also has the same shortcomings as FAB-MAP when place recognition in changing illumination conditions. Compared to the first two systems that use single image descriptor for place recognition, the SeqSLAM system uses a longer image sequence for visual place recognition and achieves better robustness, but the ability to adapt to severe illumination changes still exists major limitations.

In recent years, with the development of deep learning method and theory of computer vision, researchers have tried to use deep convolutional neural networks for autonomous learning of image features [5], and applied these features to visual SLAM localization and place recognition or loop detection. The robustness of SLAM is further improved compared to the traditional hand-crafted features. Chen, *et al.* [6] proposed a CNN-based visual place recognition algorithm, which is better than the method based on features of hand-crafted. Gao, *et al.* [7] proposed an unsupervised learning method to extract image features and used the similarity matrix between image sequences to detect closed loop, it has achieved good results in public datasets; Hou, *et al.* [8] used the PlaceCNN model to extract image features and had certain robustness under varying illumination conditions; Bai, *et al.*[9] proposed a method of SeqCNNSLAM, which combines the deep learning model AlexNet and SeqSLAM methods to simultaneously solve the robustness of view and illumination in the scene, but this method requires online adjustment of relevant parameters when place changes and the process is tedious.

Existing visual SLAM systems have done a lot of research that directly affect the accuracy of visual place recognition, such as feature extraction and image descriptor's generation. However, the robustness of the method to improve the place recognition from the perspective of enhancing the quality of candidate keyframe is seldom seen. At present, generative adversarial network(GAN) [10] is used to generate image. Isola, *et al.*[11] proposed a solution to the generation of original images to target images, which named pix2pix, requiring paired training samples in the training set; Zhu, *et al.*[12] proposed a CycleGAN model, which does not need to establish one-to-one mapping between training data, and had achieved good results. Anoosheh, *et al.* [13] proposed an unsupervised learning method-TodayGAN, which can restore night images to day images and apply them to locate a place. Although this method verifies the model's ability to recover the scene, it does not evaluate the consistency of the geographic location of the image before and after the scene is restored, and uses the VLAD [14] algorithm to encode the SIFT features. The generated image descriptor has obvious limitations in the robustness of image matching under changing environment of illumination.

3. Proposed Approach

This paper proposes a visual place recognition method with high robustness against severe illumination changes. As shown in Fig. 2, the method's design and implementation are specifically divided into four sub-modules: first, the generative adversarial network to enhance the candidate keyframe; Second, global image descriptors for assessment of geographical locations' consistency corresponding to the image before and after quality enhancement; third, construction of illumination invariant image descriptors; fourth, image matching and place recognition. A detailed description of each sub-module will be completed in Section III.A and Section III.B.

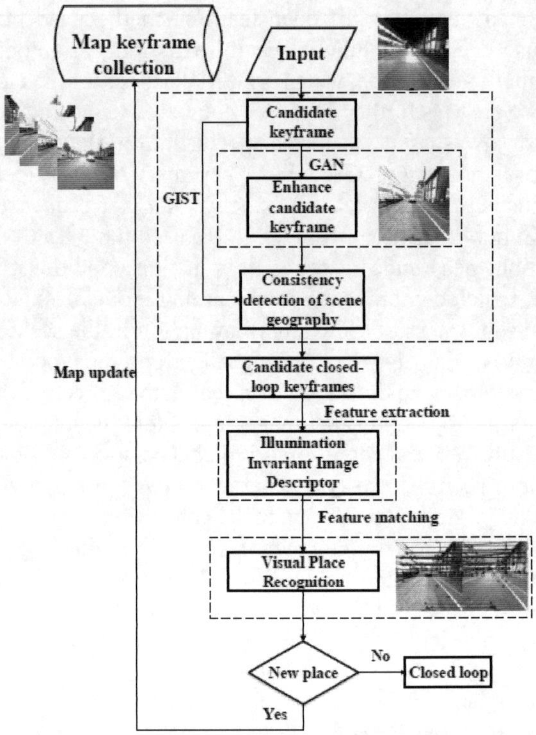

Fig. 2. The flow chart of visual place recognition method

3.1 Enhancement of Illumination Quality of Candidate Keyframes and Evaluation of Geographical Authenticity

The quality of the environment map constructed by the robot depends on the quality of the candidate keyframes. Therefore, the newly observed image sequence of the robot will be further enhanced after being filtered by the candidate keyframe method. Then the quality of subsequent maps will be guaranteed and the image matching process that plays a decisive role in place recognition is also performed between high-quality keyframes, thereby reducing the degree of interference of illumination changes on place recognition. In addition, it is a critical step whether the geographical location described by the candidate keyframe image with enhanced illumination quality is the same as the original real geographic location. The enhancement about candidate keyframe and the assessment of geographical location authenticity are described as follows:

3.1.1 The method uses the TodayGAN to enhance the illumination quality of candidate keyframes.

In the process of visual place recognition, it is inevitable to encounter scenes with strong illumination changes. This paper uses the TodayGAN model to enhance the illumination quality of candidate keyframes. As shown in Fig. 3, the generator network G performs quality enhancement on the current candidate keyframe image x, and generates an enhanced image G(x). In order to ensure that the enhanced image and the real candidate keyframe image are consistent in content, the generator network F performs an inverse process on the generated enhanced image G(x) to obtain an image F(G(x)), and ensures the image F(G(x)) is basically the same as the image x. At the same time, the discriminator network D discriminates the enhanced image G(x) and the map keyframe image y. The discriminator network are mainly composed of three parts: a discrimination network D_t about image texture, a discrimination network D_c about image color, and a discrimination network D_g about image gradient. Three parts work together to ensure that the illumination quality of candidate keyframe images are effectively enhanced.

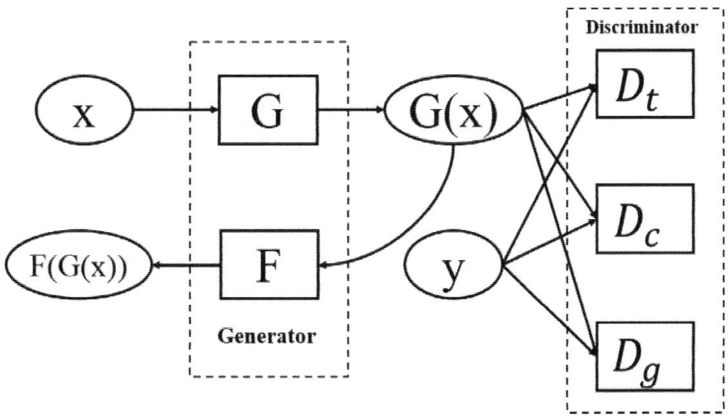

Fig. 3 Illumination quality enhancement of candidate keyframes

3.1.2 Evaluation of the authenticity of the geographic location of the candidate keyframe image expression after the enhancement of the illumination quality

If the candidate keyframe images after enhancing illumination quality can still express the true geographical location, then the images before and after the quality enhancement corresponding to this location should maintain a very high similarity. The multiple similarity curves calculated by candidate keyframe image sequences before and after enhancement should show highly consistent waveforms and peak positions. Based on this fact, the method uses the global image descriptor GIST [15] to calculate the similarity matrix of the candidate keyframe image sequence before and after the enhancement of the illumination quality. Next, the method analyzes each row of the similarity matrix to evaluate whether the candidate keyframe image can maintain the authenticity of the original geographical location after being enhanced by illumination quality. This is verified in Section IV.B of the experiment.

3.2 Illumination Invariant Image Descriptor and Place Recognition Method

The method in this paper uses the local feature package with illumination invariance to generate robust image descriptors, and then performs image matching by calculating the similarity of image descriptors to achieve the purpose of place recognition. This method extracts the DELF[16] feature with high invariance to illumination. The learning process of this feature is mainly connected to the attention mechanism module at the output of the conv4-x layer of ResNet50, so that the keypoints of the image are provided with robustness to changes in the external environment. The image descriptor is shown in Equation 1:

$$D(i) = \{f(1), f(2), \ldots, f(i)\} \tag{1}$$

Among them: *D(i)* represents the image descriptor, and *f(i)*represents the vector of the ith feature point in the image.

Image descriptor the similarity between image descriptors is equivalent to image matching and place recognition, as shown in Equation 2:

$$S(A, B) = \frac{M(A,B)}{N(A)+N(B)-M(A,B)} \tag{2}$$

Among them: *M (A, B)* represents the number of matching, *N(A)* represents the number of feature points of image *A*, *N (B)* represents the number of feature points of image *B*, the value range of similarity $S(A, B)$ is [0,1].

4. Experimental Evaluation

4.1 Datasets

In order to verify the robustness of the illumination invariance of the method in this paper, the Oxford RobotCar Dataset [17] was used as the test image for visual place recognition. The dataset contains image sequences collected by a camera for a long time along a predetermined route in the urban area of Oxford, England. The dataset is more complex, including the dramatic changes in illumination caused by changes throughout the four seasons, rain and snow, etc. Scenarios are more in line with the requirement of this experiment to verify the robustness of the illumination changes. In this paper, 100 day and night images are selected from the entire test dataset to form two subsets. The image sequences in both subsets describe the same observation path, and the frame-by-frame images in the two sequences correspond to each other.

4.2 Evaluation of the Consistency of the Geographic Locations Described before and after the Enhancement of the Illumination Quality of Candidate Keyframes

Whether the corresponding geographic locations of the candidate keyframes are consistent before and after the enhancement of the illumination quality can be achieved by using human eye observation method and image descriptor-based evaluation method. As shown in Fig. 4, the experiment performed a batch of enhancements on the image quality of the night image data to obtain a corresponding high-brightness image sequence. The research team used a random detection method to compare the images content before and after the quality enhancement and the consistency of geographical location with human eyes. There is no distortion in the description of the geographical location of the candidate keyframes. The comparison results preliminarily verify the validity of the method, and then the experiment continues to use the image descriptor-based evaluation method to further verify the validity of the method.

In fact, if the image descriptors before and after the enhancement of the illumination quality of the candidate keyframes can still maintain a high degree of similarity, it will indicate that the geographic locations described by successive images are consistent. Based on this logic, this experiment uses GIST as the global descriptor of the image, and calculates the similarity matrix of the two image sequences before and after the quality enhancement. The horizontal axis represents the night image sequence, and the vertical axis represents the night image sequence with enhanced quality. As shown in Fig. 5, the similarity matrix has a very continuous and significantly higher brightness than the diagonals of the outer area, which indicates that the geographic location described by the night image after illumination enhancement still maintains a high consistency with the previous one.

Fig. 4. Before and after enhancement of illumination quality of candidate keyframes images

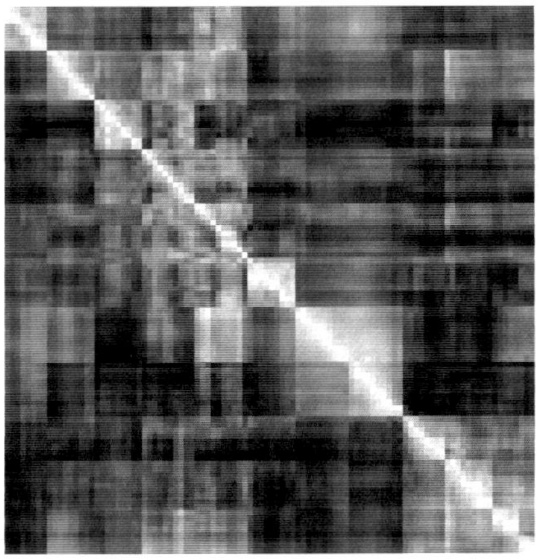

Fig. 5. Image similarity matrix before and after candidate keyframe's enhancement

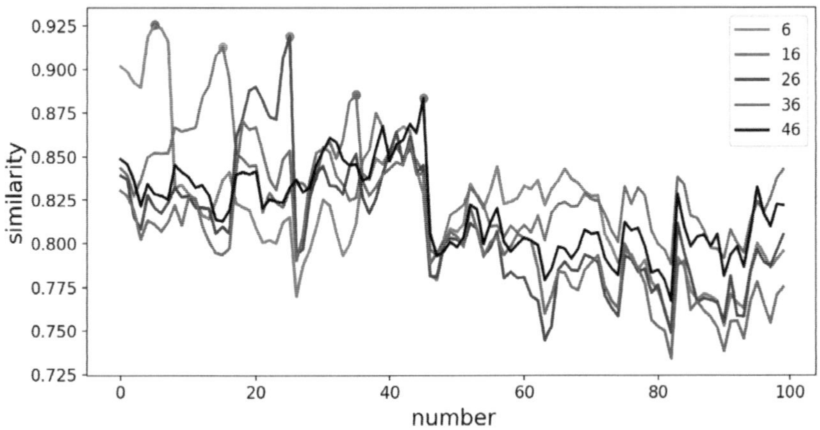

(a) The images matching similarity curves of 6,16,26,36,46

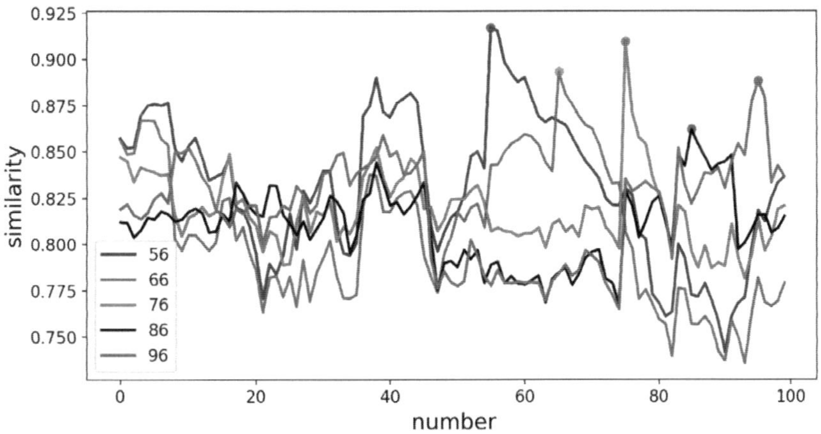

(b) The images matching similarity curves of 56,66,76,86,96

Fig. 6. Similarity matrix quantitative sampling analysis: geographical authenticity evaluation of candidate keyframes with enhanced illumination quality

Journal of Physics: Conference Series **2209** (2022) 012001 doi:10.1088/1742-6596/2209/1/012001

4.3 Verification of the Method's Robustness

In the previous Section, it was verified that the candidate keyframes would not have a distorted description of the geographical position after the enhancement of the illumination quality in this paper, which provides a prerequisite guarantee for the further visual place recognition. This Section continues to verify the robustness of the proposed method in visual place recognition, and performs necessary experiments. First, verifying the robustness improvement with enhanced candidate keyframe quality on visual place recognition. Second, verifying that the proposed method performs better than other representative methods in robust place recognition.

In the experiment, the illumination-invariant image descriptor proposed in Section III.B was used to quantify and express the night image sequence A, the enhanced night image sequence A', and the day image sequence B. Then, the original feature package based on the image matching method calculates A and B, A 'and B, and obtains an image similarity matrix, that is, judgment matrices M1 (as shown in Fig. 7) and M2 (as shown in Fig. 8) for visual place recognition. The horizontal axis is the day image sequence, and the vertical axis is the enhanced night image sequence.

It can be concluded from Fig. 7 and Fig. 8 that after the illumination quality enhancement of the candidate keyframes, a higher judgment matrix can be obtained when used for visual place recognition calculation. Compared with the intermittent diagonal of matrix M1, the diagonal of matrix M2 is more continuous and has higher brightness, which shows the method with a generative adversarial network introduced into visual SLAM for candidate keyframe's enhancement has a positive effect on improving the robustness of visual place recognition.

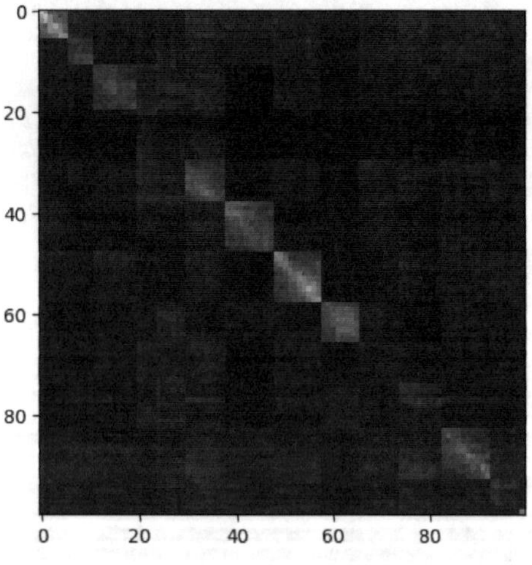

Fig. 7. Similarity matrix M1 of the original candidate keyframe sequence

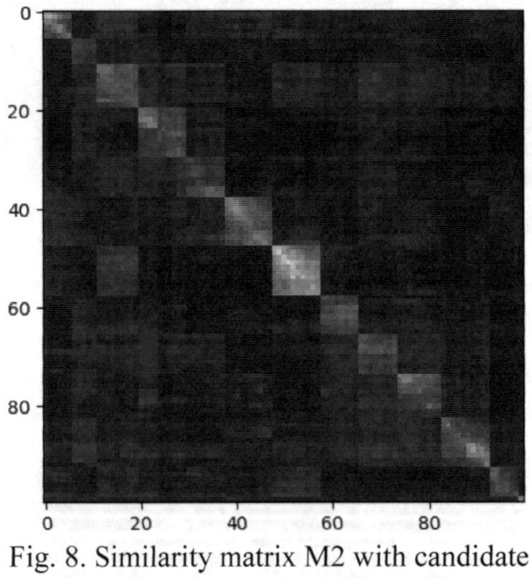

Fig. 8. Similarity matrix M2 with candidate keyframe sequence enhanced

In order to verify the robustness of the method in visual place recognition, the experiment compared the method in this paper with the representative method. First, the candidate keyframes are subjected to the calculation process described in Section III.A to enhance the illumination quality, and then the image descriptors are constructed based on the existing typical hand-crafted features SIFT and SURF, and the existing representative deep learning feature SuperPoint [18]. According to the generation process of the discrimination matrix M2, similar discriminant matrices for visual place recognition are sequentially calculated.

It can be seen from Fig. 9 that the identification degree of the place recognition shown by the matrix DELF_GAN and the matrix SUPERPOINT_GAN calculated based on the learning feature package is much better than the previous two matrices SFIT_GAN and SURF_GAN based on the hand-crafted feature package. When the illumination changes drastically, the place recognition's ability shown by Fig.

9(a) and Fig. 9(b) is almost ineffective. Moreover, the matrix DELF_GAN has continuous highlighted diagonal area than the matrix SUPERPOINT_GAN, which means that the proposed method in this paper performs better in robust visual place recognition.

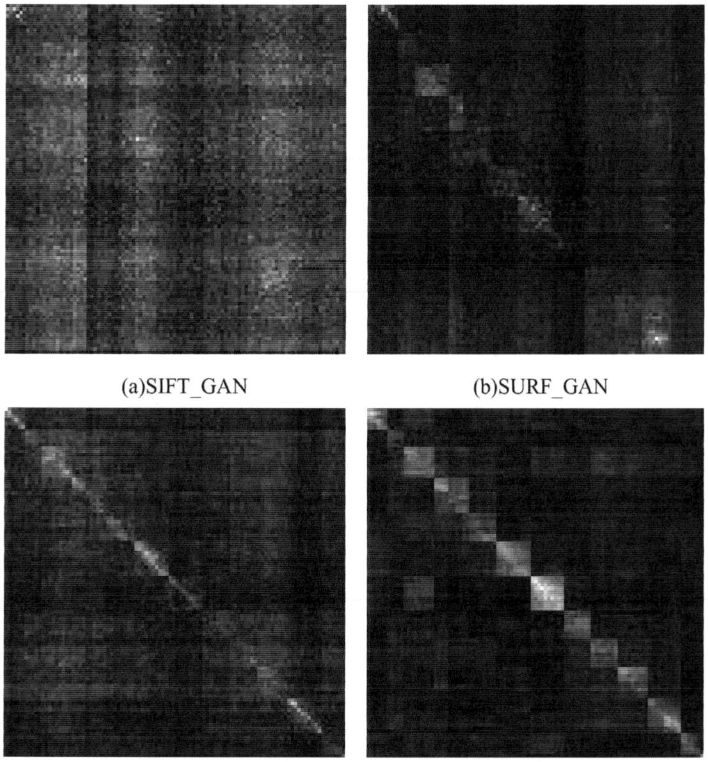

(a)SIFT_GAN (b)SURF_GAN

(c)SUPERPOINT_GAN (d)DELF_GAN

Fig. 9. Comparative analysis of similarity matrix generated by our method and other methods

The experiment has quantitatively analyzed and compared the row data of the above four matrices. As shown in Fig. 10, it shows that the method of this paper has obvious advantages over other methods in visual place recognition at a certain location.

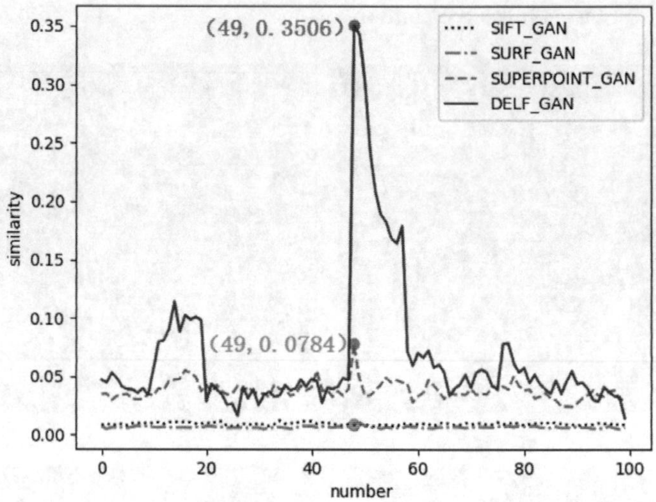

Fig. 10. Sampling for comparative analysis of the similarity matrix for visual place recognition: take the location number 49 as an example.

5. Conclusion

In view of the limitations of the robustness of the method for visual place recognition of mobile robots under severe illumination changes, this paper proposes a robust place recognition method based on the enhancement of candidate keyframes' illumination quality. First, the candidate keyframes are enhanced with illumination quality and authenticity assessment of geographical location, followed by similarity calculations using the proposed invariant image descriptors as the basis for place recognition. Finally, on the dataset of complex environment with drastic changes of illumination, it is verified that the effectiveness of the image after illumination quality enhancement to the real geographical location expression and the significance of the enhancement of illumination quality to the robustness of place recognition. At last, confirming the robustness of the method in this paper is better than other typical methods. This paper provides a valuable solution for the robust place recognition of mobile robot visual navigation systems in the face of drastic illumination changes.

The future work will focus on how to obtain a new generative adversarial network model with better performance based on NAS technology to enhance the keyframe's quality of the map constructed by the robot. In addition, semantic information will also be integrated into the map to enhance the ability of human-computer interaction.

Acknowledgements

Science and Technology Projects in Guangzhou(202102080591);Characteristic Innovation Project of Colleges and Universities in Guangdong Province(2020KTSCX254);Guangdong Medical Research Fund Project(A2020550);Natural Science Foundation of Guangdong Food and Drug Vocational College(2019ZR04);The Natural Science Foundation of Guangdong under Grant(2020A1515110255);

References

[1] Lowry, S. , Sunderhauf, N. , Newman, P. , et al. (2016) Visual place recognition: a survey. IEEE Transactions on Robotics, 32(1), 1-19.

[2] Cummins, M. , & Newman, P. . (2008). Fab-map: probabilistic localization and mapping in the space of appearance. Int.j.robot.res, 27(6), 647-665.

[3] Mur-Artal, R. , Montiel, J. M. M. , & Tardos, J. D. . (2015) Orb-slam: a versatile and accurate monocular slam system. IEEE Transactions on Robotics, 31(5), 1147-1163.

[4] Milford, M. J. , & Wyeth, G. F. . (2012). SeqSLAM: Visual route-based navigation for sunny summer days and stormy winter nights. In:Robotics and Automation (ICRA). IEEE International Conference on. IEEE.North Rhine-Westphalia.pp. 1643-1649.
[5] Zheng, L. , Yang, Y. , & Tian, Q. . (2017). Sift meets cnn: a decade survey of instance retrieval. IEEE Transactions on Pattern Analysis and Machine Intelligence, 40(5): 1224-1244.
[6] Chen, Z. , Lam, O. , Jacobson, A. , & Milford, M. . (2014). Convolutional neural network-based place recognition. Computer Science
[7] Xiang, G. , & Tao, Z. . (2017). Unsupervised learning to detect loops using deep neural networks for visual slam system. Autonomous Robots,41:1-18.
[8] Hou, Y. , Zhang, H. , & Zhou, S. . (2015). Convolutional neural network-based image representation for visual loop closure detection.In:2015 IEEE international conference on information and automation, IEEE.LiJiang .pp. 2238-2245.
[9] BAI, Dongdong, WANG, Chaoqun, ZHANG, & Bo, et al. (2018). Cnn feature boosted seqslam for real-time loop closure detection. Chinese Journal of Electronics, v.27(03), 48-59.
[10] I. Goodfellow, J. Pouget-Abadie, M. Mirza, B. Xu, D. Warde-Farley, S. Ozair, A. Courville, Y. Bengio. (2014) Generative adversarial nets. In:Advances in neural information processing systems.Montreal. pp. 2672-2680.
[11] Isola, P. , Zhu, J. Y. , Zhou, T. , & Efros, A. A. . (2016). Image-to-image translation with conditional adversarial networks. In:IEEE Conference on Computer Vision & Pattern Recognition. Las Vegas.pp. 1125-1134.
[12] Zhu, J. Y. , Park, T. , Isola, P. , & Efros, A. A. . (2017). Unpaired image-to-image translation using cycle-consistent adversarial networks. In:Proceedings of the IEEE international conference on computer vision.Venice. pp. 2223-2232.
[13] Anoosheh, A. , Sattler, T. , Timofte, R. , Pollefeys, M. , & Gool, L. V. . (2019). Night-to-Day Image Translation for Retrieval-based Localization. In:2019 International Conference on Robotics and Automation (ICRA).Montreal.pp. 5958-5964.
[14] Jegou, H. , Douze, M. , Schmid, C. , & Perez, P. . (2010). Aggregating local descriptors into a compact image representation.In: Computer Vision and Pattern Recognition (CVPR), 2010 IEEE Conference on. IEEE.San Francisco.pp.3304-3311.
[15] Oliva, A. . (2006). Building the gist of a scene: the role of global image features in recognition. Progress in Brain Research, 155(2), 23-36.
[16] Noh, H. , Araujo, A. , Sim, J. , Weyand, T. , & Han, B. . (2017). Large-Scale Image Retrieval with Attentive Deep Local Features. Ln:2017 IEEE International Conference on Computer Vision (ICCV). IEEE.Venice.pp. 3456-3465.
[17] Sattler, T. , Maddern, W. , Toft, C. , Torii, A. , Hammarstrand, L. , & Stenborg, E. , et al. (2018). Benchmarking 6DOF Outdoor Visual Localization in Changing Conditions. In:2018 IEEE/CVF Conference on Computer Vision and Pattern Recognition (CVPR). IEEE.Salt Lake City.pp.8601-8610.
[18] Detone, D. , Malisiewicz, T. , & Rabinovich, A. . (2018). Superpoint: self-supervised interest point detection and description.In:Proceedings of the IEEE Conference on Computer Vision and Pattern Recognition Workshops. Salt Lake City.pp. 224-236.

An Ultra-low Power Off Chip Capacitor-less LDO With High Transient Response

Zihang Zhang[1*],Xiaoning Xin[2],Rongyan Chuai[3] and Meng Han[4]

[1]College of Information Science and Engineering, Shenyang University of Technology, Shenyang,Liaoning,110020,China.

[2]College of Information Science and Engineering, Shenyang University of Technology, Shenyang,Liaoning,110020,China.

[3]College of Information Science and Engineering, Shenyang University of Technology, Shenyang,Liaoning,110020,China.

[4]College of Information Science and Engineering, Shenyang University of Technology, Shenyang,Liaoning,110020,China.

[*]Corresponding author's e-mail:hak22@smail.sut.edu.cn

Abstract. This paper presents a low-dropout linear regulator (LDO) with ultra-low power、high transient response. This LDO adopts GAP-dynamic-bias circuit to achieve quiescent current as low as 242nA under 0-20mA load. A transient enhancement circuit based on window comparator is used to sense the output voltage change and further enhance the transient response. The circuit is designed and implemented using TSMC0.18um technology, and is verified by simulation using Spectre software. The simulation results show that when the load changes in the range of 10uA-20mA with a 1ns transition edge, the slowest overshoot and undershoot recovery times are only 2.36us and 2.15us. In the input voltage range of 2.5-3.6V, the linear regulation rate and load regulation rate are 0.479mV/V and 2.61nv/mA, respectively.

1. Introduction

Low-dropout regulator (LDO) as a power management chip has the advantages of low noise, low ripple, and small size, and is widely used in the market. Low power, high current and fast transient response are one of the development trends of LDO. The traditional LDO is usually based on the feedback loop of the error amplifier to respond to the change of the load current. When the load current jumps at a speed of ns, to improve the transient response speed, a high-bandwidth error amplifier is required. This achieves high transient response while making it difficult to reduce power consumption. There are many ways to improve the transient response of LDO. Zhan et al. [1] used a dynamic bias circuit to improve response speed; Or P.Y. et al. [2] used capacitive coupling to detect voltage peaks; Lu Y. et al. [3] regard the output pole as the main pole, which significantly increases the bandwidth and compensates the DC accuracy through another slow loop. The appeal method requires a complicated conversion circuit or compensation circuit, and it is difficult to reduce the power consumption and design complexity of the system.

In this paper, a low quiescent current and high transient response capacitance-less LDO is designed. A GAP-dynamic-bias circuit is used to achieve extremely low quiescent current; and a transient enhancement circuit based on window comparator is used to improve the transient response. There is a good performance compromise between transient response capability and static power.

2. The structure and principle of LDO

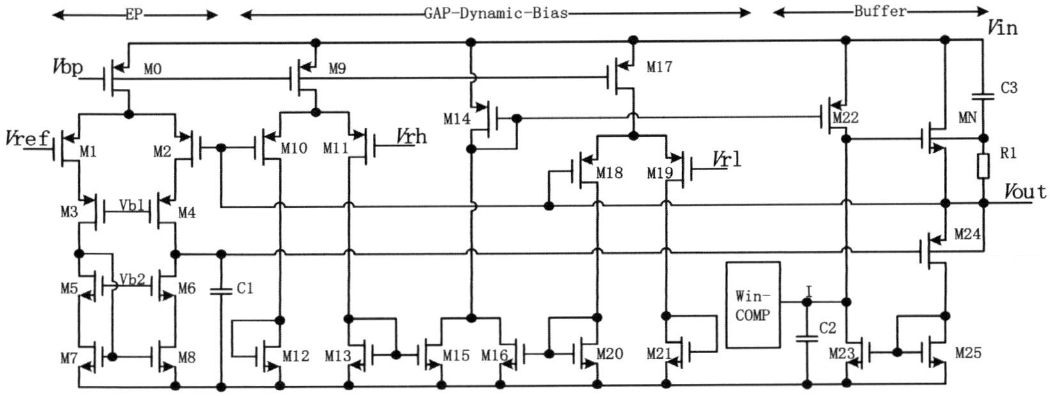

Figure 1. LDO overall architecture

The structure of the LDO circuit proposed in this paper is shown in Figure 1, including an error amplifier, a GAP-dynamic-bias circuit, and a current feedback voltage driver buffer. The LDO includes a fast path composed of M22, M23, M24, M25, MN tubes and a slow path based on a telescope cascode amplifier. The function of the fast path is to improve the transient response capability of the system. The function of the slow path is mainly used to adjust the influence of temperature, power supply voltage, and process on the output voltage to improve the static index of the system. M9, M10, M11, M12, M13, M14, M15, M16, M17, M18, M19, M20 and M21 compose a Gap-dynamic-bias circuit, which dynamically provides bias voltage to the current feedback voltage driver by detecting the change of system output voltage, so as to improve the transient response ability and greatly reduce the static power consumption of the system.

Closed-loop gain based on error amplifier feedback loop A_F:

$$A_F = \frac{V_{out}}{V_{ref}} = \frac{A_v}{1+A_v} = 1 - \frac{1}{1+A_v} \tag{1}$$

The above formula shows that there is a gain error $\frac{1}{1+A_v}$ (A_v the loop open loop gain value). To reduce the gain error and improve the static DC index of the LDO, it is necessary to increase the system open loop gain. Therefore, a telescope cascode amplifier is used to increase the gain.

The fast path is a negative feedback loop. When the load current increases, the branch current of the M24 tube decreases, and the M23 and M25 tubes form a circuit mirror. The decrease of the branch current of the M23 tube increases the grid voltage of the MN tube to generate current for the load application. Among them, the bias voltage of the M22 tube is provided by the dynamic bias circuit. The M10, M11, M18, and M19 tubes respectively constitute a differential pair, and the V_{rh} and V_{rl} voltages are set to 1.14V and 1.26V, respectively, provided by the bias voltage, and the output voltage of the system is detected at the other end. When the output voltage Vout is at the intermediate value of V_{rh} and V_{rl} at 1.2V (ie steady state), the bias current is the lowest; When the output voltage is higher than 1.2V (light load state) or lower than 1.2V (heavy load state), the highest bias current is provided to increase the bandwidth and increase the response speed.

In the case of a load of 20mA, the power tube MN will have leakage. Consider improving the current efficiency, capacitor C_3 and resistance R_1 are introduced. The introduction of C_3 suppresses the sudden change of output voltage.The function of resistance R_1 prevents the bulk from being directly connected to the output voltage(power tube source).

3. Stability analysis

In the entire LDO system, it can be seen as a two-stage circuit structure, which are the first-stage telescope cascode amplifier and the second-stage voltage driver.

The output stage of the voltage driver is composed of power tubes MN and M24, which is equivalent to a super source follower. The M24 tube is equivalent to the common drain connection for the same direction amplification. Since the Miller compensation technology requires the output stage to be amplified in the reverse direction, the parallel capacitor C_1 is used when compensating the system loop. C_1 further pushes the main pole of the system to low frequencies. The circuit diagram part of the voltage driver and its small signal model diagram are shown in Figure 2.

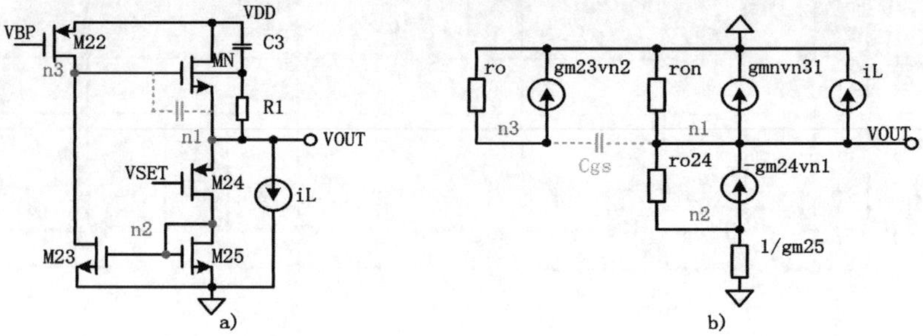

Figure 2. a) Driver circuit diagram b) Small signal diagram

According to the small signal circuit diagram above, we can know the output admittance of the output impedance of the output terminal R_{out}:

$$R_{out} = \frac{1}{1+B} \cdot \frac{1}{g_{out}} \approx \frac{g_{dn}+g_{dp}}{g_{m23}+g_{dn}+g_{dp}} \cdot \frac{1}{g_{m24}+g_{mn}} \quad (2)$$

The LDO system open loop gain A_v can be expressed as:

$$A_v = \frac{g_{m2}(g_{m6}r_{o6}r_{o8} \| g_{m4}r_{o4}r_{o2}) \cdot g_{m24} \cdot R_{out}}{1+(g_{m24}+g_{mb}) \cdot R_{out}} \approx g_{m2} \cdot (g_{m6}r_{o6}r_{o8} \parallel g_{m4}r_{o4}r_{o2}) \quad (3)$$

The main pole P_1 and the secondary point P_2 are respectively expressed as:

$$P_1 = \frac{1}{R_1 C_1} \approx \frac{1}{(g_{m6}r_{o6}r_{o8} \| g_{m4}r_{o4}r_{o2}) \cdot C_1} \quad (4)$$

$$P_2 = \frac{1}{R_{out} C_1} \approx \frac{(g_{m24}+g_{mn})(g_{m23}+g_{dn}+g_{dp})}{C_L \cdot (g_{dn}+g_{dp})} \quad (5)$$

The gain bandwidth GBW is:

$$GBW \approx A_v P_1 = \frac{g_{m2}}{C_1} \quad (6)$$

According to the above formula, it can be known that the output pole of the system P_2 will appear in the high frequency stage, far away from the origin. When the load current gradually decreases from heavy load to light load, the bandwidth of the output pole of the system gradually decreases. In order to ensure stability that the secondary point is higher than the gain bandwidth, shunt capacitor compensation technology is adopted. By designing a suitable parallel compensation capacitor C_1, the main pole of the system can be reduced, thereby reducing the gain bandwidth of the system, thereby ensuring sufficient phase margin of the system.

4. Window comparator transient enhancement circuit

The power tube in Figure 1 is an NMOS tube and has an inherent advantage of negative feedback compared to a PMOS tube. When the load changes rapidly, the transient response speed will be faster. Due to the large size of the power tube, there is also the gate-source parasitic capacitance. Because of the coupling of the parasitic capacitance, the NMOS tube's innate negative feedback advantage will be weakened, and the overshoot or undershoot values will be too large. In order to solve the appeal problem, a ground compensation capacitor C_2 is connected to the grid of the power tube. But C_2

will affect the bandwidth of the voltage driver itself, so the transient enhancement circuit based on the window comparator in Figure 3 is designed to improve the transient response speed.

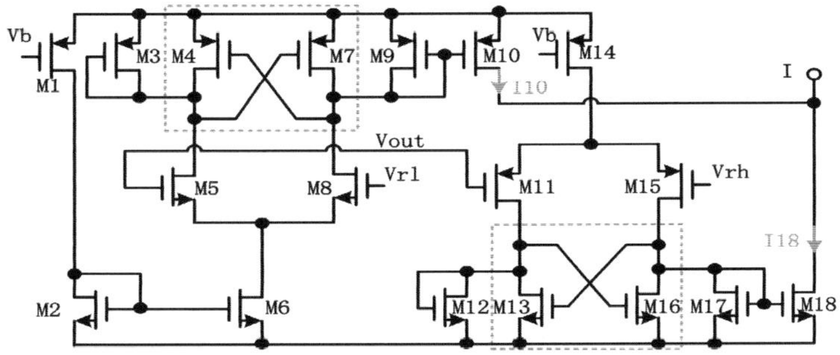

Figure 3. Window comparator transient enhancement circuit

In Figure 3, the bias voltage V_b is provided by GAP-dynamic-bias circuit, and the reference voltages V_{rh} and V_{rl} are provided by a bandgap reference voltage source. The circuit includes two positive feedback loops, the gate-drain interconnection structure composed of M4 and M7 tubes and M13 and M16 tubes respectively. When the load jumps from light load to heavy load, the output voltage Vout decreases rapidly, and the differential pair composed of M5 and M8 is used for voltage comparison to drive the M10 transistor to generate a current I_{10} to quickly charge the capacitor C_2. Similarly, when the load jumps from heavy load to light load, the output voltage V_{out} rises rapidly, and the differential pair composed of M13 and M16 is compared to drive the M18 transistor to generate a current I_{18} to quickly discharge the capacitor C_2, thereby reducing the power tube MN generate a large current.

5. Simulation results

The LDO proposed in this article supports an input voltage range of 2.5V-3.6V. The typical output voltage is 1.2V, the load current range is 100uA-20mA, and the load capacitance range is 0-100pF.

When the load current is 100uA and the load capacitance is 100pF, the output pole of frequency is the lowest, which corresponds to the worst case of stability. The amplitude-frequency characteristic curve under no-load and the load of 100uA and 100pF is shown in Figure 4. It can be seen that the phase margin is 85.9°, which can maintain the stability of the system.

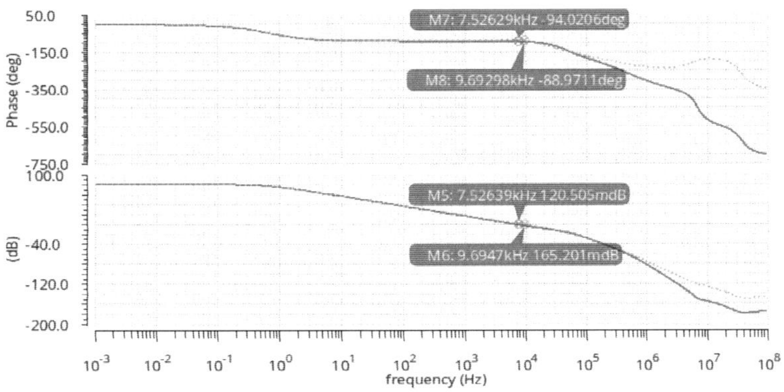

Figure 4. Amplitude-frequency characteristic curve under no-load and 100uA, 100pF load

Figure 5 shows the transient response of the LDO when the load current undergoes 10uA-20mA and 20mA-10uA jumps within 1ns(C_L = 0pF). When the 10uA-20mA jumps, the time for the output voltage to recover to VOUT - 5% is only 2.15us; when the 20mA-10uA jumps, the time for the output voltage to recover to VOUT + 5% is only 2.36us.

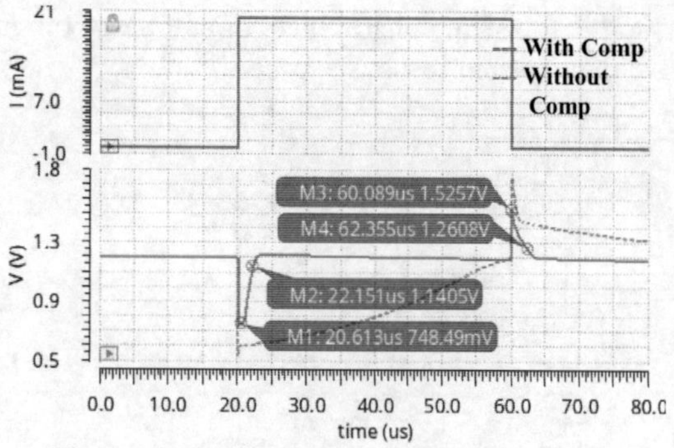

Figure 5. Transient simulation of load transient

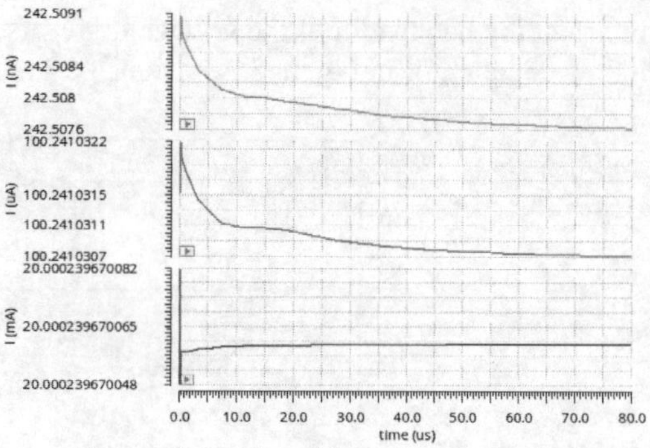

Figure 6. LDO quiescent current power

($I_L = 0\mu A\backslash 100\ \mu A\backslash 20mA$)

Figure 6 shows that under no load, 100uA load, and 20mA load, the quiescent current of 242nA remains basically unchanged.

The parameter comparison of this article and other literature heavy LDO is shown in Table1 .

Table 1 Performance comparison of LDO

Publication	[4]	[5]	[6]	This work
Technology [µm]	0.065	0.35	0.18	0.18
Input Voltage [V]	1.3	3.7	2-3.3	2.5-3.6
Output Voltage [V]	1.1	3.25	1.8	1.2
C_{laod} [pF]	0-2000	100	100	0-100
I_{laod} [mA]	0.1-50	0.1-50	0.1-200	0.1-20
Quiescent Current [µA]	50-190	26	1.2	< 0.242
Line Regulation [mV/V]	1	1.046	15.38	0.479
Load Regulation [nV/mA]	40	75.2	400	2.61
Edge Time [nS]	2	100	100	1
Transient Response [µS]	-	0.2	10	2.36

6. Conclusions

In this paper, An ultra-low power off chip capacitor-less LDO with high transient response is designed, and the static power consumption is only 242nA. The window comparator transient enhancement circuit is designed in this LDO to increase the transient response capability. The simulation results show that when the load current changes in the range of 10uA-20mA with a 1ns transition edge, the slowest time for the output voltage to recover VOUT±5% is only 2.36us.

Acknowledgments

This design has received enthusiastic help from teachers and classmates

References

[1] Zhan, C., Ki, W.H. (2012) An output-capacitor-free adaptively biased low-dropout regulator with sub-threshold under shoot reduction for SoC. J. IEEE Transactions on Circuits & Systems I Regular Papers, 59(5): 45-48.

[2] Or, P.Y., Leung, K.N. (2010) An output-capacitorless low dropout regulator with direct voltage-spike detection. J. IEEE Journal of Solid-State Circuits, 45(2): 458-466.

[3] Lu, Y., Ki, W.H., Yue, C.P. (2014) A 0.65 ns response- time 3.01ps FOM fully-integrated low-dropout regulator with full-spectrum power-supply-rejection for wideband communication systems. In: IEEE ISSCC. San Francisco, CA, USA. 306-307.

[4] Huang, M., Feng, H., Lu, Y. (2019) A fully integrated FVF-based low-dropout regulator with wide load capacitance and current ranges. J. IEEE Trans Power Elec, 34(12): 11880-11888.
[5] Amayreh, M., Leicht, J., Manoli, Y.I. (2016) A 200 ns setting time fully integrated low power LDO regulator with comparators as transient enhancement. In: IEEE International Symposium on Cireuits and Systemg (ISCAS). Motreal,QC,Canada. 494-497.
[6] Cheng, X., Zhang, Y., Xie, C.J. (2018) An Ulra-Low Power Output Capacitor-Less Low -Dropout Regulator with Slew-Rate-Enhanced Circuit. J. Journal of Semiconductors, 39(3): 70-75.

Ferroelectric Properties of $KNbO_3$ Doped with Na : First-Principles Calculations

Peng Wang[1] , Haolei Shen[1], Kui Xia[1] , Wenjun Zong[1a*] , Rongri Tan[1b*]

[1]Department of Physics, Jiangxi Science and Technology Normal University, Nanchang 330013, China

Corresponding author's email: [a]*13807065116@163.com , [b]*rogertanr@hotmail.com

Abstract—The traditional $BaTiO_3$/$PbTiO_3$ ferroelectric material has excellent performance and is widely used, but the Pb in the material has polluted the environment. Studies have found that other ferroelectric materials can also have good performance after doping and improvement. The problem is that the development is too slow. If a reasonable explanation can be given for the changes from the original structure, the research and development process will be accelerated. In this paper, we investigated the ferroelectric properties at the orthogonal-crystal $KNbO_3$ using first-principles calculations coupled with a phenomenological theory model. By replacing the K atoms with Na, studying the stability, polarization, energy band, and density of states after replacement doping, explaining the internal mechanism of its generation, and providing theoretical support for the research and development of new materials.

1. Introduction

ABO_3 perovskite oxides have been widely studied for their unique physical properties and broad technical applications. Among a large number of functional perovskite oxides, ferroelectric materials $BaTiO_3$ and $PbTiO_3$ are used in capacitors and sensors in modern electronics industry due to their unique properties[1]. Some Zr-doped $PbTiO_3$ have been widely used in random access memory and other devices[2-3]. Although of great application value, it contains lead and is toxic. It has become a challenging problem to find lead-free ferroelectric oxides to replace $PbTiO_3$ in industry. The researchers turned to other perovskite materials. Potassium niobate ($KNbO_3$) is one of the typical ferroelectric materials[4], with better ferroelectric properties, optical properties. And large electromechanical coupling coefficient, its dielectric constant is also lower, more importantly, it is environmentally friendly pollution-free.

It has been used to synthesize many piezoelectric materials, such as $K_{0.5}Na_{0.5}NbO_3$ (KNN) ceramics, and its research is also focused on doping modification[5]. Some compounds prepared by Li^+, Ta^{5+}, Sb^{5+} element doping and reaction template grain growth technology (RTGG) have significantly enhanced their properties. KNN research also includes microstructure analysis and control, phase structure analysis and control, ferroelectric, dielectric and piezoelectric properties. The research about KNN lead-free piezoelectric ceramics has reached a climax[6]. Guo et al. also studied the influence of Li^+ doping on the performance of KNN, and predicted that $(K_{0.5}Na_{0.5})_{1-x}Li_xNbO_3$ has good piezoelectric performance at x=0.06.There are also studies using computational methods to study the coexistence of ferroelectricity and magnetic properties of iron-doped tetragonal crystal $K_{0.5}Na_{0.5}NbO_3$ (KNN) at different iron concentrations. It is found that the magnetic properties increase with the increase of Fe atoms. This situation provides a method of obtaining multiferroic materials and the origin of the induced magnetic moment are given.

There have been many studies on the improved composite structure of $KNbO_3$ doping, and the expected good results have also been obtained. However, there are few studies on the original structure from the perspective of ferroelectric formation mechanism, and it fails to give a clear contrast for doping of different proportions. In this article, the ferroelectric properties of materials are studied by doping with different proportions of Na atoms, and the mechanism is analyzed from the electronic point of view. The use of first-principles for research can provide certain guiding significance for the improvement of preparation experiments.

2. Method

All the calculations are carried out by the Vienna ab initio simulation package (VASP 5.4.4)[7,8] which is based on density functional theory (DFT) and the plane-wave pseudopotential methods. The generalized gradient approximation (GGA) with the Perdew-Burke-Ernzerhofer (PBE) exchange correlation function is used[9]. $KNbO_3$ ferroelectric material was modeled with 40 atoms based 1x2x2 supercell, which has $KNbO_3$ orthorhombic phase with the lattice parameter a = 5.744 Å,b=3.980 Å and c = 5.771 Å[10]. The wave function was represented as a plane wave expansion that is truncated at a cut-off energy of 550 eV, and the Brillouin zone was sampled using a 4x4x6 Monkhorst-Pack grid[11]. The structural optimization is obtained until the energy tolerance was 1×10^{-6} eV/atom and force tolerance is 1.5×10^{-3} eV/Å. The macroscopic polarization was calculated by using the modern theory of polarization[12,13]. Then the structure of K atom substituted by Na with different concentration was optimized. The properties, energy bands and density of states diagrams of the energy double potential well of proportionally doped $K_{1-x}Na_xNbO_3$ are calculated, and finally the data are exported for comparative analysis.

3. Results and discussion

Fig.1(a) displays the $KNbO_3$ orthorhombic monocell structure, derived from parameters in the literature. In these 1x2x2 supercells, the number of K ionic is gradually replaced by more Na, as shown in Fig.1(b)-(d). With the addition of Na, the crystal volume was found to be smaller than pure KNO. For example, the crystal volume of $K_{1-x}Na_xNbO_3$ (x = 0.25, 0.5, 0.75) are 517.74 Å^3, 512.57 Å^3, 500.70 Å^3 respectively compared to 522.70 Å^3 of pure $KNbO_3$. The decrease volume is caused by the smaller ionic radius Na (102pm) than that of K (151pm). After optimization, it can be seen that the respective free energies are gradually decreasing. These can be observed in Fig.2. Therefore, it can be concluded that makes sense to change Na to K internally. In Fig.3, we can see the orderly movement of the atoms at the B site, and it can be concluded that the new KNN crystals can be strongly bonded together. All these indicate that after K is replaced by Na, the internal atoms are strongly coupled and can exist stably.

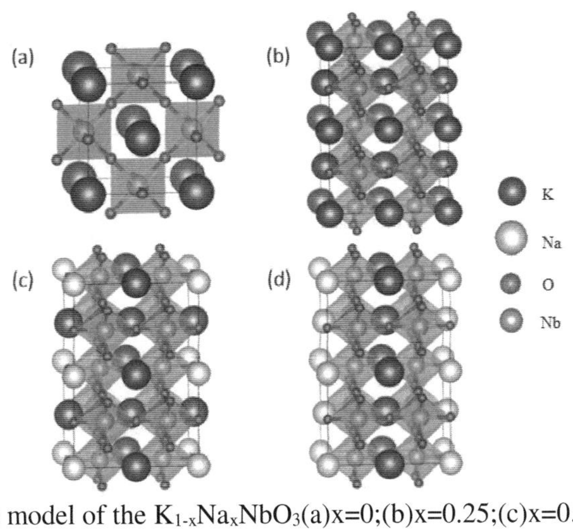

Fig.1 Lattice model of the $K_{1-x}Na_xNbO_3$(a)x=0;(b)x=0.25;(c)x=0.5;(d)x=0.75.

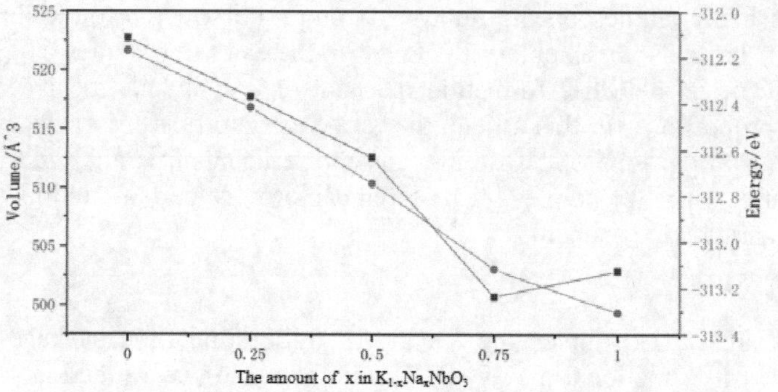

Fig.2 Volume and energy of $KNbO_3$ of different Na.

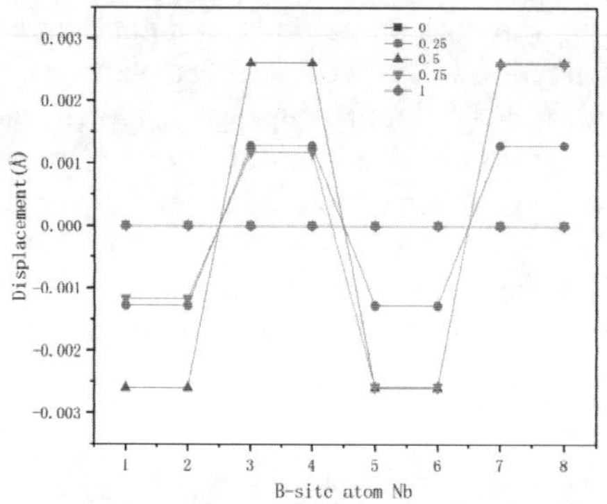

Fig.3 The relative ionic conditions of B site atom in each KNbO3 crystal.

In order to better understand the effects of B-site ionic displacement on the ferroelectricity of KNO, the macroscopic polarization was calculated by using the modern theory of polarization. The polarization intensity of the electrode phase calculated by Berry phase method is 0.40736 C/m^2. It is close to the values in literature, which shows the correctness of the method. The spontaneous polarization of ferroelectric materials can be calculated by the change in polarization of a crystal from a highly symmetrical paraelectric structure to a ferroelectric phase. As shown in Fig.4, the two minima in the double-well potential curve correspond to the two stable spontaneous polarization states, namely, the P+ and P- states. In the case of pure $KNbO_3$, the spontaneous polarization is around 0.18 C/m2. Both the values of spontaneous polarization and well-depth of $K_{1-x}Na_xNbO_3$ increase with the increase of x, indicating the intrinsic ferroelectricity of the $K_{1-x}Na_xNbO_3$ ceramics is increased with increasing x when compared with that of original $KNbO_3$. The energy-polarization curves are fitted using phenomenological Landau-Ginsburg-Devonshire (LGD) theory with the following equation,

$$\Delta G = \frac{1}{2}\alpha P^2 + \frac{1}{4}\beta P^4 + \frac{1}{6}\gamma P^6$$

where ΔG is the energy difference between the P+ and P = 0 state, α,β,γ are the coefficient constants.

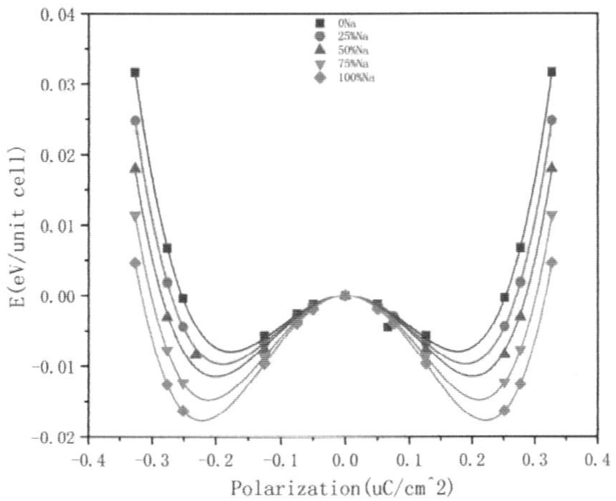

Fig.4 Double-well potential versus ferroelectric polarization of $K_{1-x}Na_xNbO_3$.

The calculated electronic band structure along the high symmetry directions of the Brillouin zone is presented in Fig.5. In all the compounds of $KNbO_3$, the valence band maximum (VBM) is located at the G point and the conduction band minimum (CBM) is also at the G point. The results show that $K_{1-x}Na_xNbO_3$ is a semiconductor with a direct band gap. All strip shapes are similar except for minor differences. It can be observed that the band gap value increases slightly with the increase of the replacement ratio. The addition of Na increases the insulation and provides a basis for improving resistivity. It can also provide an explanation for the increased polarization.

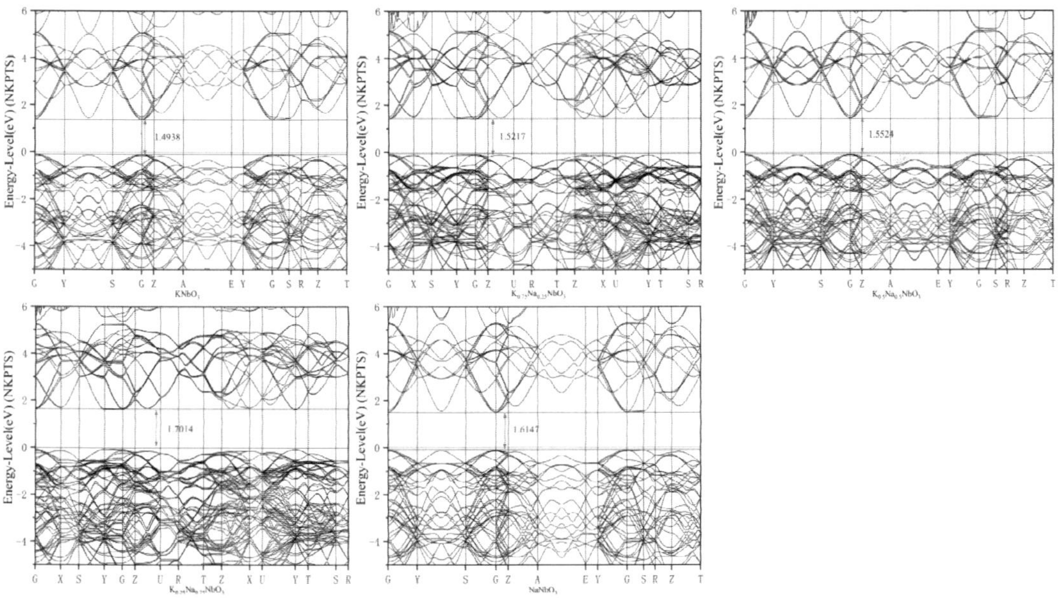

Fig.5 Band structure of $K_{1-x}Na_xNbO_3$. (1) x = 0.0; (2) x = 0.25; (3) x = 0.5; (4) x = 0.75; (5) x = 1.

The first image in Figure 6 is the initial density of states of $KNbO_3$. It can be seen from the figure that in the energy range of -6-8 eV, valence bands are mainly formed by 2p orbital of O atom and 4d orbital of Nb atom, and the region is overlapped, which indicates that Nb atom has strong interaction with nearby O atom and forms chemical bonds. The 2p electron of O atom and the 4d electron of Nb atom have strong orbital hybridization. Orbital hybridization reduces the short-range repulsion between ions determined by the Pauli exclusion principle, which allows the ferroelectric distortion to form. In this energy segment, there are also bands formed by a small number of S orbitals of K atoms. This suggests that the K and O atoms also interact weakly in the material. The following pictures show the density of states after substituting doping at different ratios. It can be seen that the shape is roughly the same as that before doping, but the only difference is that in the 0-8eV energy segment. The common bonding region to the left of Nb and O decreases, and the resulting change is that the interaction between Nb and O decreases. It shows that doping affects the formation of octahedral distortion of Nb and O, which can be explained by the mechanism of ferroelectric change. This analyzes the causes of ferroelectric changes from the electronic level.

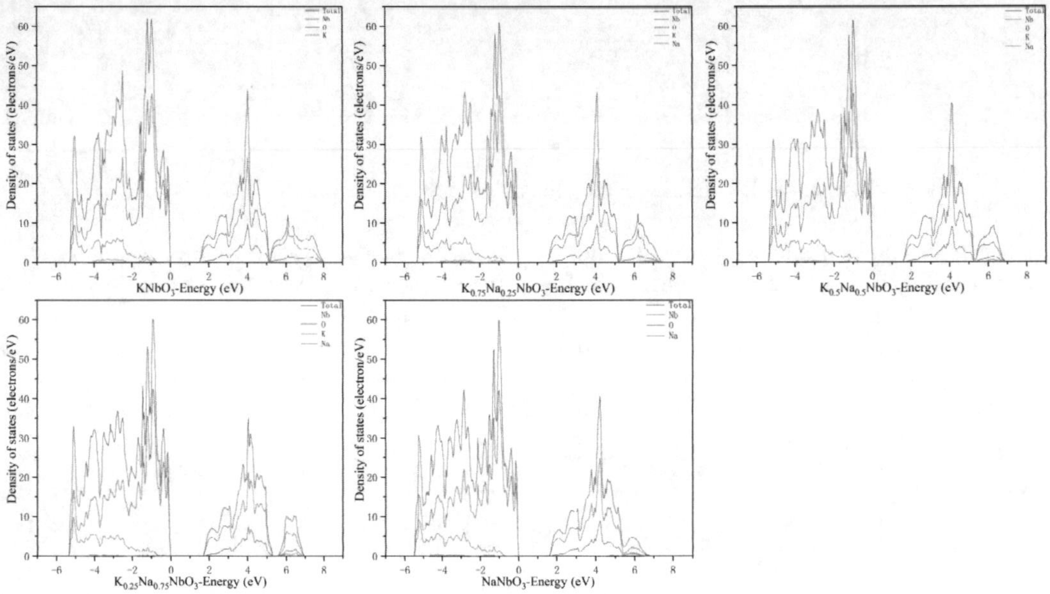

Fig.6 Density of states of $K_{1-x}Na_xNbO_3$. (1) x = 0.0; (2) x = 0.25; (3) x = 0.5; (4) x = 0.75; (5) x = 1.

4. Conclusion

In summary, according to the first-principles method, $KNbO_3$ can exist stably after replacing Na. It can be concluded from the change of polarization intensity that the ferroelectric property gradually increases with the increase of Na. In terms of band structure, the band gap becomes larger with the increase of Na. As you can see from the density of states, the energy at the right end of the conduction band is also moving to the left. From the perspective of internal electrons, the addition of Na affects the bonding of Nb-O, and results in a slight change in the structure of the niobium and oxygen octahedron, which leads to a change in ferroelectric properties. These results show that the ferroelectric properties of $KNbO_3$ can be enhanced by introducing atoms, which opens a way for the design of new materials.

Acknowledgments

This work was financially supported by the National Natural Science Foundation of China (Grant No. 11564015) , the Fund for Distinguished Young Scholars of Jiangxi Science & Technology Normal University (Grant No. 2015QN-BJRC002) and the Educational Commission of Jiangxi Province of China.

References

[1]Weber U, Greuel G, Boettger U, et al. (2010) Dielectric Properties of Ba(Zr,Ti)O3-Based Ferroelectrics for Capacitor Applications[J]. J Am Ceram Soc, 84(4): 759-766.

[2]Kohlstedt H,Mustafa Y,Gerber A,et al. (2005) Current status and challenges of ferroelectric memory devices [J]. Microelection Eng , 80(1): 296-304.
[3] Norga G J, Fe L, Wouters D J, et al. (2000) Effect of RuO2 growth temperature on ferroelectric properties of RuO2/Pb(Zr,Ti)O3/RuO2/Pt capacitors [J]. Appl Phys Lett, 76(10): 1318-1320.
[4] Hayashi H, HaKuta Y, Kurata Y. (2005) Hydrothermal synthesis of KNbO3 powders in supercritical water and its nonlinear optical properties[J]. J Mater Chem,14:2046.
[5] Abrahams S C. Ferroelasticity [J]. (1971) Materials Reseaarch Bulletin, 10(6): 881-890.
[6] E. Gross. Lead-free at last [J]. (2004) Nature, 432(4): 24-25.
[7] G. Kresse, J. Furthmuller, (1996) Efficient iterative schemes for ab initio total-energy calculations using a plane-wave basis set, Phys. Rev. B Condens. Matter 54 11169–11186.
[8] G. Kresse, J. Furthmüller, (1996) Efficiency of ab-initio total energy calculations for metals and semiconductors using a plane-wave basis set, Comput. Mater. Sci. 6 15–50.
[9] J.P. Perdew, K. Burke, M. (1996) Ernzerhof, Generalized gradient approximation made simple, Phys. Rev. Lett. 77 3865–3868.
[10] Wan L F, Nishimatsu T, Beckman S P. (2012) The structural, dielectric, elastic, and piezoelectric properties of KNbO3 from first-principles methods[J]. Journal of Applied Physics, 111(10): 476-38.
[11] H.J. Monkhorst, J.D. Pack, (1976) Special points for Brillouin-zone integrations, Phys. Rev. B 13 5188–5192.
[12] R.D. King-Smith, D. Vanderbilt, (1993) Theory of polarization of crystalline solids, Phys. Rev. B 47 1651–1654.
[13] R. Resta, (1992) Theory of the electric polarization in crystals, Ferroelectrics 136 51–55.

Research on Non-uniform Interrupted Sampling Repeater Jamming for Phase Coded Radar

Zhihao Luo[1a], Jianbing Li[1b*], Xueyu Dong[1c], Jianbang Sun[1d]

[1] School of Information System Engineering, PLA Strategic Support Force Information Engineering University, Zhengzhou, 450000, China

[a]email: lzh10260709@163.com, [c]email: 1766485350@qq.com

[d]email: xyqm818@163.com

*Corresponding author: [b]email: 1043105469@plaieu.edu.cn

Abstract. Because of using correlation processing technology, it is difficult for traditional jamming patterns to achieve the desired jamming effect on phase coded radar. At present, the forward deception jamming of phase coded radar is mainly realized by uniform interrupted sampling repeater jamming (UISRJ). In order to solve the problem that the UISRJ can only produce a single and delayed false target, a kind of non-uniform interrupted sampling repeater jamming (NUISRJ) is proposed. The non-uniform sampling pulse train is generated through pseudo random sequence, and then sampled and forwarded to generate jamming signal, which is processed by radar pulse compression to form dense false targets. Combining the theoretical analysis and simulation results, it can be seen that the NUISRJ has the dual effects of suppressing jamming and spoofing jamming, and the distribution of false target string is more random, including the leading and lagging false target. The amplitude and coverage area of false targets group can be controlled by adjusting sampling pulse width and forwarding times to effectively implement jamming.

1. Introduction

With the rapid progress of modern radar detection technology, phase coded radar [1] is widely used because it can not only ensure the detection range of radar but also obtain high range resolution. The interference that does not match phase coded signal cannot gain corresponding gain during the pulse compression processing, so it has strong anti-interference performance. Therefore, effective jamming of phase coded radar becomes an important research direction in the field of radar countermeasures.

In recent years, the digital radio frequency memory (DRFM) [2] has become a common way of modern radar to counter coherent interference because it can sample, quantify, encode, store and delay the transmission of the received signal. There are many researches on jamming for linear frequency modulation (LFM) pulse compression radar at home and abroad. Multiple jamming modes of LFM signals have been studied in [3]-[6], which can produce sound spoofing or suppressing jamming effects. However, the interference of phase coded signal is seldom studied. The simulation analysis of UISRJ of phase coded radar in [7] shows that only single and delayed false target can be generated. Lai et al. [8] firstly propose a dense dummy target jamming method based on random symbol modulation, but the dummy target density is not controllable. As mentioned in [9]-[10], interrupted sampling predictive repeater jamming can form the leading false target. However, the effectiveness of this jamming method is based on the accurate interception of radar signal and the successful prediction of the complete coding sequence, so it is difficult to implement.

Therefore, the sampling method of intermittent sampling is changed in this paper. The interference is generated by NUISRJ, and the side lobe output will be deteriorated after pulse compression treatment. Multiple false targets will be generated through the elevation of side lobe amplitude, which can effectively implement interference.

2. Phase coded signal characteristics

The phase modulation function of LFM is continuous, while the phase coded signal is completely different and its phase modulation function is discrete. Phase coded signals are usually modulated by binary pseudo random sequences for high frequency and wide pulse. When the code value is only 0 and 1, it is the binary phase-coded signal which is easy to realize, and the commonly used mainly includes Barker code, m-sequence and so on.

The general binary phase-coded signal can be expressed as:

$$s(t)=a(t)e^{j\varphi(t)}e^{j2\pi f_0 t}(0\le t\le T) \tag{1}$$

Where $a(t)e^{j\varphi(t)}$ is the signal complex envelope function, $e^{j\varphi(t)}$ denotes the phase coding value, 1 or -1. $e^{j2\pi f_0 t}$ is the carrier frequency signal, when T denotes the signal time length. If the envelope of the binary phase-coded signal is rectangular, the complex envelope can be written as:

$$u(t)=\frac{1}{\sqrt{N}}\sum_{k=0}^{N-1}c_k v(t-k\tau) \tag{2}$$

Where $v(t)$ is defined as the complex envelope of the signal sub-impulse function. τ and N are the time widths and numbers of signal sub-impulse respectively when $c_k=e^{j\varphi(k)}=+1,-1$ .Using the properties of function δ, Equation (2) can be rewritten as:

$$u(t)=v(t)*\frac{1}{\sqrt{N}}\sum_{k=0}^{N-1}c_k\delta(t-k\tau)=u_1(t)*u_2(t) \tag{3}$$

Which satisfies $u_1(t)=1/\sqrt{\tau}$ and $u_2(t)=\frac{1}{\sqrt{N}}\sum_{k=0}^{N-1}c_k\delta(t-k\tau)$ .The fuzzy function of the binary phase-coded signal is its time-frequency compound autocorrelation function; it can be obtained:

$$\begin{aligned}\chi(\tau_d,\xi_d)&=\int_{-\infty}^{+\infty}u(t)u^*(t+\tau_d)e^{j2\pi\xi_d t}d_t\\&=\chi_1(\tau_d,\xi_d)*\chi_2(\tau_d,\xi_d)\\&=\sum_{m=-(N-1)}^{N-1}\chi_1(\tau_d-m\tau,\xi_d)\chi_2(m\tau,\xi_d)\end{aligned} \tag{4}$$

Where $\chi_1(\tau_d,\xi_d)$ and $\chi_2(\tau_d,\xi_d)$ are the fuzzy functions of sub-pulse signals $u_1(t)$ and $u_2(t)$ respectively. If τ_d determined, m is the uniquely certain integer. If τ_d is greater than or equal to 0, then $m=\lfloor t_d/\tau\rfloor$, otherwise $m=\lceil t_d/\tau\rceil$.

According to the definition of fuzzy function, it can be expressed as:

$$\chi_1(\tau_d-m\tau,\xi_d)=e^{j\pi\xi_d(\tau-\tau_d+m\tau)}\sin c(\xi_d(\tau-|\tau_d-m\tau|))\cdot\frac{\tau-|\tau_d-m\tau|}{\tau}(|\tau_d-m\tau|\le\tau) \tag{5}$$

$$\chi_2(m\tau,\xi_d)=\frac{1}{N}\sum_{k=0}^{N-1-m}c_k c_{k+m}e^{j2\pi\xi_d k\tau}(0\le m\le N-1) \tag{6}$$

According to the symmetry of the fuzzy function, the above equation also applies when $-(N-1)\le m\le 0$.By substituting equations (5) and (6) into (4), the corresponding fuzzy function of the binary phase-coded signal can be calculated.

3. Analysis of interrupted sampling repeater jamming

3.1 The principle of UISRJ

The principle of UISRJ based on DRFM is shown in Figure 1. Radar signals are sampled and forwarded, and the above actions are repeated until the last radar signal ends.

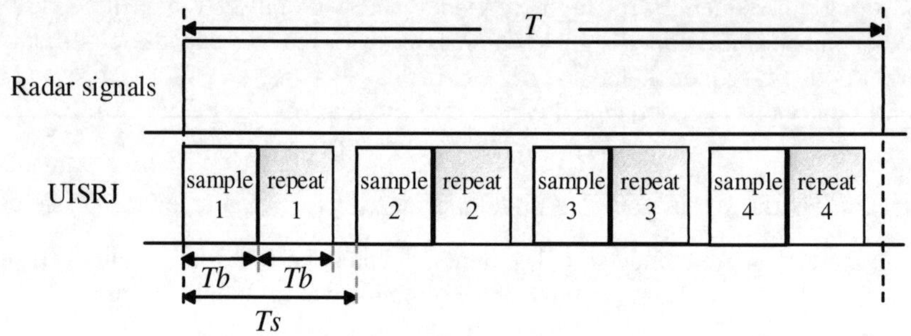

Figure 1. Schematic diagram of UISRJ.

Interrupted sampling of radar signal is equivalent to multiplying a rectangular envelope pulse train on the base of the original radar signal. If N is sampling times, T_b represents the sampling pulse width and forwarding time, and satisfies sampling pulse period $T_s\ge 2T_b$. The sampled pulse signal is denoted as $p(t)$, and its expression by Fourier transform is

$$p(t)=rect\left(\frac{t}{T_b}\right)*\sum_{n=-\infty}^{+\infty}\delta\left(t-nT_s\right)=\sum_{n=-\infty}^{+\infty}a_n e^{j2\pi nf_s t} \tag{7}$$

Where satisfies $a_n=T_b f_s\sin c(nf_sT_b)$, and T_b/T_s denotes the duty cycle. When the jammer receives radar signal $u(t)$, its matching filter satisfies $h(t)=u^*(t_0-t)$.Without loss of generality, t_0 can be 0, then $h(t)=u^*(-t)$. If the signal after sampling is $s(t)$ and the pulse pressure is

$$y(t)=s(t)*h(t)=p(t)u(t)*h(t)=\sum_{n=-\infty}^{+\infty}a_n u(t)e^{j2\pi nf_s t}*h(t)=\sum_{n=-\infty}^{+\infty}a_n u(t)*u^*(-t)e^{j2\pi nf_s t} \tag{8}$$

If $\xi_d=nf_s$, according to fuzzy function definition, then

$$y(t)=\sum_{n=-\infty}^{+\infty}a_n\chi(\tau_d,-\xi_d)=\sum_{n=-\infty}^{+\infty}a_n\chi(\tau_d,-nf_s) \tag{9}$$

From equations (9), it can be seen that $y(t)$ is equivalent to the superposition of pulse pressure results of transmitted signals after countless times of different Doppler frequency shifts, so the UISRJ is the result of joint action of signals of different frequency shifts. However, the phase coded signal is sensitive to Doppler shifts, and except for the zero-order component, the weighting coefficient of other orders is far less than 1, so it can also be written as:

$$y(t)\approx a_0\chi(\tau_d,0)=\frac{T_b}{T_s}\chi(\tau_d,0) \tag{10}$$

The generated false target amplitude is T_b/T_s of the true target and is related to the duty cycle. Because there is a delay between the forwarded interference and the sampled signal, the false target generated by the UISRJ is single and lags behind the true target.

3.2 The principle of NUISRJ

The UISRJ results in a delay between the jamming signal and the sampled signal, so the false target with lag will be formed after the pulse compression. Therefore, we consider the NUISRJ of phase coded radar signal, whose essence is to disturb the received signal element, so that the generated jamming signal element is mismatched with the radar matching filter. In the case, the side lobe output of the jamming signal will be deteriorated after pulse compression processing, and multiple false targets are formed before and after the real target by the elevation of the side lobe amplitude. We can also use the pseudo random sequence for the selection of non-uniform sampling pulse train (NUSPT).

Pseudo random sequence is a deterministic sequence with properties similar to random sequence and its correlation function is close to white noise. The most common binary pseudo random sequence is the m-sequence, which is a code sequence of period 2^n-1 generated by the n-order shift register of the linear feedback. It has the following three features:

(1) The number of zeros and ones in the sequence is roughly equal.
(2) The proportion of consecutive 0 and 1 subsequences in the sequence is determined.
(3) The sequence has good autocorrelation.

The random sampling pulse width and relatively more sampling times can be guaranteed by using the continuous sub-sequence of pseudo random sequence as each sampling pulse width. It is regular and easy to produce. Supposing T_m is the symbol width, the number of symbols of successive sub-sequence is N_n, and $\tau_n = T_m N_n$ denotes the pulse width of each sample, then the pulse train of non-uniform intermittent sampling is shown in Figure 2.

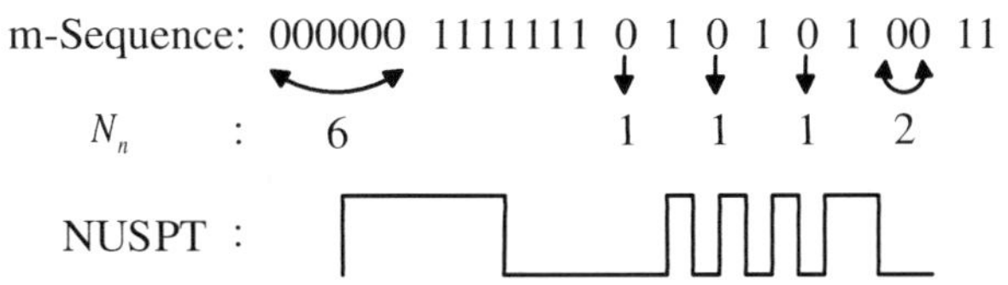

Figure 2. The generation of non-uniform intermittent sampling.

4. Analysis of simulation results

To further demonstrate the jamming characteristics presented in the former analysis, some simulations are carried out to verify its effectiveness and correctness. Different jamming styles will produce different jamming effects on false targets. The following is the analysis of simulation results and comparison of two different ways of sampling repeater jamming. The main parameters are given in Table 1.

Table 1. The simulation parameters.

Parameters	Value
Encoding type	m-sequence
Series	7
Feedback coefficient	[1000001]
symbol width	$0.5\mu s$
signal length	$63.5\mu s$

4.1 Analysis of UISRJ

When the duty cycle is determined, the jamming effect can be changed by changing the sampling pulse width. In order to facilitate analysis, the sampling method is integral sampling and integral forwarding, and different parameter combinations of sampling period and sampling pulse width are :(a) $T_b = 2\mu s, T_s = 4\mu s$ (b) $T_b = 4\mu s, T_s = 8\mu s$ (c) $T_b = 8\mu s, T_s = 16\mu s$.

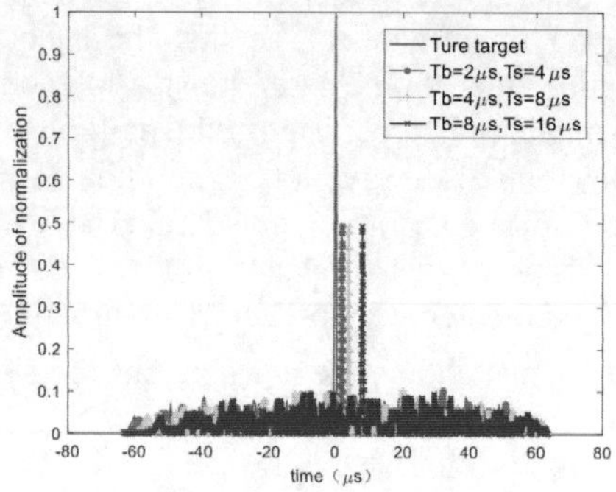

Figure 3. Influence of sampling pulse width on UISRJ.

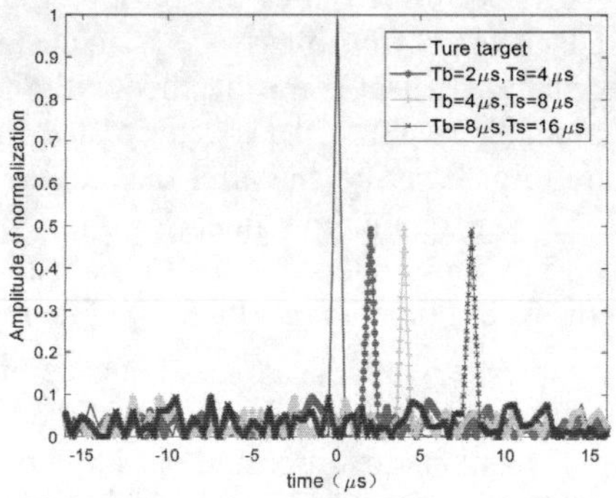

Figure 4.A larger view of local detail of Figure 3.

The simulation results are shown in Figure 3 which shows the experimental results of UISRJ with different sampling pulse widths. Figure 4 is a detailed enlarged image. It can be seen that the forwarding interference of uniform intermittent sampling can only form a false target with lag, and when the sampling duty ratio remains constant, the amplitude of the false target is basically unchanged, and the wider the sampling pulse width, the longer the lag time of the false target.

When the sampling pulse width is determined, the jamming effect can be changed by changing the duty cycle. For the sake of analysis, the sampling method is sampling and forwarding, and different parameter combinations of sampling period and sampling pulse width are :(a) $T_b = 2\mu s, T_s = 8\mu s$ (b) $T_b = 4\mu s, T_s = 12\mu s$ (c) $T_b = 8\mu s, T_s = 16\mu s$.

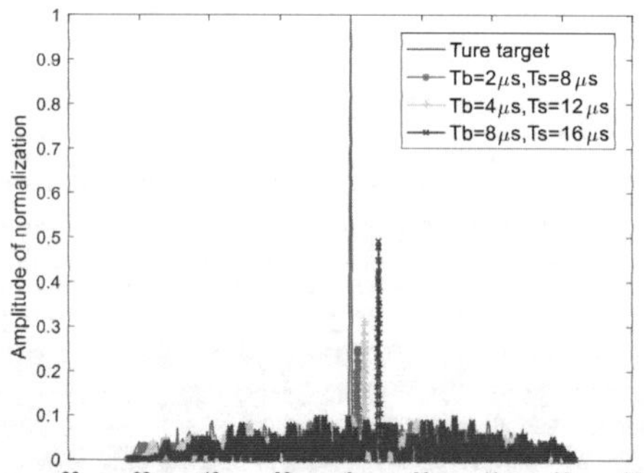

Figure 5. Influence of the duty cycles on UISRJ.

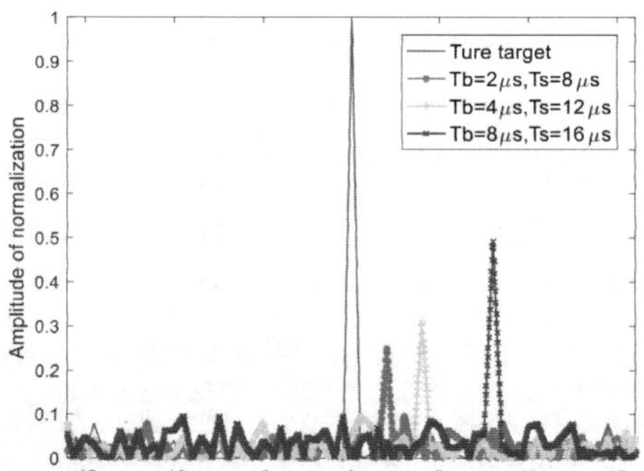

Figure 6.A larger view of local detail of Figure 5.

The simulation results are shown in Figure 5 which shows the experimental results of UISRJ with different duty cycles. Figure 6 is a detailed enlarged image. It can be seen that the forwarding interference of uniform intermittent sampling can only form a false target with lag, and when the sampling pulse width remains constant, the delay of the false target is consistent, and the amplitude of the false target formed by the direct forwarding of intermittent sampling is related to the sampling duty ratio, and the larger the sampling duty ratio is, the larger the amplitude of the false target is.

Therefore, the amplitude of the false target is proportional to the duty cycle of the sampled signal, and the lag time of the false target is proportional to the sampling pulse width, and only one false target is formed in UISRJ.

4.2 Analysis of NUISRJ

In the case of NUISRJ, m sequence, series 7 and feedback coefficient [1000001] are adopted to interfere phase coded signals, and the interference effect is shown in Figure 7. When the forwarding gain is 16dB, the interference effect is observed again as shown in Figure 8.

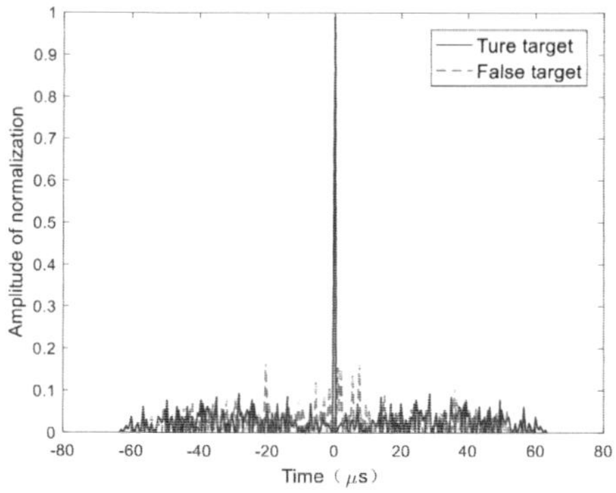

Figure 7. The rendering of NUISRJ for the symbol width $T_b = 1\mu s$.

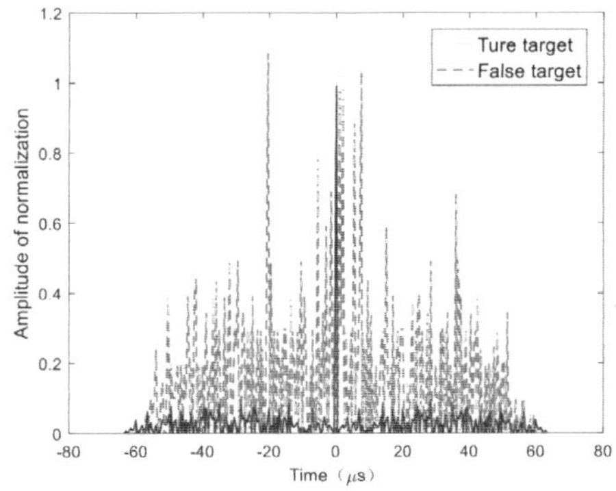

Figure 8. The jamming effect when the forwarding gain is 16dB.

Combined with Figure 7 and Figure 8, it can be seen that the number of false targets generated by NUISRJ increases significantly, with both leading and lagging false targets, and the distribution of false targets is random. Although the amplitude is reduced, the false targets can suppress the real target by appropriate compensation of forwarding gain, which has the dual effects of suppressing jamming and spoofing jamming.

When the symbol width of the sampled pulse train changes, the jamming effect also changes.

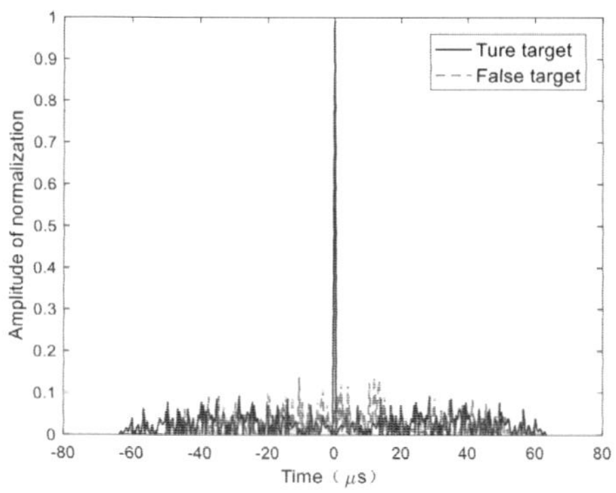

Figure 9. When the symbol width of sampling pulse string $T_b = 2\mu s$, the jamming effect of NUISRJ.

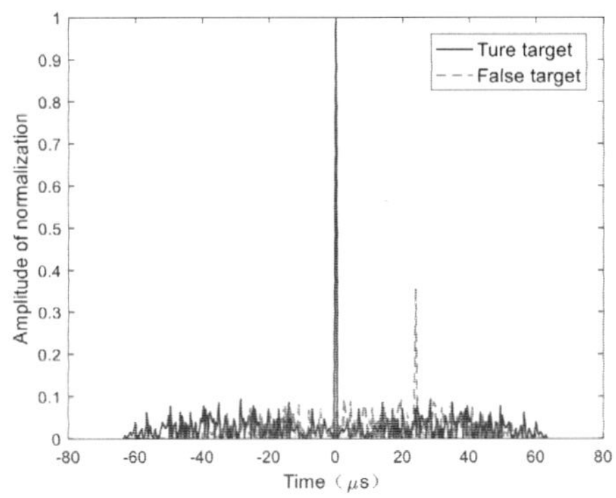

Figure 10. When the symbol width of sampling pulse string $T_b = 4\mu s$, the jamming effect of NUISRJ.

By comparing Figure 7, 9 and 10, it can be seen that with the increase of the width of the sampling code element, the false target group gradually moves to the right, and when it reaches a certain extent, it presents the interference effect of UISRJ. This is because the increase of symbol width leads to the decrease of sampling times. After non-uniform sampling of the received message number element, the mutual correlation of the original symbol element is still maintained. In the case, the output of the pulse compression of the interference signal is similar to the UISRJ.

Therefore, a better jamming effect can be achieved by selecting appropriate forwarding gain and sampling symbol width.

5. Conclusion

In this paper, the characteristics of interference signals generated by different sampling methods are studied by analyzing the sampling methods of interrupted sampling repeater jamming. In order to solve the shortcomings of single and lag false target in UISRJ, the NUISRJ based on pseudo random sequence is proposed, which has the following advantages:

(1) The NUISRJ can generate the leading and lagging false targets with more random distribution and wide suppression range.

(2) The amplitude and coverage area of false targets group can be controlled by adjusting sampling pulse width and forwarding times to effectively implement jamming.

Combining the theoretical analysis and simulation results, it can be seen that the NUISRJ can generate false targets before and after the real targets by deteriorating the side lobe output after pulse compression, and the positions of false targets are more dispersed. Meanwhile, it can adapt to the interference requirements under different conditions by controlling the width of symbols and appropriate forwarding gain. Compared with UISRJ, it has stronger jamming effect on phase coded radar. However, the generated false target has a low amplitude, which is the next problem to be solved in the future.

References

[1] GRECO M., GINI F., FARINA A. (2008) Radar detection and classification of jamming signals belonging to a cone class. IEEE Transactions on Signal Processing, 5(5): 1984—1993.

[2] Lv, H.T. (2009) Design method of digital radio frequency memory (DRFM). *Fire Control Radar Technology*, 38(3): 34-37.

[3] Han, W.B, Li X.Y, Guo J.L, et al. (2014) Analysis on suppression effect of interrupted sampling repeater jamming. *Electronic Information Countermeasure Technology*, 29(2): 47-50, 69.

[4] Shi, L.F., Zhou, Y., Li, D., et al. (2005) Interference of multiple false targets in constant false alarm detection of LFM pulsed radar. *Systems Engineering and Electronics*, (5): 818-822.

[5] LIU, Z., WANG, X.S., LIU, J.C., et al. (2008) Jamming technique of interrupted-sampling and periodic repeater based on digital radio frequency memory. *Acta Armamentaril,* 29(4): 405—410.

[6] Cheng, L., ZHANG, J.M, Yang A.P. (2020) Simulation of frequency shift jamming based on LFM pulse compression radar. *Ship Electronic Countermeasures*, 43(6): 53-56.

[7] Li, J.F., Sheng, J.S. (2009) Study of phase-coded radar jamming technology. *Modern Electronic Technology*, 13(21) : 16 —18.

[8] Lai, Z.j., Gao Y.L., Wan J.G., et al. (2016) Research on dense false target jamming of phase coded radar. *Journal of Air Force Early Warning Academy*, 30(4): 267-270.

[9] Xu, L.T., Feng D.J., Zhao, J., et al. (2013) Leading false target jamming for phase-coded radar. *Journal of Astronautics,* 34(1): 133-138.

[10] Li, J., WANG, H., Wang, W. (2019) Analysis of interrupted sampling repeater jamming of multi-carrier phase coded radar. Signal Processing, 35(1): 49-56.

Research on Equalization Technology of Broadband Satellite Communication Channel

Jie Jiang[1a*], Jian Zuo[2b], Hao Feng[1c]

[1] Chongqing Aerospace Polytechnic, Chongqing 400021, China

[2]China Mobile Group Chongqing Co., Ltd., Chongqing 401121, China

[a*]Jiangjie@cqht.edu.cn, [b]13808304812@139.com, [c]282269726@qq.com

Abstract. Satellite channel is the representative non-linear channel with limited frequency band. The random change of the channel causes the communication stability to be affected. With the development of satellite communication technology, the bandwidth of satellite communication has increased and the data rate has also increased significantly, resulting in more serious code interference, which has become a key factor affecting communication quality. Finding an efficient equalization technology to eliminate channel inter-code interference is the key to ensuring satellite broadband communications. This article first briefly discusses the principle of eliminating inter-symbol interference and common equalizer algorithms, and summarizes the main factors that affect the performance of channel equalization algorithms. On this basis, an improved LSTM blind equalization algorithm is proposed.

1. Introduction

With the development of aerospace and communication technology, the development and utilization of foreign space in the world has been increasing. These include free communication satellites, earth observation satellites, data relay satellites, multimedia satellites and military satellites on low earth orbit. The satellite communication band includes multiple wavelengths, with a maximum capacity of 200Gbps[1]. Current satellite communications require real-time processing and transmission of massive data, and the data rate of the star is even higher, and the data transmission rate is getting higher and higher. The broadband satellite communication system has many types of broadband filters to limit signal bandwidth and suppress noise. The liter cosine filter is the most representative, and the resulting cosine filter pulse response will keep the length of the length of the drag tail to obtain a better filter effect. If there is a sequence mismatch, it increases the interference between the rear and rear, thus forming the inter symbol interference. In addition, the nonlinear phase of the band-pass filter is also the main factor in the interference of the resulting code. The interference of the elimination code is the key to the high speed data transmission in the satellite channel. The equilibrium technique can greatly reduce the interference of the code, improve the channel distortion caused by the spectrum extension and nonlinear distortion, and improve the transmission rate. Therefore, the increase of the equalizer can balance the digital pulse of the distortion and effectively reduce the interference of the code. Therefore, the research on equilibrium technology is of positive significance.

2. Principle of inter symbol interference elimination

In baseband transmission, there is always interference between codes because of the non-ideal characteristics of the channel. Therefore, it is necessary to insert some type of tunable filter in the baseband system to resist interference, i.e., equalizer. In other words, an equalizer is added between the receiving filter and the decider[2], as shown in Figure 1.

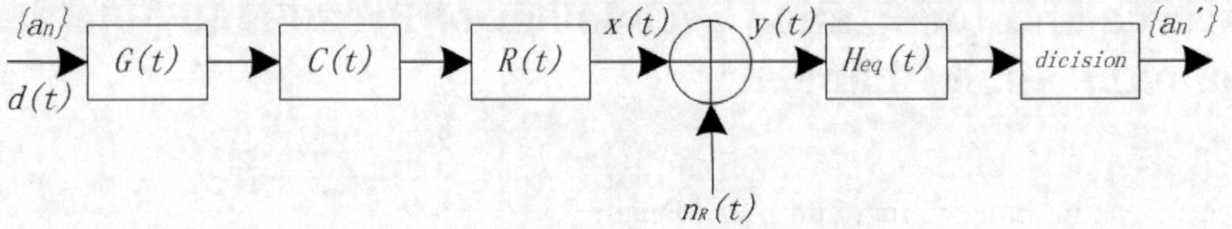

Fig.1 Baseband system with equalizer

If $d(t)$ is the original signal and $x(t)$ is the complex baseband impulse response, that is, $x(t)$ fully reflects the overall transmission characteristics of the receiving/sending filter and channel, and the equalizer is actually the inverse filter of the transmission channel. If the transmission channel is frequency selective, the equalizer improves the frequency attenuation of the spectrum so that the attenuation region of each part of the received spectrum is flat and the phase tends to be linear. For time-varying channels, the adaptive equalizer can track the time-varying characteristics of the channel in real time.

3. Common equalizer technologies

Equalization techniques are usually divided into two categories: linear equalization and nonlinear equalization. The difference between the two types is how the output of an adaptive equalizer is used to control a partial equalizer sequence. Typically, analog signal $y^{'}(t)$ performs amplitude limiting or threshold operations through the adjudicator in the receiver to determine the digital logic value $y(t)$ of the signal. If $y(t)$ is not applied to the feedback logic of the equalizer, the equalizer is linear. Conversely, if $y(t)$ is applied to feedback logic and helps change the equalizer's subsequent output, the equalizer is non-linear. There are many types of filter structures to achieve equalization, and many algorithms for each structure. Common equalizer technologies can be classified according to the type, structure and algorithm of equalizer, as shown in Figure 2.

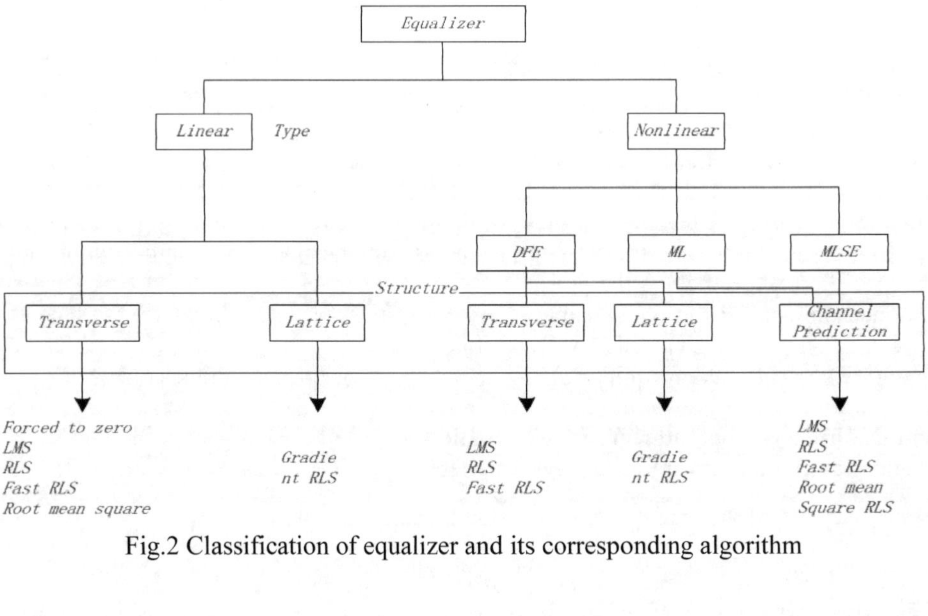

Fig.2 Classification of equalizer and its corresponding algorithm

3.1. Linear equalizer
Linear equalizers can be implemented with FIR filters, which are the simplest type. The current and past values of the received signal are linearly superimposed according to the filter coefficients, and the summation after superposition is used as the output. If the delay element and stage gain are analog signals, the continuous signal waveform output by the equalizer will be sampled at symbol rate and sent to the decision device. However, equalizers are usually implemented in the digital domain, and their sampled signals are stored in shift registers. Linear equalizers can also be implemented using raster filters. The advantages of grating equalizer are: good numerical stability and faster convergence rate. Because of the special structure of the grating equalizer, the most effective length can be dynamically adjusted. Therefore, if the time expansion feature of the channel is not obvious, it only takes a few steps to achieve, and if the time expansion feature of the channel is improved, the number of steps of the equalizer can be automatically increased by this algorithm, and the equalizer process does not need to be interrupted. However, the structure of grating equalizer is more complex than linear FIR filter.

3.2. Decision feedback equalizer
The basic principle of decision feedback equalization is that once an information symbol is recognized and determined, it can predict and eliminate the inter-symbol interference caused by the information symbol before recognizing the subsequent symbols. Decision feedback equalization can be realized directly by transverse filter or grid filter. The transverse filter consists of feed-forward filter and feedback filter. The feedback filter is driven by the output of the detector, and its coefficients can be adjusted to eliminate the interference between the previous symbol and the current symbol. Like traditional decision feedback equalization, predictive decision feedback equalization also has a feed-forward filter. However, the feedback filter is driven by the difference between the output of the detector and the output of the feed-forward filter. Because it predicts residual interference between noise and symbols contained in the feed-forward filter and subtracts the detector's output after the feedback delay, the feedback filter is referred to here as a noise predictor. The performance of predictive decision feedback equalization is similar to that of conventional decision feedback equalization due to the finite number of taps in feed-forward filter and the infinite size of feedback filter. The feedback filter in predictive feedback equalizer can also be realized by grating structure. At this point, RLS grid algorithm can be used to quickly form convergence.

3.3. Adaptive equalizer
During satellite communication, due to noise factors, communication bandwidth is limited and communication capacity is reduced. Because of the group delay characteristics, inter-symbol interference is more serious than ground communication, which greatly reduces communication throughput and data quality. Researchers propose equalization techniques to reduce noise and inter symbol interference in data transmission quality. The traditional method is to use the training sequence to transmit in the channel based on the channel parameters, and compare the training sequence with the received sequence deviation, so as to update the tap coefficient to reduce the deviation. This iterative calculation makes the receiving sequence in the middle of the equalizer the best estimate of the transmitting sequence, and the final tap coefficient is stored for actual communication. The adaptive equalizer that changes the weight belongs to the adaptive equalizer. Adaptive equalizer to produce good input signal, the output signal of the error is minimized, and tap coefficient can adaptive setting, variable in the channel performance is better, but the adaptive equalizer has a disadvantage, in order to send actual information it must be in the actual signal before sending training sequences, and through learning weights training sequences, which reduces the frequency utilization[3].

3.4. Blind equalizer
Adaptive equalizers must repeatedly send training sequences to recover and equalize data with time-varying channel characteristics. Blind equalization technology to realize the data need for

training, effectively made up for the shortcomings of adaptive equalization technology, blind equalization algorithm is the first to use the input signal of prior information to design the function value, and then through learning algorithm to minimize cost function to obtain the best solution, so as to realize the function of convergence, its purpose is to make the output sequence is close to the best estimate of transport order column.

4. Blind equilibrium algorithm with RL feedback

4.1. Blind balanced system framework

In the broadband satellite communication system, the ground modulation signal is transmitted through the satellite channel to the satellite to reduce the signal-to-noise ratio, and the blind equilibrium technique is used to realize the channel blind equilibrium without training and to recover the initial signal as much as possible. In order to balance the transmission characteristics of the normal channel, improve the balance ability, make the modulation recognition signal more reliable, and improve the detection rate of the modulation detection network. The text proposes a blind equilibrium system with integrated transport load (RL) feedback. In the process of transmission of the link to the ground station, the signal is transmitted through the blind balanced network to improve the signal-to-noise comparison, then send it to the Saprk distributed parallel processing platform, the corresponding signal is processed, and the signal pattern identification network is trained, and the final training network is sent to the satellite through the upward link, and the subsequent modulation signal pattern recognition is carried out. The interference of the signal in the signal of the satellite channel reduces the detection effect of the modulation signal. In order to reset the modulation signal with code interference as the original modulation signal, the blind equilibrium algorithm with RL feedback is used, and before the modulation signal goes into the deep convolution neural network (DCNN), the detection accuracy of the modulation signal is improved by reducing the interference of the original signal by means of equilibrium technology[4].

When the modulation signal pattern identifies the blind equilibrium, the modulation signal is disturbed by the satellite channel generation code. The blind equilibrium algorithm with RL feedback is used to eliminate the interference between codes. The signal is compressed and stored in the storage, and the distributed spark computer platform is used for parallel computing and training deep convolution neural network DCNN. Until the condition is fulfilled, the best training model is saved, and then the modulation mode of the signal modulation and the accuracy of the evaluation model is submitted. In order to improve the quality of the signal, it is necessary to restore the original signal as much as possible so that the signal can be obtained through the RL feedback algorithm. The blind equalizer that has RL feedback is a equalizer based on the network of long and short memory networks (LSTM) network. LSTM network not only can learn the characteristics of current information and input information, but also adjust the characteristics relationship between current information and previous information, and better restore the original signal to achieve a better equilibrium effect. By entering the data at the time of learning, the enhanced learning can be applied to the blind equilibrium technology, and the purpose of the training sequence is not required. After retrieving the input data, the current model is updated and the performance of the current network is evaluated using the model and the next update direction is determined, and the iteration is updated until the model converses. Enhanced learning algorithms require explicit goals. For example, the goal used in the equalizer is to restore the equilibrium data to obtain the best estimate of the transmission sequence, and then run the network update based on this target. Therefore, enhance learning to treat the network and the uncertain environment as a whole[5].

4.2. The blind equilibrium algorithm with RL feedback is designed

Based on the known channel parameters, the training sequence and the received sequence deviation are compared, so that the stroke coefficient is updated to reduce the deviation, the optimal estimate of the sending sequence is obtained, and the final extraction coefficient is stored for actual communication, which must be sent in advance to send the training sequence, waste a lot of bandwidth, and reduce the efficiency of the transmission of information. In addition, the algorithm requires frequent transmission of training sequences for communication systems that are used in real-time adjustment of channel conditions. Therefore, the blind balanced technology that does not require training sequences is widely concerned. The blind equilibrium technique containing RL feedback can effectively improve the equalization, make the modulation signal in the modulation detection network more reliable, and the detection accuracy of the high profile detection network.

After the transmission of the satellite channel, the signal can be run by the blind equilibrium algorithm with RL feedback, and the difference between the modulation signal output and the original signal value of the current network recovery is evaluated, and the difference is constantly reduced to improve the balance effect. In the case of the equalizer, the signal is compressed and stored in the storage, and the DCNN network is calculated and evaluated on the spark platform, until the completion condition is reached, the accuracy of the current model is predicted and the forecast modulation mode is output. The modulation signal of the satellite channel is used to obtain the recovery modulation signal, compressed the modulation signal and put it in the storage, and then USES the flow calculation module to handle the flow, and sends the modulation signal to the DCNN network to carry out the iterative training, until the termination condition is reached, the best model is saved and the new modulation signal is predicted[6]. The pseudo-code for blind equilibrium algorithm containing RL feedback is shown in figure 3.

```
img=open('/path/Pictures');
forecastImg =open('/path/ forecastPictures');
newImg=img.shuffle();
epoch=50000;
NL=4; lr=0.1;
for i in range(0,epoch) do
  c(0)=newImg;
  for j in range(0,NC) do
        c(j)=LSTME(c(j-1));
  end for
  x̄_n = c(NL)
  J(m)=Σ_n(x̄_n − x_n)^2/2n
  min(J(i))=min(Σ_m w^{1/m} · J(m))
  model=save(α, β);
end for
Img=model(forecastImg)
return Img
```

Fig.3 Blind equilibrium algorithm containing RL feedback

As input of broadband satellite modulation signal, it is divided into test set, training set, initialization filter, learning rate and other parameters. The signal first passes the 4 layer LSTM network, and then evaluates the current number of iterations based on the loss function. Also calculate the value function of the current result and save the current model. Update the weights and iterate the offset to minimize the value of the loss function and set the maximum iteration times. When the maximum iteration number is reached, the test set is used to predict the value function of the current model, the difference between the signal and the recovery signal and the difference between the final output signal of the modulation. According to the simulation analysis of error curve, the signal $s(k)$ is the independent and distributed sequence of $\{\pm 1\}$(BPSK), the total of N=3000 data points, the finite length coefficient is the first of the $Lf = 41$, $R2 = 1$, the channel order is $Lh = 5$, and the adaptive learning rate is $\mu = 0.001$. The error curve is shown in figure 4.

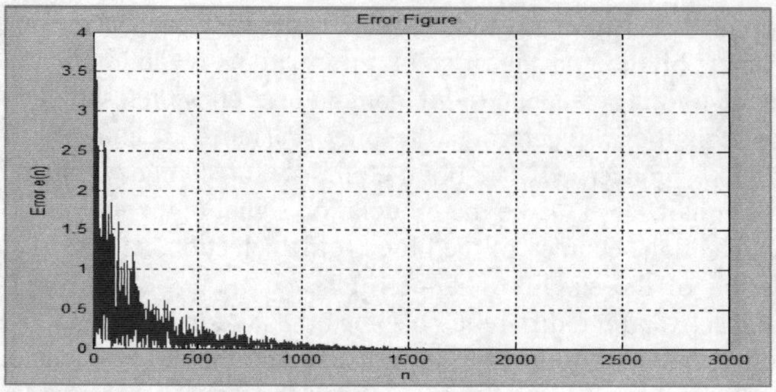

Fig.4 Error figure

Results:

The noiseless channel estimation: $h = \begin{pmatrix} -0.07809 \\ 0.11818 \\ -0.21688 \\ 1.2021 \\ 0.43782 \\ -0.16754 \end{pmatrix}$, error rate = 0.

The noise channel estimation: $h = \begin{pmatrix} -0.07778 \\ 0.11942 \\ -0.21419 \\ 1.2053 \\ 0.43829 \\ -0.16937 \end{pmatrix}$, error rate = 0.

5. Conclusion

In conclusion, the poor and changeable satellite communication channel conditions affect the effect of the channel equilibrium and reduce the accuracy of the modulation signal pattern recognition. In this paper, several major equalizer is analyzed, and then the blind equilibrium algorithm with RL feedback is proposed to reduce the bandwidth of the information transmission by using the normal equilibrium in advance.

References

[1] Taiyu Chen. (2016)The OFDM broadband satellite is based on the communication system channel estimation and equilibrium technology research. Beijing University of Post Telecommunication.

[2] Zhuzhen He, Zhipeng Qie, Jiang Zhu. (2013)The adaptive equilibrium correction study of broadband satellite channel group. Communication technology., 11:14-19.

[3] Hehuan Liu. (2011)High order modulation and efficient coding technology research in broadband satellite communication. Xi'an University of Electronic Science and technology.

[4] Rong Yang. (2019)Research on the blind equilibrium algorithm based on extreme learning machine. Lanzhou University.

[5] Huorong Chen. (2019)Research on the control technology of ship communication network," Marine science and technology., 41(02):170-172.

[6] Zhi Sun. (2014)D The satellite channel judgment feedback fuzzy control multi-mode blind equilibrium algorithm. Technical bulletin., (8):173-17.

Measuring the orbital angular momentum mode of vortex beam based on phase matching sequence

Zhaohui Li[1a*], Wenying Xue[1b*]

[1]College of Communication and Information Engineering, Xi'an University of Science and Technology, Xi'an 710054，China

[a*]lzh791001@163.com, [b*]1033687097@qq.com

Abstract—Measuring the orbital angular momentum(OAM) mode of vortex beams is of great significance in applications based on vortex beams.we propose a method of phase matching characteristics of vortex beams to measure the OAM mode. The method first use a high-speed spatial light modulator(SLM) to sequentially load a set of helical phase sequence images, so that the vortex beam is modulated by this SLM. Then obtain the modulated optical field through pinhole filtering, and then the synchronized tilt phase modulation is performed by the 4f system and high-speed SLM. When the OAM mode of vortex beam is opposite to the topological charge of helical phase image, that is, the incident vortex beam degenerates to the fundamental Gaussian mode beam,and through our optical system, the output plane obtained a indicating spot with the position related to the OAM mode. The Simulation and experimental results show that the method obtains a clear OAM mode indicating spot on the final output plane, which verifies the theoretical derivation.

1. Introduction

The vortex beam is a hollow ring beam with a continuous spiral phase wavefront. In 1992, Allen et al. first discovered that such beams carry OAM[1].

Because OAM has an infinite discrete value space, it can be used in large-capacity optical communications[2],and the spiral phase structure and a dark hollow ring light intensity distribution makes it applicable in the field of optical micro-manipulation[3].Especially it used in OAM multiplexing communication, when the channel terminal demultiplexs the information, and it needs to determine the OAM mode in the current optical field.Therefore,measuring the OAM mode is the key to the application based on vortex beam. Scholars have completed a lot of research on the detection of OAM mode,currently there are two main types of detection methods,one is the Interference method [4-6], which interferes with the vortex beam and another reference beam, and judges the OAM mode by the interference pattern characteristics obtained, however, any slight fluctuations related to the beam can easily cause the interference pattern to be unstable, this kind of method has high requirements on the experimental optical path; Another type of detection method is based on diffraction optical elements method , which is to pass the vortex beam through the aperture such as double slits[7], triangular holes[8], the ring gratings[9,10], then judge the OAM mode according to the characteristics of the diffracted spot. However, the traditional diffraction optical device is not flexible in application.

In recent years, scholars have proposed more OAM mode detection methods [11-16]. One of the methods is us the optical device make the log-polar coordinate transformation to the vortex beam , so that the spiral wave-fronts of different OAM modes are converted into tilted plane wave-fronts with different phase gradients,then focus through the lens to obtain the focused spot, and it spatial position distribution is related to the tilt phase gradient, so as to judge the OAM mode.However, the experimental optical system of this method is not only complicated and difficult to align, which brings difficulties to OAM mode recognition.

In order to simplify the optical system, this paper experimentally verifies an OAM mode detection method based on the phase matching of vortex beam. By using a high-speed SLM to sequentially load a set of helical phase modulation sequence images, so that the vortex beam to be modulated by this SLM,When the OAM mode of vortex beam is opposite to the topological charge of helical phase image, then the phase matching is succeed, at this time the incident vortex beam degenerates to the fundamental Gaussian mode beam,that is,the hollow ring beam is transformed into a solid beam;When the phase is not matched, there is still a hollow vortex beam on the optical field. After the pinhole filtering, the bright spot is obtained under the condition of phase matched, otherwise, the pinhole has no bright spot passed through.After synchronized tilt phase modulation by the 4f system and high-speed SLM,there can obtain a tilted plane wave related to the OAM mode. Finally,after focusing by the lens, the indicating spot related to the OAM mode is obtained on the output plane.It provides a new idea for OAM mode detection.

2. Measurement principle

Generally, without considering the amplitude, the complex optical field of the beam carrying OAM propagating along the z-axis is expressed as:

$$U_0=\exp(il\theta) \tag{1}$$

Where θ is the azimuth angle on the beam section , l is the spiral topology charge, its value represents the number of 2π cycles experienced by one rotation of the azimuth angle, and the symbol represents the direction of rotation, $\exp(il\theta)$ represents the spiral phase factor of the beam. For vortex beams, the measurement of the OAM mode is the measurement of the spiral phase factor of the vortex beam.

Due to the existence of the phase singularity in the spiral phase center, the beam carrying OAM forms a hollow ring-shaped structure during the propagation process,which is vortex beam.The phase matching method uses this ring-shaped light intensity distribution characteristic for OAM detection, and its principle is: When the OAM mode of vortex beam is opposite to the topological charge of the phase matching sequence,that means the incident vortex beam degenerates to the fundamental Gaussian mode beam,that is,the phase matching is succeed.If the OAM mode of vortex beam is opposite to the topological charge of the phase matching sequence,The spiral phase factor of the vortex beam is not canceled, and the output optical field is still a hollow ring-shaped vortex beam, that is, the phase is not matched. The mathematical expression of this process is:

$$\begin{aligned} U_1&=U_0\exp(iL\theta)=\exp(il\theta)\exp(iL\theta) \\ &=\begin{cases} 1 & l+L=0 \\ \exp[i(l+L)\theta] & l+L\neq 0 \end{cases} \end{aligned} \tag{2}$$

Where,the phase matching sequence is represented by $\exp(iL\theta)$,and the OAM mode of the beam to be measured is set within the range of L, that is, $L=-n\sim+n(|l|<|n|)$.

Placing a pinhole $P\left(x_p,y_p\right)$ at the center of the output optical field U_1 after matching, The mathematical expression of this process is:

$$U_2=U_1P\left(x_p,y_p\right),\ P\left(x_p,y_p\right)=\begin{cases} 1 & \left(x_p^2\right)+\left(y_p^2\right)<r_p^2 \\ 0 & \text{other} \end{cases} \tag{3}$$

Where x_p and y_p represent the coordinates on the plane where the pinhole is located, and r_p represents the radius of the pinhole.

In the case of phase matching succeed,there is a bright spot passing though from the pinhole, in the case of phase unmatched, only the central dark spot can pass through the pinhole, then U_2 is:

$$U_2=\begin{cases} \delta(x,y) & l+L=0 \\ 0 & l+L\neq 0 \end{cases} \tag{4}$$

So far, it can be judged whether the phase matching is succeed according to whether there is a bright spot passing through the pinhole, so as to identify the OAM mode.However, due to the fixed position of the pinhole, the spatial position of the bright spot passing through the pinhole is also fixed, and the OAM mode cannot be obtained directly through the spatial position here. If the vortex beams of different OAM modes correspond to the output spots of different spatial positions, and the OAM mode can be directly judged according to the spatial positions.

When the phase matching is succeed, the bright light spot passing through the pinhole can be regarded as a point light source,using a Fourier lens to perform a Fourier transform on U_2,we will obtained a plane wave parallel to the z-axis:

$$\mathscr{F}\{U_2\}=\mathscr{F}[\delta(x,y)]=1 \tag{5}$$

Where $\mathscr{F}$ represents the Fourier transform operator.

Using the tilt modulation grating to modulate the plane wave, and then it will transform a tilted plane wave after modulation, which is represented by U_3 as:

$$U_3=1\cdot\exp\left[-i2\pi\left(f_x a\right)\right],a=l\cdot(D_x/N_L) \tag{6}$$

Where the tilt modulation grating is represented by $\exp\left[-i2\pi\left(f_x a\right)\right]$,$a$ is the parameter of tilt modulation, D_x is the length of the horizontal interval of the light field space, and N_L is the number of spiral phase sequence images, divide the optical field space into N_L equal parts in the horizontal direction, and the fixed interval of each part is D_x/N_L, treat the fixed interval as a constant C ,then the tilt modulation parameter is $a=l\cdot C$,that is, the parameter a takes the topological charge l as a variable and a is proportional to l .

After the tilted plane wave passes through the Fourier lens, that is, performing a Fourier transform on U_3, and the output optical field U_4 is :

$$U_4=\mathscr{F}^{-1}(U_3)=\mathscr{F}^{-1}\left\{\exp\left[-i2\pi\left(f_x a\right)\right]\right\}=\delta(x-a,y) \tag{7}$$

U_4 is the focus spot with the lateral position offset related to l, so that the OAM mode of the vortex beam can be directly judged according to the lateral position of the focus spot.

3. Experiment

In order to verify the theoretical derivation,the method proposed in this paper was verified experimentally.

The experimental setup is shown in Fig.1;Schematic diagram of experimental optical system is shown in Fig.1(a), the beam emitted by the He-Ne laser becomes a quasi-plane wave after passing through the beam expand system. After passing through the SLM1 loaded with the helical phase image, the vortex beam is generated and irradiated on the SLM2. Here, SLM2 and SLM3 are connected to the same computer PC2, the purpose is to achieve synchronization between them two; PC2 load on a set of helical phase sequence image dynamically (Set the speed of loading images to switch one per second and play in loop),as shown in Fig.1(b), the right half is the tilt modulation grating image, which will be used in the Contents of subsequent experimental links; the left half is the helical phase sequence image which is loaded by SLM2. The outgoing beam from SLM1 irradiates the center of the left half of the SLM2 liquid crystal display(LCD) screen to perform phase matching. The output optical field of SLM2 is filtered by a pinhole to obtain the bright spot in the case of phase matching succeed , and then output a plane wave after passing through the Fourier lens1, this plane wave irradiates the center of the right half of the SLM3 LCD screen to achieve tilt modulation of the plane wave.The tilted plane wave emitted from SLM3 focused by Fourier lens 2, and then there will obtained an offset spot on the focal plane of lens2, in the last, it received and recorded as intensity image by charge coupled device(CCD).There, the exposure time of CCD is not less than the cycle time of loading sequence images on PC2.

First, verify the phase matching result of the vortex beam and the spiral phase sequence image.As shown in Fig.2.When the vortex beam with l=5 passes through the SLM loaded with the spiral phase image of L=-5, obtained the fundamental Gaussian mode beam with l=0, that is,obtained the solid beam;when the vortex beam with l=5 passes through the SLM loaded with the spiral phase image of L=-3, obtained the vortex beam with l=2 , which is still a ring-shaped beam with a dark area at the center. This is consistent with the theoretical derivation.

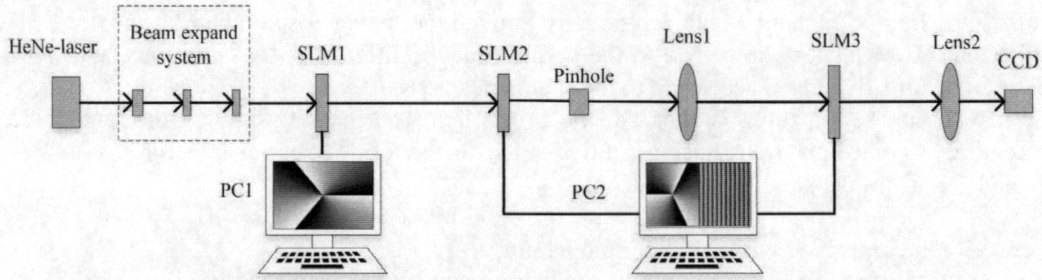

(a) Experimental optical system

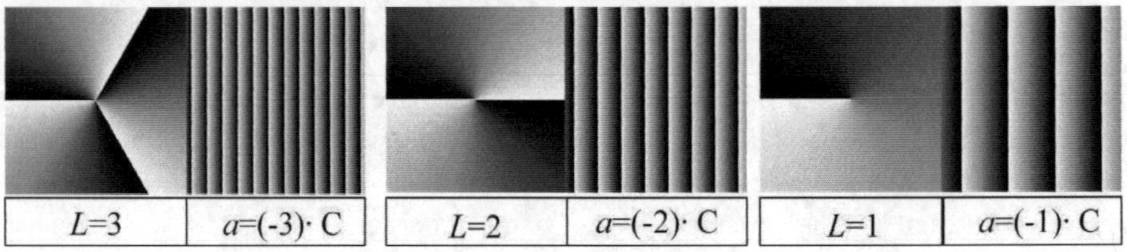

(b)Part of the spliced sequence images loaded by PC2

Fig.1 Experimental setup

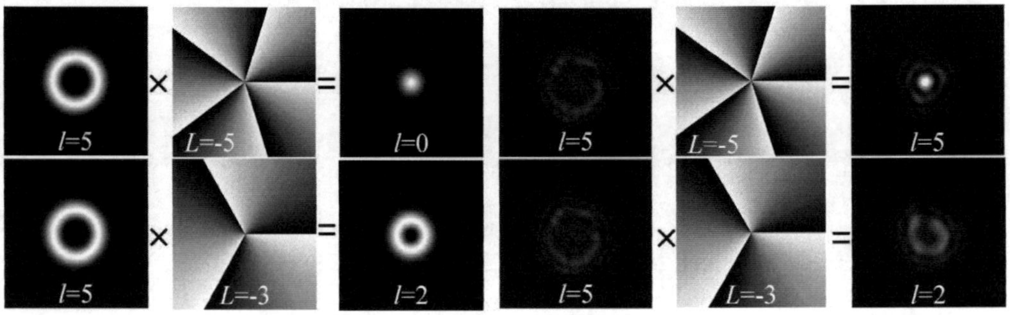

(a)Simulation results (b) Optical experiment results

Fig.2 Phase matching experiment results

Experiment to verify the OAM mode detection method proposed in this paper . In this experiment, the range of topological charge of helical phase image loaded on SLM2 is L=-5~5, this is also the scope of the detectable OAM mode. The simulation and optical experiment are the detection of the OAM mode of the vortex beam with l=-5~5.

The simulation result is shown in Fig.3(a), the position of the bright spot in the figure indicates the OAM mode of the vortex beam currently. When the bright spot is at the center of the figure, it means OAM mode $l = 0$,when it is in the left half of the center position, the OAM mode number is negative,when it is in the right half of the center position indicates that the OAM mode number is positive,and the farther away from the center position, the greater the absolute value of the OAM mode number.If the pinhole is in an ideal state, use one pixel to characterize the pinhole in the simulation, and we will obtain a light spot with one pixel the on the output plane. However, in order to consider the actual effect, the pinhole is enlarged, and we use a pinhole of $2*2$ pixels in the simulation. Therefore, it can be seen from the simulation results that, in addition to the light spot that appears at the correct indication position, there is still weak light intensity at other positions,This is because the pinhole aperture is larger than the unmatched vortex light ring light intensity, causing the pinhole to not only pass through the dark area of the ring light intensity center, but also part of the light energy leaks to the output plane to form a weak light intensity.

The optical experimental results are shown in Fig.3.(b), It can be seen that the indicator light spot appears at the correct OAM position, and In the non-indicating position, there is still weak light leakage. This is also caused by the larger size of the pinhole hole, which is mutually verified with the simulation results. The light leakage caused by the pinholes brings about a decrease in the contrast of the output plane, which affects the final interpretation result, but within a certain threshold range, correct interpretation can be performed. Except for the indicator spot indicating the OAM mode, there is always zero-order diffraction in the detection result ,this is because optical experiments have high requirements for equipment, and the SLM3 we used in the experiment is not well on the performance of phase modulation, so the amplitude and phase modulation exists at the same time.

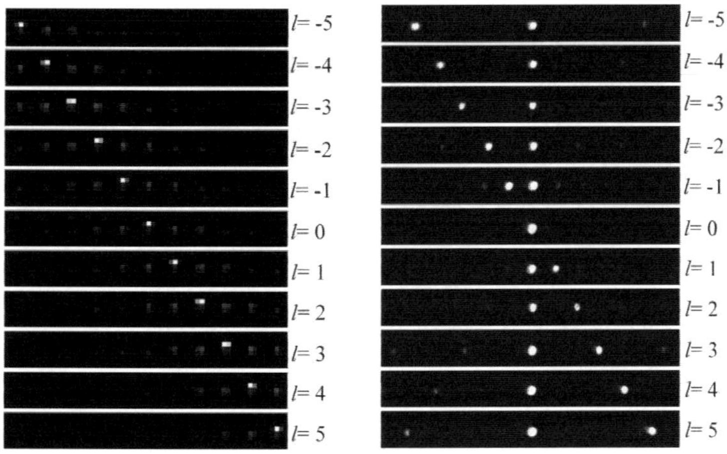

(a)Simulation results (b)Optical experimental results

Fig.3 OAM measurement results

4. Conclusion

Based on the above discussion, the following conclusions are drawn:

(1)We studies the detection method of OAM mode, and proposes an OAM mode detection method based on phase matching sequence. This method modulates the vortex beam by loading the SLM of the spiral phase sequence images,Then, obtained the light spot in the case the phase matching succeed from the pinhole filtering, after passing through the optical system , we obtained an indicator light spot whose position is related to the OAM mode on the output plane. The results show that this method can judge the OAM mode of the vortex beam according to the position of the indicated light spot.

(2)The scope of the detection of OAM mode is limited by the range of topological charge of the helical phase sequence images. During the experiment detection process, the larger the range of helical phase sequence, the larger the detection range of the OAM mode, that means the more sequence images need to be prepared in advance, and the amount of work to be prepared in the early stage is complicated.the method to detect the OAM mode is by a time sequential manner,the helical phase images with different topological charges loaded on the SLM dynamically, that is equivalent to different signals that do not overlap each other on the time axis.Even if there are vortex beams with different OAM modes in the optical field,they can be phase-matched with different signals on the time axis one by one ,then to obtain the OAM modes.Therefore, this method is also suitable for multi-mode OAM detection.

(3)The disadvantage of this paper is that there exists zero-order diffraction in the results of optical experiments, which is related to the phase modulation performance of the SLM equipment. If SLM with better phase modulation performance can be used in the experiment, the zero-order diffraction can be reduced or eliminated.

Acknowledgments
This work was financially supported by the National Natural Science Foundation of China(61705178).

References

[1] Allen L, Beijersbergen M W, Spreeuw R J C, et al. Orbital angular momentum of light and the transformation of Laguerre-Gaussian laser modes[J]. Physical review A, 1992, 45(11): 8185-8189.
[2] YANG Yuanjie, ZHAO Qi, LIU Linli, et al. Manipulation of orbital-angular-momentum spectrum using pinhole plates[J]. Physical Review Applied, 2019, 12(6): 064007.
[3] ZHANG Hao, LI Xinzhong, MA Haixiang, et al. Grafted optical vortex with controllable orbital angular momentum distribution[J]. Optics express, 2019, 27(16): 22930-22938.
[4] Ke X, Xu J. Interference and detection of vortex beams with orbital angular momentum[J]. Chinese Journal of Lasers, 2018, 43(9): 0905003.
[5] Pan S, Pei C, Liu S, et al. Measuring orbital angular momentums of light based on petal interference patterns[J]. OSA continuum, 2018, 1(2): 451-461.
[6] Feng Z, Wang X, Dedo M I, et al. High-density Orbital Angular Momentum mode analyzer based on the mode converters combining with the modified Mach–Zehnder interferometer[J]. Optics Communications, 2019, 435: 441-448.
[7] Emile O, Emile J. Young's double-slit interference pattern from a twisted beam[J]. Applied Physics B, 2014, 117(1): 487-491.
[8] Stahl C, Gbur G. Analytic calculation of vortex diffraction by a triangular aperture[J]. Journal of the Optical Society of America A, 2016, 33(6): 1175-1180.
[9] Dai K, Gao C, Zhong L, et al. Measuring OAM states of light beams with gradually-changing-period gratings[J]. Optics letters, 2015, 40(4): 562-565.
[10] Liu Z, Gao S, Xiao W, et al. Measuring high-order optical orbital angular momentum with a hyperbolic gradually changing period pure-phase grating[J]. Optics letters, 2018,43(13): 3076-3079.
[11] Berkhout G C G, Lavery M P J, Courtial J, et al. Efficient sorting of orbital angular momentum states of light [J]. P Physical Review Letters, 2010, 105(15): 153601.
[12] Li C, Zhao S. Efficient separating orbital angular momentum mode with radial varying phase[J]. Photonics Research, 2017, 5(4): 267-270.
[13] Wen Y, Chremmos I, Chen Y, et al. Spiral transformation for high-resolution and efficient sorting of optical vortex modes[J]. Physical review letters, 2018, 120(19): 193904.
[14] Yang Y, Niu L, Yang Z, et al. Measuring the topological charge of terahertz vortex beams with a focal hyperbolic lens[J]. Applied Optics, 2020, 59(15): 4685-4691.
[15] Ding L, Meng Z, Feng S, et al. Measuring the topological charge of optical vortices using elliptical airy phase mask[J]. IEEE Photonics Technology Letters, 2020, 32(12): 741-744.
[16] Paroli B, Siano M, Potenza M A C. Measuring the topological charge of orbital angular momentum radiation in single-shot by means of the wavefront intrinsic curvature[J]. Applied Optics, 2020, 59(17): 5258-5264.

Study of Spacecraft Low Frequency Cable Installation Technology Oriented to CSB Bus

Xiaohui SONG [1,2,a,*], Wei WANG [1,2], Xianing Xing [1], Qian ZHANG [1], Xu LIU [1], Huaqing LIAN [1], Lei GU [1]

[1]Beijing Institute of Spacecraft Environment Engineering, Beijing, China

[2]Beijing Engineering Research Centre of the Intelligent Assembly Technology and Equipment for Aerospace Product, Beijing, China

[a]songhsiaohui@126.com

Abstract. In response to the needs of spacecraft cable three-dimensional crossover layout, research on cable installation methods based on CSB bus has been carried out. Aiming at the status of using CSB bus cable in the satellite communication cabin of DFH-4 platform for the first time, on the basis of detailed analysis of the performance of CSB bus, the laying and binding methods are analyzed and tested, and the cable installation method on the carbon fiber bracket is proposed. Technical solutions. It has been proved through practice that this method meets the actual needs of satellites, solves the problems of laying and tying the bus cables of the communication cabin, and effectively meets the needs of the spacecraft cable three-dimensional crossover layout, and greatly improves the spacecraft assembly efficiency.

1. Introduction

Communication satellites are developing in the direction of large capacity, high power, and high heat consumption, which will inevitably bring about the continuous increase in the number and weight of the low-frequency cable network of the entire satellite.

How to effectively use the layout space and reduce the weight increase and reliability decrease of the whole satellite caused by the limitation of the low-frequency cable network has become a problem that has attracted more and more attention. A spacecraft used CSB bus cable for the first time in the load bay layout. The CSB bus cable is connected to 56 devices in the communication cabin. The cable direction is complicated and the plugs are widely distributed so that the entire bundle of cables shuttles between the north and south boards and the north-south partitions of the communication cabin[1].

Because the bus cable is very fragile, the requirements for laying and binding are very strict. And when laying the CSB bus cable, the satellite low-frequency cable has been laid and bound in place. The equipment has passed the installation test, and the low-frequency cable plug has basically been inserted in place, and it is very difficult to lay the cable.

Compared with the platform's 1553B bus cable, the CSB cable has a longer length and more plugs, which brings great challenges to operations such as laying and binding.

2. CSB Bus Cable

2.1 Brief introduction

According to the results of the communication cabin equipment layout and cable network layout, the communication cabin is equipped with two independent CSB (C-serial Bus) bus cables, one of which is arranged on the north instrument board of the communication cabin, and the other is arranged on the south instrument board of the communication cabin. Each bus connects 28 TWTA components and 1 PLISU control port. The CSB bus cable shuttles through the entire communication cabin and needs to connect 56 devices in the communication cabin. The length of a single cable is greater than 20 meters, and the equipment and plugs on the satellite have been checked in place when laying, which brings great difficulties to the laying of the cable. There are many branches and plug terminals of the CSB bus cable, which need to be worn when laying, or between high and low frequency cables. The operation space is narrow, and the laying and binding are very difficult. The SPLICE (splits) on the bus cable is unevenly distributed, and there are at least two splitters at the location of the cable branch, and there are special binding requirements for the splitters. In order to facilitate the binding and fixing, the binding position of the splitter should be determined when laying the cable[2].

The total number of plugs is large. When laying the cable branches and plugs, measures should be taken in advance to prevent the cable branches from being bent under force or the plug tails from being damaged. The plug type is mainly divided into PLISU plug model is D-sub25 core connector, LCAMP plug model is MDM37P.

2.2 Installation conditions

Due to the long length of the CSB bus cable, 56 devices in the communication cabin are involved, and the plugs are all over the communication cabin. When laying, the equipment in the communication cabin has been installed, the low-frequency cable has been laid and bound in place, the semi-rigid high-frequency cable has been installed and connected in place, some plugs have been inserted, the operating space is very limited, and the CSB cable needs to be shuttled between the high and low frequency cables. At the same time, the cabin needs to be worn, which brings huge difficulties to the installation work.

2.3 Cable splitter

Compared with the 1553B bus cable, the coupler on the 1553B bus cable is a key point that requires special attention in binding[3]. There are 3 types of SPLICE (splicer, similar to 1553B coupler) of CSB bus cable, namely "L" type, "R" type and "T" type. Therefore, the lashing must be targeted, as shown in Figure 1.

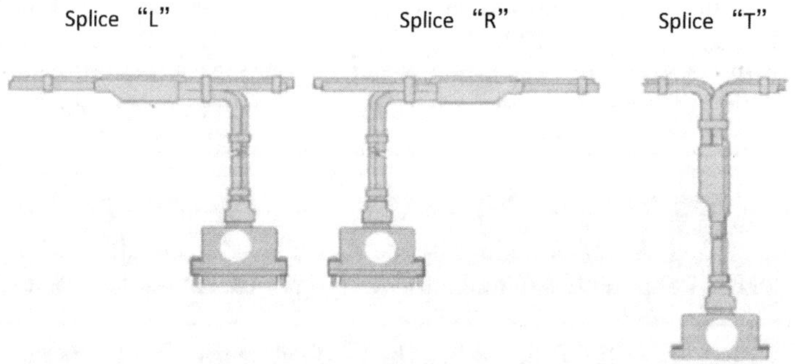

Figure 1. Three forms of SPLICE branch

3. Carbon fiber bracket

The composite cable bracket can be glued to the carbon fiber panel honeycomb panel, aluminum panel honeycomb panel, equipment housing with an identification temperature of not more than 40℃, and tied between multiple cable bundles. Supports of different heights can be flexibly selected to bind and fix the transmission line according to requirements, thereby optimizing the cable layout, reducing the dependence of the cable installation on the satellite cabin wall panels and the fixed installation holes of the equipment, and greatly increasing the flexibility of the final assembly design. The composite bracket is a thin-walled structure and needs to bear a certain mechanical load. The application temperature range is -10℃～50℃. The composite stent molding process must not only ensure the structural size, but also ensure the performance of the product[4].

According to the load-bearing conditions and dimensions of the support, the order of the plying angles and the thickness of the single layer are designed, and the mechanical simulation analysis of the support structure is carried out to ensure the best mechanical performance. To ensure that the composite cable support has a high bearing capacity, the compactness and fiber volume content of the product must be ensured, and there must be no defects such as large pores and delamination. The preparation of prepreg must ensure the glue content. In order to ensure that the glue content is within a reasonable range, the fabric prepreg controls the glue density and brushing amount, and the weftless fabric prepreg controls the winding tension, winding speed, glue density and other process parameters. The prepreg layer should ensure that the layers are dense, without porosity, and fibers without buckling. On the one hand, determine the best layup pot life according to the viscosity characteristics of the resin system, and on the other hand, increase the pre-compaction process appropriately during the layering process. . Pre-compaction should be based on the curing characteristics of the resin, not only to ensure that the product is pre-compacted and compact, but also to ensure that the resin cannot be completely cured.

The surface layer of the carbon fiber composite cable support rod is T300-1k carbon cloth, and the inner layer is laminated with no weft cloth. Therefore, the carbon fiber composite bracket adopts manual latitude-free cloth layering and press curing molding process. The carbon cloth prepreg adopts manual brushing process, and the latitudeless cloth prepreg adopts wet roller molding. The prepreg strictly controls the surface density of the fiber and the amount of glue. During the laying process, vacuum pre-compaction is performed after every 3 to 4 layers are laid. After laying, press and cure according to the curing curve of epoxy resin. After the support rod is demolded, it is processed by cutting equipment according to the external dimensions to form the final product.

Figure 2. Carbon fiber composite cable bracket

The surface density of the carbon cloth of the cable support is 120g/m2, and the number of layers is 3; the surface density of the weftless cloth is 106 g/m2, and the number of layers is 8 layers. The amount of carbon fiber is as follows.

$$Wf = 120 \times 0.08 \times 0.013 \times 3 + 106 \times 0.08 \times 0.013 \times 8 = 1.26g$$

The calculated fiber weight content is 1.26g/1.9g=66.3%. The fiber density is 1.76g/cm3 and the resin density is 1.26g/cm3. The fiber volume content is as follows.

$$Vf = \frac{0.663/1.76}{0.663/1.76 + (1-0.663)/1.26} = 58.5\%$$

4. Test verification

4.1 Static test

The static test uses two different bonding installation datum planes of aluminum alloy and carbon fiber composite materials for static test. According to the static test results, a kind of bonding surface is determined for dynamic test. The static test includes the body static test and the glued static test[5].

The static test is carried out on the electronic universal testing machine. The static test of the main body tests the static compressive resistance of the cable bracket body in the X and Y directions, and the static test of the glue joint tests the failure load of the glue connection in 3 directions. Gluing load test The cable bracket is glued to the L-shaped mounting board with normal temperature curing adhesive, the mounting board is fastened to the beam of the testing machine with screws, one side of the cable bracket is glued to the mounting surface, and the other side is used for processing the mounting holes Load tooling for installation.

It can be seen that the lowest destructive load of the carbon fiber composite cable support (1) is 727N, the lowest destructive load of the cable support (3) in the X direction is 979N, and the lowest destructive load of the cable support (3) in the Y direction is 1990N.

4.2 Bonding load test

The composite cable bracket adopts 420 room temperature curing adhesive with high viscosity and not easy to flow. Before bonding, the bonding surface of the carbon fiber composite bracket is polished and cleaned, and the aluminum alloy bonding surface is phosphoric acid anodized. The test results of carbon fiber composite cable bracket 1 and cable bracket 3 on the aluminum bonding surface and the carbon bonding surface are basically the same. The minimum load of the aluminum bonding surface is 61N, and the minimum load of the carbon bonding surface is 60N.

Figure 3. Bonding test of carbon fiber composite scaffold

4.3 Kinetic test

The dynamic test focuses on verifying the dynamic load-bearing capacity of the composite cable bracket, and adopts the same installation form as the static test. Since the X-direction (perpendicular to the bonding surface) of the cable bracket is much stronger than the Y-direction and Z-direction (parallel to the bonding surface), the dynamic test only performs the Y-direction and Z-direction tests. According to the cable bundling suitable spacing and cable bending requirements provided by the Ministry of Communications, the cable bracket spacing is determined to be 300mm. The vibration tooling is designed according to the Y-direction and Z-direction simultaneous testing of the cable support, and it is L-shaped when installing the cable.

After the partial vibration test, the maximum response frequency became 66.5 Hz. After the test, it was confirmed that after the mechanical vibration test, the cable support body and the adhesive surface were not damaged, the cable was installed firmly and there was no slipping phenomenon, and the test results met the requirements of technical indicators.

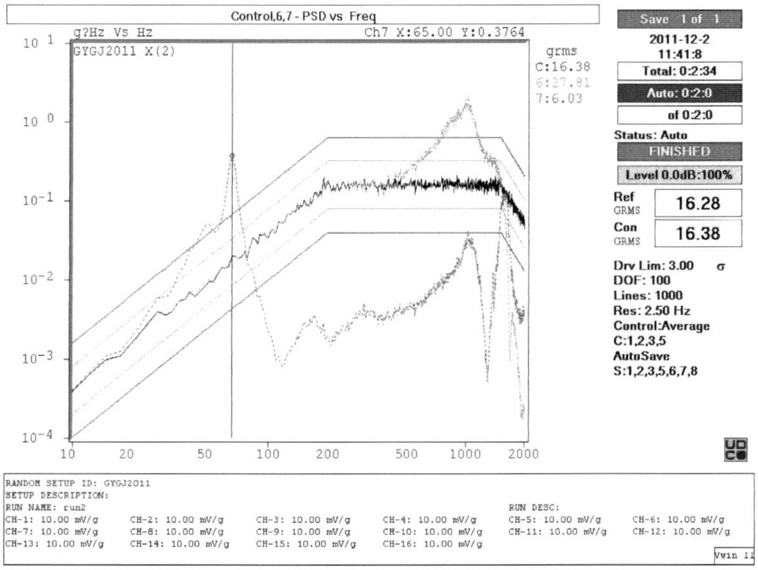

Figure 4. Full-scale random vibration test for weighted cable identification

5. CSB bus installation method

5.1 Preparation before laying

First, spread the CSB bus cable flat, and at the same time, ensure that the cable is in a natural state, and the plug and splitter are not stressed. Determine the start and end of the CSB cable laying according to the process documents. Temporarily fix the plug with the same direction and the main beam cable with 3M. Slowly turn into a circle from the laying end of the cable to make the cable slack and fasten it with temporary straps so that when laying on the satellite, the cable is too long to cause compression and pulling to cause damage to the cable.

5.2 Laying and binding method

When laying the CSB cable, the satellite should be in a state where the low-frequency cable is laid and bound in place, so the laying of the CSB cable should shuttle between the T-shaped brackets that bind the low-frequency cable. When laying from the beginning of the cable, make sure that the cable that turns into a circle is not bent, scratched, or stressed during the shuttle. Use anti-static bubble film to do the necessary protection when in the cabin. When laying, make sure that the plug number is consistent with the socket end of the device and fix it with a temporary cable tie, and then continue laying. This not only ensures the correctness of the laying, but also determines the length of the cable laying. After all the laying is completed in accordance with the process requirements, the laying direction, turning angle and insertion position of the cable need to be checked. The inspector determines that the fixing conditions are met before lashing. Before binding, wrap 3-4 layers of nylon tape on the T-shaped bracket where it needs to be fixed, and at the same time wrap 3-4 layers of nylon tape on the position where the CSB cable needs to be fixed. Teflon cable ties are the preferred choice for fixing, which can be fixed in a crisscross buckle mode, which can be tightened by hand. Tighten the cable tie on the bracket first when binding, and then tighten the cable tie on the fixed cable. In actual operation, you can choose a bracket that can be bound and fixed according to the cable direction and cable length[6].

5.3 CSB bus cable binding technology

The surface layer of the carbon fiber composite cable support rod is T300-1k carbon cloth, and the inner layer is laminated with no weft cloth. Therefore, the carbon fiber composite bracket adopts manual latitude-free cloth layering and press curing molding process. The carbon cloth prepreg adopts manual brushing process, and the latitudeless cloth prepreg adopts wet roller molding. The prepreg strictly controls the surface density of the fiber and the amount of glue. During the laying process, vacuum pre-compaction is performed after every 3 to 4 layers are laid. After laying, press and cure according to the curing curve of epoxy resin.

According to the requirements of the CSB bus assembly, the main cable bundles need to be fixed 10mm before and after the splitter. Therefore, a carbon fiber bracket is attached to the upper surface of each multi-function module to fix the CSB cable. The strength of the bonding surface of each carbon fiber bracket can bear a weight of not less than 1.22kg in the direction perpendicular to the mounting surface, and it can weigh not less than 0.96kg in parallel to the mounting surface.

When binding, pay attention to whether there are sharp-edged devices or high-frequency brackets around the cables. In this case, you need to protect the cables after binding and fixing them. You can choose to wrap 3 to 4 layers of nylon tape where the cable is in contact or where it is likely to be contacted, and tie it with a Teflon cable tie (3.4mm is preferred).

In the position where the CSB bus cable passes through the cabin, a low-frequency cable can be used as a binding point for attachment. Before lashing, you also need to wrap nylon tape and use a cable tie to make a crisscross buckle. Special attention should be paid to the cabin position, and the cables in the cabin position should be protected. It can be wound with heat-shrinkable film, the number of winding layers is more than 5, and the heat-shrinkable film at both ends of the cabin should exceed the cabin plate by 20mm to 25mm. In the absence of T-shaped brackets and other lashing points, nylon bases can be pasted on the deck for fixing with the approval of design and workmanship. Pay attention to the secondary insulation. It is important to note that the CSB bus cable is a signal cable, and it is not allowed to directly use GD414 to paste and fix it on the cabin board.

6. Conclusion

Spacecraft cables should avoid metal components such as spacecraft equipment, grounding stakes, pipelines, heat pipes, etc., and a safety distance of 20mm should be left between them. Low-frequency cables cannot be bent, and the inner bending radius should not be less than 4 times the cable diameter. Low-frequency cables should not touch sharp, rough, and hard edges and corners. Cables near sharp edges and corners and at the hole should be protected by 5-6 layers of heat shrinkable film, and the protective parts should grow at least 20mm from the protected part. At the same time, the sharp edges and corners should be passivated by insulating coating by pasting polyimide single-sided pressure-sensitive tape.

In response to the needs of the spacecraft cable three-dimensional crossover layout, the CSB bus cable installation plan was designed, which can adapt to the complex cable routing conditions. Various experimental results prove that the method is practical and effective. This is the first time that the CSB bus cable has been applied in the development of a positive sample satellite, achieving a three-dimensional, spatial network installation mode, greatly improving the overall assembly efficiency, ensuring product quality, and having a good promotion and application prospect.

References

[1] Tan Weichi, Hu Jingang, Spacectaft System Engineering, China Science& Technology Press, 2009:131.

[2] Song Xiaohui. Experimental Study on Thermal Characteristics of Spacecraft. Cylindrical Roller Bearings[J]. LUBRICATI0N ENGINEERING，2016，V01．41（8）：121-125.

[3] Song Xiaohui. Research on Cable Assembly Technology Facing Tridimention Layout in Spacecraft[C]. 2015 International Conference on Mechanical Engineering and Control System，2015：59-64.

[4] Song Xiaohui. Research of Bolt Pre-Tightening Technology in Narrow Space for Spacecraft[J]. AERONAUTICAL MANUFACTURING TECHNOLOGY，2020，63(13)：97-101.
[5] Song Xiaohui. Research on Torque Loading Technology based on L Lever in Spacecraft Bolt [C]. 2015 International Conference on Engineering Technology and Application，2015：224-228.
[6] Song Xiaohui. Research on Application of Digital Assembly Technology based on MBD in Spacecraft Field[C]. 2018 2nd International Conference on Advanced Technologies in Design, Mechanical and Aeronautical Engineering，2018：163-166.

An implantable loop antenna with Hilbert curve structures

Jinfeng Huang[1a], **Aoran Li**[2b]

[1] College of Electronic and Information Engineering, Nanjing University of Aeronautics and Astronautics, Nanjing, Jiangsu, 210000, China.

[2] College of Computer Science and Technology, Nanjing University of Aeronautics and Astronautics, Nanjing, Jiangsu, 210000, China.

[*]Corresponding author's e-mail:[a] huangjf@nuaa.edu.cn

Abstract. In this work, an implantable loop antenna with Hilbert curve structures is proposed. This antenna uses Hilbert curves to achieve extremely small electrical dimensions. In addition, we use magnesium instead of copper in the antenna structure, as it is healthier than copper. The proposed antenna covers a frequency band ranging from 765 to 1143 MHz (S11< -10 dB), a high efficiency (1.5%), and a maximum realized gain of -15.42 dBi. It is a promising candidate for implantable capsule antenna in medical devices.

1. Introduction

Powered by advanced medical wireless communication systems, implantable antennas attract increasing attention in recent years. Implantable antennas are designed for purposes: (1) monitoring information inside the body of a human or animal, such as stomach endoscopy, heartbeat monitoring, blood glucose level monitoring; (2) wireless power supply for implanted medical devices, such as cochlear implants, pacemakers; (3) locators for protecting animals or studying their behavioral patterns. The wireless transmission of data or energy, reducing the limitations of traditional equipment, has brought great convenience to today's medical treatment. As a key part of the wireless transmission system, implantable antennas become a popular area of research.

Implantable antennas mainly include capsule antennas [1-3] and plane antennas [4-5]. Due to the special application environment, implantable antennas have certain requirements for size and shape. To achieve miniaturization, miniaturized antenna structures such as PIFA (Planar Inverted-F Antennas), loop antennas [1], dielectric resonant antennas [2], spiral antennas [6], dielectric loaded antennas [7], and slit antennas [8] are widely used.

Copper is a commonly used material for metal structures in antennas due to its ductility, low cost and high electrical conductivity. However, according to the International Union for Nutritional Sciences (IUNS), the daily intake of copper for adults is 1.6 - 3 mg, while the daily intake of magnesium is 315-360 mg. Excessive intake of copper can cause liver toxicity and other symptoms. Magnesium, with its high daily intake and its ability to be dissolved and absorbed in the body, is a new alternative to metal materials for implantable antennas [9].

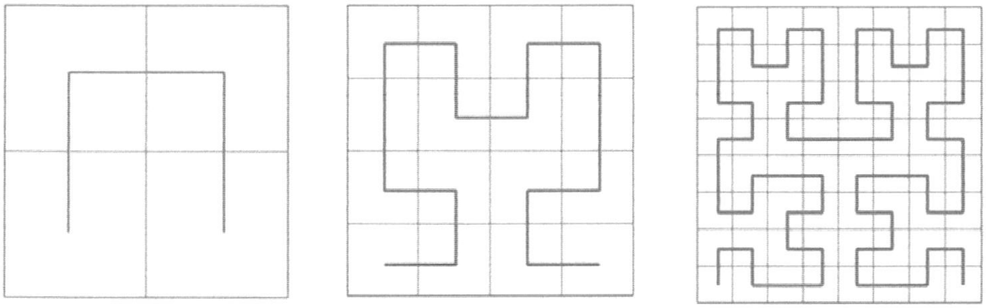

(a) (b) (c)
Figure 1. Hilbert curve diagram: (a) Order 1; (b) Order 2; (c) Order 3.

We propose an implantable loop antenna with Hilbert curve structures. Compared to the proposed implantable antenna, this design has the following advantages.

- The antenna achieves a very small electrical size by utilizing the Hilbert curve.
- The antenna achieves a large bandwidth, high gain and efficiency.
- The antenna design is made of magnesium, which can be absorbed by the human body and is non-toxic and non-hazardous.

2. Antenna Design and Analysis

2.1 Hilbert Curve

The Hilbert curve is a fractal pattern of space-filling curves [10] as shown in Figure 1. The plane space is subdivided into 4^n squares and the center of each square is connected by a curved line, "n" is the order of the curve.

A U-shaped structure of order 1 can be regarded as the smallest unit of the Hilbert curve. To get a Hilbert curve, we execute the following steps:

- Dividing the square equally into 4 × 4 small squares;
- Dividing 16 small squares the center line into 4 quadrants: top left, bottom left, top right, and bottom right;
- Plotting order 1 Hilbert curve in each quadrant, as shown in Figure 1(a);
- Keeping the curve of the top left and top right quadrants unchanged, rotate the curve in the bottom left quadrant 90 degrees clockwise and the curve in the bottom right quadrant 90 degrees counterclockwise;
- Connecting curves in four quadrants.

By analogy, Hilbert curves of order 3, order 4 and even higher can be obtained. Theoretically, Hilbert curves can fill the whole space in two, three, and even higher dimensions.

Interpreting a Hilbert curve in terms of a mapping, it can be seen as a mapping of the plane region [0,1] X [0,1] to the interval [0,1]. Taking the order 1 Hilbert curve as an example, using binary, the four small squares are coded as 0.00, 0.01, 0.10, 0.11. And each subsequent increment in order is increased by 2 binary decimal places. The first square (0.00) is subdivided into four squares: 0.0000, 0.0001, 0.0010, 0.0011. By continuously subdividing and encoding each point in the two-dimensional square space, a one-to-one mapping from the plane to the interval is accomplished.

The proposed antenna structure takes advantage of the fractals of the Hilbert curve to reduce the antenna size.

2.2 Antenna configuration

The antenna consists of three order 3 Hilbert curves in series, as shown in Figure 2. The antenna is carved from a 1 mm thick magnesium board with a total length of D_1 and a width of H. The distance

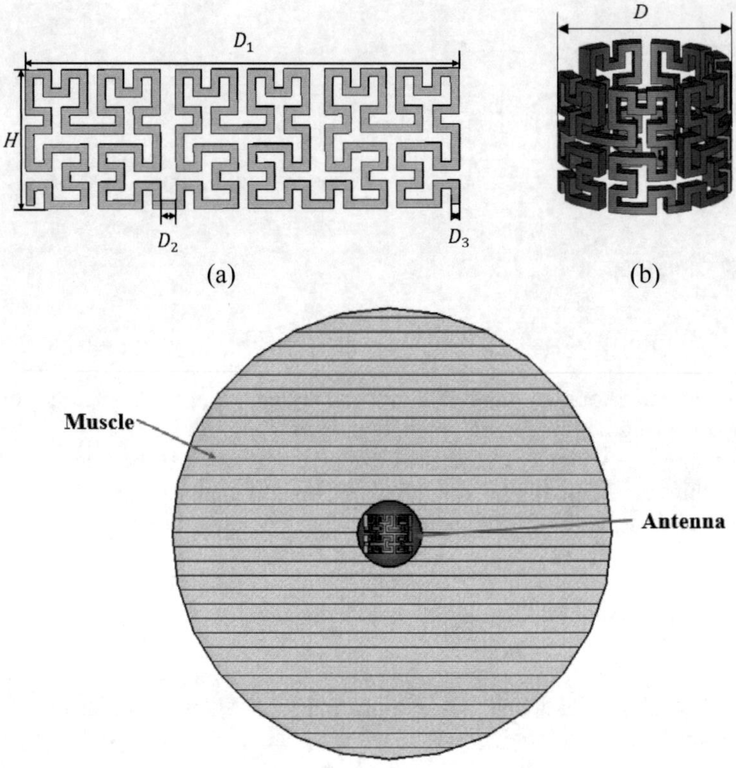

(c)
Figure 2. The proposed antenna structure: (a) Plan sketch; (b) Spatial structure; (c) Simulation environment.

between each order 3 Hilbert curve is D_2 and the width of the curve is D_3. After bending, a circular antenna structure with a diameter of D is formed. Therefore, the bent antenna is placed in a ball muscle tissue ($\varepsilon_r \approx 55$, $\sigma \approx 0.948$ @915 MHz) with a radius of 50 mm.

By optimizing the parameters of the antenna, the optimum dimensions are: D_1=27.5 mm, D_2=27.5 mm, D_3 =1 mm, D =11 mm, and H =8.5 mm.

3. Simulation Results and Discussions

The simulation results for the reflection coefficient (S_{11}), total efficiency, and gain of the antenna inside the muscle tissue are given in Figure 3. It can be seen that the antenna has a bandwidth of 765 - 1143 MHz (S_{11} < -10 dB) and can cover the ISM band (902 - 928 MHz) for medical implants. 41.3% relative bandwidth ensures that the antenna can operate even when the environment changes or processing errors occur. The efficiency of the antenna in the operating band is 1- 1.6%. The maximum gain of the antenna is -15.42 dBi. Figure 3(d) shows an omnidirectional pattern of this antenna.

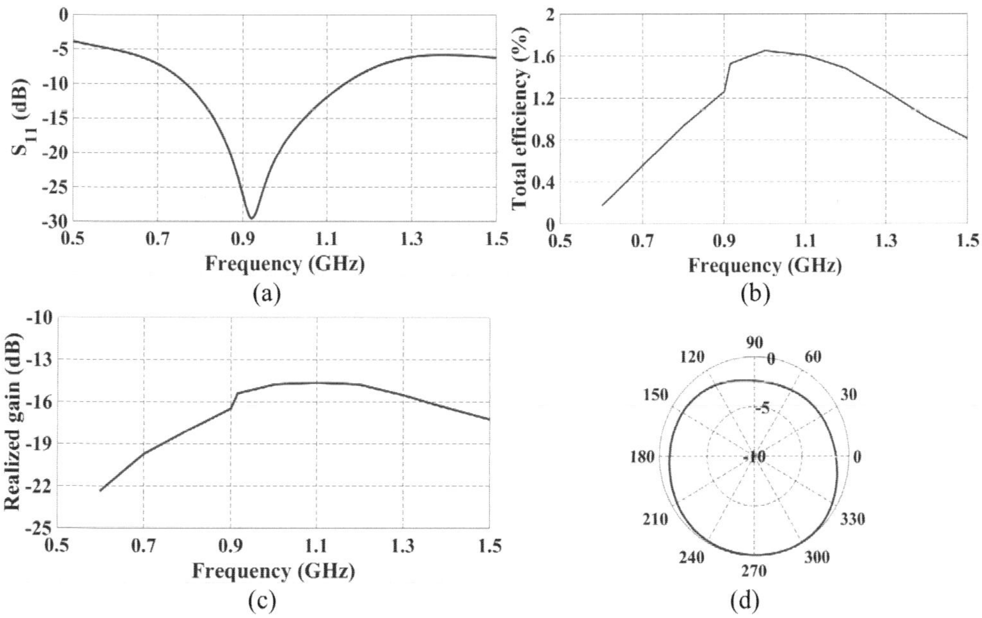

Figure 3. Simulation results of the proposed antenna structure: (a) Reflection coefficient; (b) Total efficiency;(c) Realized gain; (d) Radiation pattern at 915 MHz.

To calculate the antenna SAR values, the antenna is placed in spherical muscle tissue with a radius of 50 mm in the simulation, using the standard IEEE/IEC 62704-1 and an input power of 1 W. The simulation SAR results for the antenna SAR values at 915 MHz for an average mass of 1 g and 10 g are shown in Figure 4. From the figure, we can see that the highest SAR values of the antenna are 216 W/kg and 61.8 W/kg respectively.

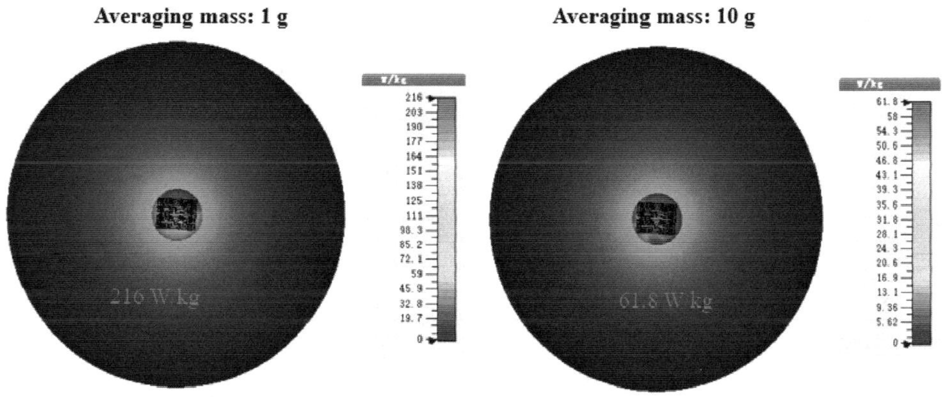

Figure 4. Simulation SAR of the proposed antenna structure.

Table 1. Performance comparison with related design.

Ref	Relative bandwidth (%)	Size (mm^3)	Gain (dBi)	Efficiency (%)
[11]	8.9	Ø10 × 5.8	-21.0	\
[12]	13.0	Ø10 × 15	-24.9	0.1
[13]	19.0	Ø11 × 22	-22.1	0.3
This design	**41.3**	**Ø11 × 9.5**	**-15.4**	**1.6**

Furthermore, the comparison between the related antennas and this design is also given, as shown in Table 1. It is obvious that the performance of this design offers considerable advantages.

4. Conclusion

This paper proposes a broadband, high gain, high efficiency implantable loop antenna with Hilbert curves. The antenna design uses magnesium that is harmless and can be dissolved in the body, making it suitable for implantable medical applications.

Simulation results show that the antenna can work in the ISM band with a bandwidth of 765 - 1143 MHz and cope with the environmental changes that may occur during the implantation process. The antenna gain in the operating band is -15.42 dBi and the efficiency is greater than 1.2 %. Simulation results of the antenna in the muscle tissue model indicate that the design is suitable for implantable antenna applications.

References

[1] Shang, J.L., Yu, Y. (2019) An Ultra-Wideband Capsule Antenna for Biomedical Applications. IEEE Antennas and Wireless Propagation Letters.,18:2548-2551.

[2] Lai, J., Wang, J., Zhao, K., Jiang, H. and Liu, J. (2021) Design of a Dual-Polarized Omnidirectional Dielectric Resonator Antenna for Capsule Endoscopy System. IEEE Access.,9:14779-14786.

[3] Zhu, D., Xu, L.J., Gao, S. and Wen, G.Y. (2018) Integrated Design of Wideband Omnidirectional Antenna and Electronic Components for Wireless Capsule Endoscopy Systems. IEEE Access.6:29626-29636.

[4] Hamouda, Z., Wojkiewicz, J.L., Pud, A.A. et al. (2018) Flexible UWB organic antenna for wearable technologies application. Iet Microwaves Antennas & Propagation.,12:160-166.

[5] Nguyen, V.T., Jung, C.W. (2016) Radiation-Pattern Reconfigurable Antenna for Medical Implants in MedRadio Band. IEEE Antennas and Wireless Propagation Letters.,15:106-109.

[6] Peng, Y.X., Saito K., Ito K. (2019) Dual-Band Antenna Design for Wireless Capsule Endoscopic Image Transmission in the MHz Band Based on Impulse Radio Technology. IEEE Journal of Electromagnetics, RF and Microwaves in Medicine and Biology.,3:158-164.

[7] Nikolayev, D., Joseph, W., Skrivervik, A.K., et al. (2019) Dielectric-Loaded Conformal Microstrip Antennas for Versatile In-Body Applications. IEEE Antennas and Wireless Propagation Letters.,18:2686-2690.

[8] Bao, Z., Guo, Y.X., Mittra, R. (2017) An Ultrawideband Conformal Capsule Antenna With Stable Impedance Matching. IEEE Transactions on Antennas and Propagation.,65:5086-5094.

[9] Koo, J., MacEwan, M.R., Kang, S.K. et al. (2018) Wireless bioresorbable electronic system enables sustained nonpharmacological neuroregenerative therapy. Nature Medicine.,24:1830–1836.

[10] Séébold, P. (2007) Tag-systems for the Hilbert curve. Discrete Mathematics and Theoretical Computer Science.,9:213-226.

[11] Ke, Z., Liu, C., Liu, X., et al. (2018) A Conformal Differentially-Fed Antenna for Ingestible Capsule System. IEEE Transactions on Antennas and Propagation.,66:1695-1703.

[12] Bao, Z., Guo, Y.X., Mittra, R. (2017) Single-Layer Dual-Tri-Band Inverted-F Antennas for Conformal Capsule Type of Applications. IEEE Transactions on Antennas & Propagation.,65:7257-7265.

[13] Duan, Z., Xu, H., Gao, S.S. and Geyi, W. (2021) A Circularly Polarized Omnidirectional Antenna for Wireless Capsule Endoscope System. IEEE Transactions on Antennas and Propagation., 69: 1896-1907.

Design of Intelligent Anti-suffocation System in Vehicle

ZHENG LI[1,a] , YUN CAO[2,b*], ZHULE JIN[3,c]

[1,2] Institute :School of Mechanical Engineering,Nanjing University of Science and Technology. School of Mechanical Engineering, Nanjing,China

[3] Institute :School of Mechanical Engineering,Nanjing University of Science and Technology. School of Automation, Nanjing,China

[a]2711171685@qq.com, [b]cy1990@njust.edu.cn, [c]718963471@qq.com

Abstract—With the progress of society, more and more families take cars as the main means of transportation. At the same time, due to the negligence of some parents, there are more and more suffocation accidents caused by children left in the car and adults resting in the car. At present, the research can only detect people who are moving around in the car, but not those who are resting in the car. At present, many systems are detecting and alerting, rather than directly solving the problem. The people left in the car with engine cut off and all the doors closed.The capacity of the car battery is very small, which requires the system operates with minimal power consumption. This problem is not considered in the current relevant research. To solve the above problems, this paper designs a system based on low power consumption dual microprocessor, combined with power management program, equipped with carbon dioxide sensor, temperature sensor and camera, which has low power consumption and can detect people resting and staying in the car at the same time. The system can effectively solve the shortcomings of the current research in view of the two situations of people resting and staying in the car at the same time.

1. INTRODUCTION

With the continuous development of society, cars have become the choice for more and more families to travel, which greatly facilitates people's life. But at the same time, the incidence of traffic accidents is also increasing. In recent years, there are more and more accidents in which children are left in the car to suffocate and die, and adults suffocate when they rest in the car, which has brought harm to countless families. The closed environment in the car will lead to heatstroke, overheating, dehydration, suffocation and death of people staying in the car. Therefore, a system that can detect and execute is particularly important.

At present, the relevant research on the detection of people in the car can only detect the people moving in the car, not the people resting. In the paper *A Mechatronic System to Prevent Death due to Suffocation in a Locked Car* ,motion sensor and temperature sensor are used for detection, which can not deal with the situation that someone is asleep in the car[1]. In the paper *Child Presence Detection and Alerting System in an Unmanned Car,* the message is sent to adults by detecting the activities in the car, without considering the sleeping people[2].

In addition, the current research is only to detect the issue and wait for rescue, which does not directly solve the problem. In the paper *Design of Anti-suffocation Monitoring System in Vehicle,* Although the system has an executive module, it can only give an alarm when the vehicle has no sunroof, which also has shortcomings[3]. In the paper *Alert System for Child in Car Using NodeMCU,* parents are informed by sending a message that they can remotely control the steering gear to open the window to prevent suffocation, but there are also situations where parents can't receive the message[4].

In order to solve the above problems, the system designed in this paper is composed of detection module, control module and execution module. The system can make corresponding judgment according to the situation, so as to make the window open and close, rather than only detect whether there are people without solving problems. For safety and low power consumption, the system adopts dual MCU design. MCU 1 controls the working mode of MCU 2. Only when the interior environment meets the preliminary preset conditions, MCU 1 will wake up MCU 2 through the serial port, perform further detection and output the corresponding results. The system can effectively reduce accidents.

2. System Design Scheme

The system consists of detection module, main control module and execution module. The three parts are connected through control bus,data bus . The detection module is composed of carbon dioxide sensor, temperature sensor, sound sensor and camera, the control module is composed of MCU 1 and MCU 2 and their peripheral circuits, and the executive part is composed of relay and motor. The detection module can be installed at any position in the vehicle, and the execution module needs to be installed inside the car door . The opening and closing of the car window can be achieved by controlling the opening and closing of the corresponding relay through the control signal sent by the control module.

The system designed in this paper takes dual MCU as the control core and sensors for in car environment detection. MCU 1 is a low-power processor. MCU 1 and its peripheral circuit are mainly responsible for the detection of in car temperature and carbon dioxide content, and has been in operation state during the operation of the system; MCU 2 and its peripheral circuit are responsible for the detection of people left in the vehicle. MCU 2 communicates with MCU 1 through the serial port. During the system operation, MCU 2 is in the sleep state. When the preset conditions are met, MCU 1 wakes up MCU 2 through the serial port to release the sleep state of MCU 2. After MCU 2 enters the working state, it carries out the detection of people left in the vehicle, and outputs corresponding instructions in combination with the data detected by MCU 1, Make the relay perform corresponding action. This design can not only meet the normal function, but also achieve low power consumption design. The overall design structure of the system is shown in the figure.

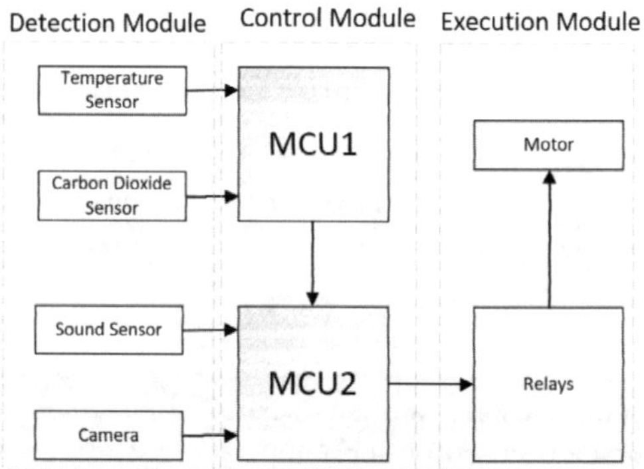

Figure 1. System design scheme

3. Hardware Design

3.1.Environment Monitoring Module

The interior environment monitoring module consists of temperature sensor, carbon dioxide sensor, sound sensor and camera. The temperature sensor and carbon dioxide sensor are respectively connected with MCU 1, and the sound sensor and camera are connected with MCU2[5].

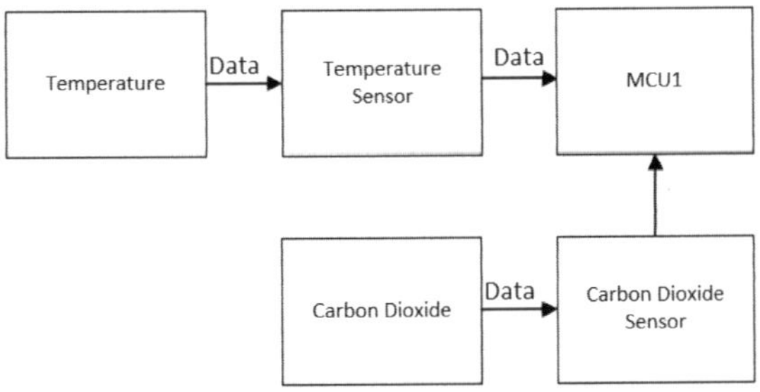

Figure 2. Environmental monitoring scheme

*1) Temperature sensor:*Temperature sensor is a sensor that can sense the temperature change and convert the temperature into output voltage signal.

The temperature sensor used in this paper is DS18B20 digital temperature sensor. The measuring temperature range of DS18B20 sensor is between minus 55 ℃ - 125 ℃, and the sensor has high precision, low energy consumption, high stability, strong anti-electromagnetic interference ability and standard digital output[6].

*2) Carbon dioxide sensor:*Mh-z14a carbon dioxide gas sensor is a general intelligent small sensor. It uses the principle of non dispersive infrared to detect CO2 in the air, and has good performance.

With built-in temperature compensation, digital output, analog output and PWM output, it is convenient to use.

The sensor is a high-performance sensor made by combining mature infrared absorption gas detection technology with precise optical path design and excellent circuit design.

3) Sound detection module:Ky-037 sound sensor can detect the sound of the surrounding environment, with adjustable sensitivity, working voltage of 3.3v-5v, and the output form is digital switching output.

4) Camera:OV5640 camera adopts 1.4um × 1.4um pixel size, with high sensitivity, low crosstalk and low noise; automatic image control functions; automatic exposure, automatic white balance, automatic elimination of light fringes, automatic black level calibration, automatic band-pass filtering, etc.; the port supports image quality control: color saturation adjustment, tone adjustment, gamma calibration, sharpness and lens calibration, etc., Flash, auto focus, image zoom, translation and window settings support image compression, JPEG image data can be output, digital video interface and Mipi interface are supported, embedded microprocessor is provided, active crystal oscillator is integrated, LDO is integrated without external clock, and only 3.3V power source is required to work normally[7].

3.2.Main Control Module

The main control module is controlled by dual MCU. MCU 1 selects stm32f103zet6. The central processor of the chip is 32-bit Cortex-M3 based on arm, embedded with 8MHz RC oscillator, and the maximum working frequency is 72MHz. Flash memory between 256 and 512 bytes, 64K byte SRAM, 3 analog-to-digital converters, 3 line peripheral interfaces, 2 single and two-way two-wire synchronous serial bus interfaces, 4 FSMC interfaces, 11 timers, 5 serial ports, 1 controller area network interface and 112 GPIO ports are integrated on the chip[8].

MCU2 selects imx6ull from NXP company, which is a low-power, high-performance and low-cost application processor based on ARM cortex A7 core. The core can reach 900 MHz, parallel 24bit RGB LCD interface, 1366 × 768 resolution, 8 / 10 / 16 bit parallel camera sensor interface and up to 8 UART interfaces.

3.3.Execution Module

The execution module consists of two relays and a motor. The control signal from MCU 2 controls the relay. When the relay is closed, motor M is connected to the power supply[9]. The two relays control the forward and reverse rotation of the motor respectively, so as to control the lifting and lowering of the window.

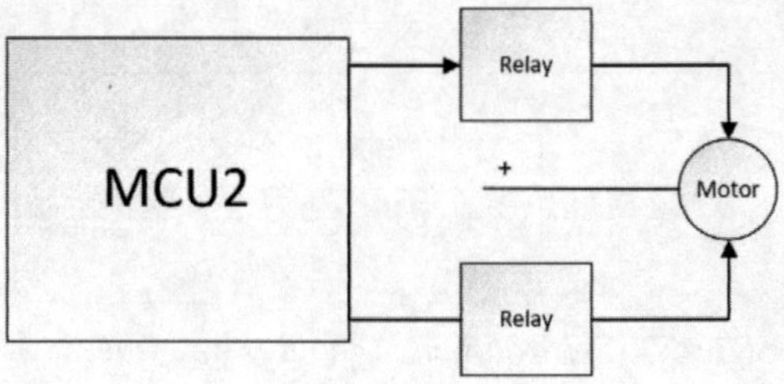

Figure 3. Execution module

4. SOFTWARE DESIGN

In the pictures below,we use gray fill indicates that the carbon dioxide concentration and temperature is out of range .The system based on Dual MCU designed in this paper has 5 working conditions:

Case 1: when the carbon dioxide concentration in the car is lower than 3% and the temperature is lower than 30 ℃, MCU 1 works normally and MCU 2 is in sleep state. At this time, the relay does not act, as shown in Figure 4;

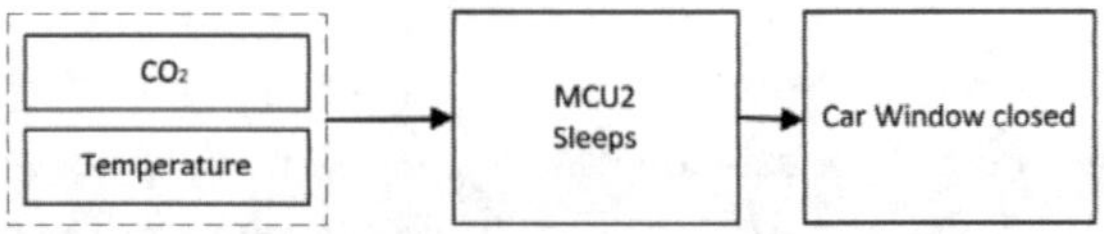

Figure 4. Case 1

Case 2: when the carbon dioxide concentration in the car is lower than 3% and the temperature is higher than 30 ℃, MCU 1 works normally and wakes up MCU 2. When no person is detected, the relay does not act, as shown in Figure 5;

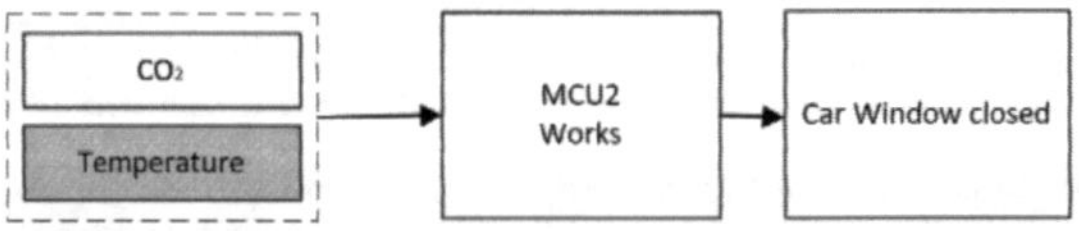

Figure 5. Case 2

Case 3: when the carbon dioxide concentration in the vehicle is lower than 3% and the temperature is higher than 30℃, MCU 1 works normally and wakes up MCU 2. When a person is detected, the relay acts to open the car window one -fifth, as shown in Figure 6;

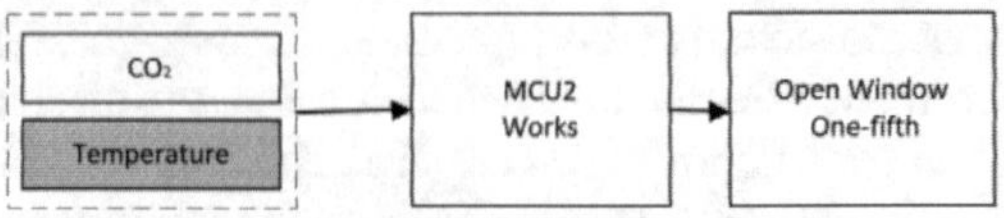

Figure 6. Case 3

Case 4: when the carbon dioxide concentration in the vehicle is higher than 3%, MCU 1 works normally and wakes up MCU 2. When no person is detected, it is also possible that someone is asleep in the vehicle, so the carbon dioxide concentration will. The relay will act to open the car window one -fifth, as shown in Figure 7;

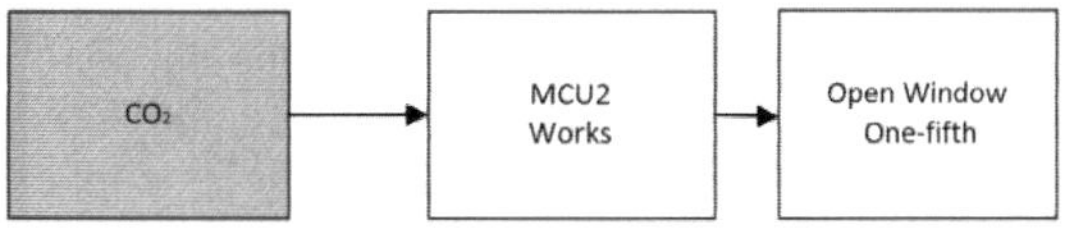

Figure 7. Case 4

Case 5: when the carbon dioxide concentration in the vehicle is higher than 3% and the temperature is lower than 30℃, MCU 1 works normally and wakes up MCU 2. When a person is detected, the relay works to make the performs action to half open the car window, as shown in Figure 8;

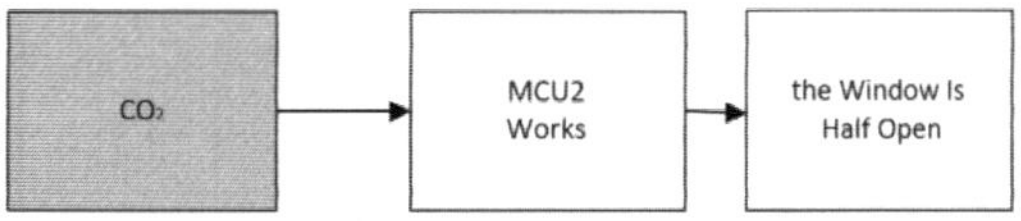

Figure 8. Case 5

The system also adds a feedback function. In the above cases, when the window is opened, if the temperature and carbon dioxide concentration in the car do not tend to the normal value, the MCU 2 control relay continues to make corresponding actions to make the temperature and carbon dioxide concentration tend to the normal value.

5. Conclusion

The system designed in this paper relates to the field of automobile safety. It mainly prevents the occurrence of high temperature and suffocation accidents caused by children left in the car and adults sleeping in the car.

The system uses dual MCU design and MCU 1 low-power chip as the main control to control the work and sleep of MCU 2, so as to save battery power. Assuming that the car battery is 12V and 60ah, the working voltage of MCU 1 is 3.3V and the normal working current is 20mA, so the working time is 454 days. Therefore, the battery consumption is very small, which can completely avoid the battery loss caused by working during parking. Most of the related similar products use human body infrared sensors, gas sensors and other means to detect the personnel in the car, which requires a process, which may cause harm to the human body, and it is impossible to accurately judge whether the detained personnel in the vehicle are only children. Therefore, the system is more efficient and safe, and the combination of the system and the tachograph which has a wide market foundation makes it more convenient and practical, and has high popularization and application value. On the premise of wide application of the system, it can be expected that the accidents of children in the car due to asphyxia and dehydration will be improved.

References

[1] Dodderi M D N, Raghavendra R S, Hegde N, et al. A Mechatronic System to Prevent Death due to Suffocation in a Locked Car[J]. IRA-International Journal of Technology & Engineering, 2016, 4(3): 126-134.
[2] Bharadwaj S, Raashi A S, Prajwala A N, et al. Child Presence Detection and Alerting System in an Unmanned Car[J]. 2021.
[3] Wang Z, Wang S, Shi Z. Design of Anti-suffocation Monitoring System in Vehicle[C]//Journal of Physics: Conference Series. IOP Publishing, 2021, 1744(2): 022110.

[4] Asben D, Roslan I, Zakaria N A, et al. Alert System for Child in Car Using NodeMCU[C]//Proceedings of the 3rd International Conference on Intelligent and Interactive Computing 2021 (UTeM Press). UTeM Press, 2021: 116.
[5] STM32 ARM[J]. Nuts & volts,2012,33(7):28-29.
[6] NING XU, YONG REN. Design of a Pedestrian Detection System Based on OpenCV[C]. //International conference on education, management, commerce and society: EMCS 2015, Shenyang, China, 29-31 January 2015.:Curran Associates, Inc., 2015:66-70.
[7] CHEN JIN, WANG HUILING, QIN QIN, et al. The Development of a Performance Test Equipment of Embed Process Control System[C]. //Proceedings of 2009 international conference on information, electronic and computer science. vol. 2.:World Academic Union (World Academic Press), 2009:582-585.
[8] SUDHA R. RAMAN, MICHEL D. LANDRY, C. ANDREA OTTENSMEYER, et al. Keeping our children safe in motor vehicles: knowledge, attitudes and practice among parents in Kuwait regarding child car safety[J]. International journal of injury control and safety promotion,2013,20(4):358-367.
[9] YU JIA.SUN, XIAO MING.WANG, FANG XIU.JIA, et al. Design of a Wireless Sensor Network Node Based on STM32[J]. Applied Mechanics and Materials,2013,2560(697):1920-1923. DOI:10.4028/www.scientific.net/AMM.347-350.1920.

Application of Information Processing System Based on DSP in Electronic Information Engineering

Jingbo Huang*

School of Physics, Changchun University of Science and Technology, Changchun, 130022, China

*Corresponding author's e-mail: jhuang19@students.desu.edu

Abstract. On the one hand, target tracking technology is widely used in navigation, guidance, alarm and monitoring and other military fields, on the other hand, it is also widely used in civil fields such as traffic information monitoring and security. This paper mainly studies the application of information processing system based on DSP in electronic information engineering. This paper introduces DSP information processing system in detail. Aiming at the comprehensive requirements of real-time and tracking performance of the target tracking system, multi-feature fusion optimization method is adopted to improve the particle filter tracking algorithm to improve the accuracy of the tracking algorithm. In this paper, the simulation results show that the improved algorithm has a certain scale and occlusion robustness, and can adapt to long-term stable real-time tracking.

1. Introduction

Image tracking system is through a variety of sensors such as visible TV, infrared thermal imager, video image sequence, and then automatically extract and track the target, in the military field, image tracking system in missile, aircraft and other target early warning, precision guided weapons, fire control system plays a vital role. The image tracking system first captures the target from the acquired video image sequence containing the target, then calculates the coordinates of the target in the image, and sends the coordinate information to the tracking servo system to complete the target tracking function [1]. In the civil field, the image tracking system can target track and locate the vehicles on the specific road in the real-time monitoring of traffic road, and can monitor the road traffic condition in real time, and provide available information for the rapid transportation of congested road traffic. Play a key role in forest monitoring, through the establishment of data collection, data management, flexible monitoring, forecast and prediction, spread analysis, location tracking, identification and alarm, fire command integration platform, timely and accurate forecast of forest fire and other disasters. Due to the lack of bus bandwidth and processing speed, the application of DSP processor is limited to signal control and frequency domain analysis. In recent years, with the development of process manufacturing technology, the bandwidth and performance of DSP processor have been significantly enhanced. DSP based system has gradually entered into more complex video processing field, and embedded video processing system has been rapidly developed [2-3].

In the history of target tracking, tracking models can be roughly divided into three types, namely generative model, discriminant model and other models. Since the CORRELATION filter was introduced into the tracking field by THE MOSSE tracker, the tracker based on correlation filter was once concerned by researchers due to its advantages of good accuracy and fast speed, and gradually became the mainstream of discriminant model research [4]. KCF obtains a conclusion similar to

MOSSE from the point of view of training linear classifier with cyclic samples, and introduces kernel method and multi-channel feature to improve tracking performance significantly. In KCF, the use of multi-channel features can significantly improve the tracking performance more than the core skills, but the multi-channel features will increase the computational complexity and reduce the real-time performance of the tracker [5]. DSST constructs multi-scale space on the basis of KCF and searches for the best scale to improve the scale adaptability of tracker. In DSST, the construction of multi-scale space is realized by creating image pyramids. The more layers of the pyramid, the higher the accuracy of scale prediction and the greater the amount of calculation [6].

With the continuous improvement of microprocessor chips and computer computing power, target tracking algorithm has been gradually developed, and further promote the development of computer vision (CV), machine learning and deep learning, and further deepen the application of target tracking algorithm.

2. Target Tracking Based on DSP

2.1. DSP Information Processing System

(1) System structure and characteristics

In recent years, with the more and more in-depth research on DSP, DSP chip stability, speed of operation, integration and other aspects of performance have been greatly improved, and DSP chip is widely used in digital signal processing. At present, DSP is increasingly tending to real-time signal processing, and its performance in video and image processing is getting better and better [7-8].

The features of DSP hardware system include:

The kernel of DSP is a super-long instruction word structure with 8 functional units and 64*32 bit general purpose register. The execution speed of instructions can reach 8 instructions per clock cycle, and its operation speed is 4800MIPS.

DM642 uses two levels of super high speed buffer memory, the memory performance is high, the data can meet the requirements of high speed DSP core.

In order to improve the efficiency of instruction set, DSP system adds an instruction, which can directly package the data, so as to achieve seamless data flow.

The CORE CPU of the DM642 chip uses VelociTI Advanced Ultra-long instruction Word technology, which is faster than other chips in the series and can process up to eight 32-bit instructions simultaneously. The CPU of the chip contains two groups of functional units, each of which is composed of four small functional units and a register. The DM642 not only has fixed-point instructions, but also includes many 8-bit, 16-bit extended instruction sets. As a result, DM642 works efficiently. The loading memory structure of chip DM642 is a very key technology, in the loading memory structure, the instruction register is completely different from the memory register.

(2) Overall process

In the research on the recognition and tracking of target images, we need to carry out pre-processing, image segmentation, target recognition and target tracking algorithms in sequence [9]. In this system, the selection of DM642 chip, video signal acquisition, storage, analysis, signal output processing.

The CCD camera transmits the moving target image signal to the video decoder, converts the analog signal of the image into digital signal and then sends it to DSP. The digital signal is stored in SDRAM by VP port of DM642 chip. DM642 chip processes the video digital signal by recognition and tracking algorithm. The processed digital signals are converted into analog signals by the video encoder, and finally output in the form of images on the display.

As for the construction of DSP hardware platform, the parameters of each indicator are defined according to the performance indicators required by the system, and then the appropriate DSP chip is selected, the program is written and burned, and the debugging program is run at last.

2.2. Optimization and Implementation of Tracking Based on DSP

In the particle filter-based target tracking algorithm, the prior knowledge of a single feature to describe the target usually has limitations. For example, the texture feature of the image is filtered out when the histogram information reflects the grayscale feature of the target [10]. There is a large error in comparing the similarity between the sample particle and the target template by using a single feature, and the inaccurate updating of the particle weight leads to the phenomenon of tracking drift.

(1) Gray features

The comparison method of gray features is based on the pixel information in the target region and the candidate sample, and the formula is shown below.

$$w^T = \frac{\sum_{i=0}^{M}\sum_{j=0}^{N}[f(i,j)\cdot g(i,j)]}{\sqrt{\sum_{i=0}^{M}\sum_{j=0}^{N}[f(i,j)]^2}\cdot\sqrt{\sum_{i=0}^{M}\sum_{j=0}^{N}[g(i,j)]^2}} \tag{1}$$

In the formula, wT represents normalized similarity, while M and N represent the width and height of the image. Similarity measurement by comparison of pixel regions can make full use of the information in the image, make the correlation peak of the target and sample more sharp, solve the problem of non-convergence in similarity comparison, and better improve the tracking accuracy and robustness.

(2) Histogram features

As a common feature in images, the histogram feature is a simple and efficient digital image description feature, which has been widely applied in the field of machine vision [11]. First of all, the calculation of histogram features is efficient, and it is very fast to obtain histogram information by using target template. Secondly, the feature of the histogram is to map the 2d image digital information into the 1d model by using the statistical method, which reduces the dimension of comparison, extracts the image features, and has a certain rotational robustness to the target template. The commonly used gray histogram statistics formula is as follows:

$$H(i) = \frac{n_i}{N} \quad i = 0,1,\ldots L-1 \tag{2}$$

Where, i represents gray level, ni represents the number of pixels whose gray level is I, N represents the total number of pixels, and L represents the total gray level type, which is generally 255.

(3) DSP algorithm optimization

In the implementation of the algorithm based on the DSP processor, the program needs to be optimized to ensure the correct operation of the algorithm and the maximum efficiency of the hardware computing resources. The particle filter algorithm in this paper is first developed on the computer using OpenCV environment to ensure the correct realization of the logic, and then the algorithm is implemented in THE DSP processor [12]. The DSP processor used in this paper is high performance TMS320C6657 processor produced by TI Company. The algorithm optimization goes through three stages, and the timing function is calculated through the TSCH/TSCL register.

EMCV is a simplified version of OpenCV library functions transplanted in embedded processor, which has built-in interface and class definition of many functions in OpenCV1.0 version. This stage is used for algorithm transplantation, and the program code is basically unchanged, which ensures the logical correctness of algorithm transplantation.

In the last stage, using the EMCV library migration algorithm, the program target tracking results are correct, but the single frame processing time is long, which can not meet the real-time demand of processing time, and the program code is optimized at the language level.

Language level optimization, the first use of C language code to adapt the program, out of OpenCV library function support. Second, use operations that cost less to replace operations that cost more, and use shift operations to replace multiplication and division operations. After that, the functions that require less computation are modified to inline functions, such as size comparison functions, using inline functions. Finally, you can use the DSP processor support library VLIB to implement the barbardian distance calculation with library function calls.

In THE DSP processor, the HAver structure memory is used in the DSP, and the three-level memory memory mode is set, that is, L1, L2 and DDR3 memory architecture. In this article, the program is placed in THE L2 cache area, and the compiler is optimized at -3.

3. System Test Simulation Experiment

3.1. Experimental Scheme

In order to evaluate the tracking algorithm proposed in this paper, the performance of the algorithm is evaluated experimentally. Firstly, the algorithm was tested in THE DSP development environment, and the tracking performance of the algorithm was verified in the laboratory environment from the perspective of intuitive and qualitative.

The ideal test environment for this experiment is to embed the development system into the moving platform and carry out tracking and verification on the moving platform. However, due to the high cost of moving platform equipment, this test was completed in the laboratory simulation test. The camera used in this test was black and white, so the video collected was in black and white mode.

3.2. Algorithm Comparison

In this paper, the experimental data set Carscale and David of OTB-100 and the data set Pedestrian algorithm in VOT-2013 are selected to test the tracking effect. Clutter, partial occlusion and object deformation exist in the Data set of Carscale, while the data set of David is mainly influenced by illumination. Pedestrian1 scored background wobble, and scored MF and KCF respectively.

4. Simulation Results

4.1. Video Frame Recognition

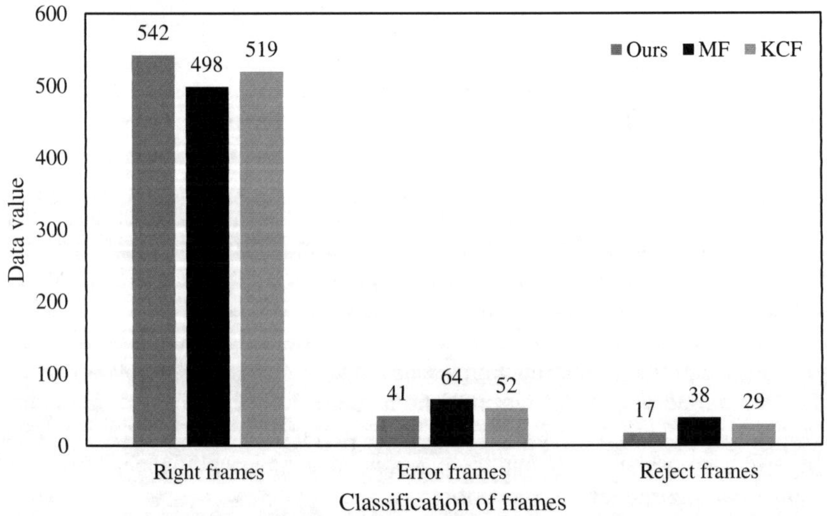

Figure1.Video frame recognition accuracy

As shown in Figure 1, the improved algorithm proposed in this paper can correctly identify 542 frames, incorrectly identify 41 frames, and reject 17 frames. The correct identification frame number of MF algorithm is 498, the wrong identification frame number is 64, and the rejected identification frame number is 38. The KCF algorithm correctly identifies 519 frames, incorrectly identifies 52 frames, and rejects 29 frames. The above results show that the algorithm proposed in this paper has the highest accuracy in video frame recognition, and the recognition rate reaches 90%, which meets the performance requirements of the algorithm.

4.2. Tracking Success Rate of Different Data Set Algorithms

Table1. Success rate table of tracking results of different algorithms

	Carscale	David	Pedestrian
Ours	93	44	35
PF	84	42	21
KCF	91	98	26

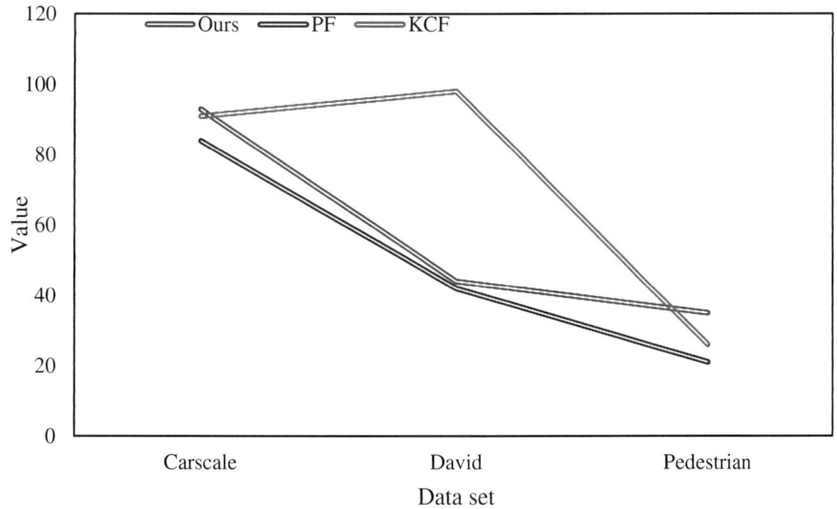

Figure 2. The success rate of tracking results of different algorithms

As shown in Table 1 and Figure 2, on Carscale data set, the success rate of the improved algorithm proposed in this paper reaches 93%, which is better than PF algorithm (84%) and KCF algorithm (91%). On David data set, the success rate of the improved algorithm proposed in this paper is only 44%, far behind the 98% success rate of the KCF algorithm. On Pedestrian data set, the success rate of the algorithm in this paper reaches 35%, leading PF algorithm 14%, and KCF algorithm 9%. The improved algorithm optimizes the target scale changes, and the tracking box can be scaled appropriately to adapt to the target scale, which reduces the center position offset error and significantly improves the success rate on Pedestrian data set. At the same time, the algorithm has good processing speed, can run in real time and adapt to DSP engineering applications.

5. Conclusions

Target tracking technology has been widely used in military fields such as navigation, guidance, alarm and monitoring, and has great potential in civil fields such as vehicle autonomous driving and robot vision. This paper focuses on the software design of HIGH definition image tracking system based on DSP, and designs the frequency domain tracking and correlation tracking algorithm, which can realize the target stable tracking function under complex background. The experiment shows that the improved tracking algorithm can run in real time and adapt to DSP engineering application.

References

[1] Sobhani B, Paolini E, Giorgetti A, et al. Target Tracking for UWB Multistatic Radar Sensor Networks[J]. IEEE Journal of Selected Topics in Signal Processing, 2017, 8(1):125-136.

[2] Perach B, Weiss S. SiMT-DSP: A Massively Multithreaded DSP Architecture[J]. IEEE Transactions on Very Large Scale Integration (VLSI) Systems, 2018, PP(99):1-14.

[3] Li J, Choutko V, Xiao L. Single Event Upset Analysis: On-orbit performance of the Alpha Magnetic Spectrometer Digital Signal Processor Memory aboard the International Space Station[J]. Nuclear Instruments & Methods in Physics Research, 2018, 885(MAR.21):98-104.

[4] Meyer F, Kropfreiter T, Williams J L, et al. Message Passing Algorithms for Scalable Multitarget Tracking[J]. Proceedings of the IEEE, 2018, 106(2):221-259.

[5] Khakpour S, Pazzi R W, El-Khatib K. Using clustering for target tracking in vehicular ad hoc networks [J]. Vehicular Communications, 2017, 9(jul.):83-96.

[6] Shojaei K, Dolatshahi M. Line-of-sight target tracking control of underactuated autonomous underwater vehicles[J]. Ocean Engineering, 2017, 133(MAR.15):244-252.

[7] Bahri N, Grandpierre T, Ayed M, et al. Embedded Real-Time H264/AVC High Definition Video Encoder on TI's KeyStone Multicore DSP[J]. Journal of Signal Processing Systems, 2017, 86(1):67-84.

[8] Xiu-Ling Z, Wu-Yang G, Yong-Jin L , et al. Flatness predictive model based on T-S cloud reasoning network implemented by DSP[J]. Journal of Central South University, 2017, 24(010): 2222-2230.

[9] Kong L, Zhu L, Zhang L, et al. Real-Time Controller Based on FPGA and DSP for Solar Ground Layer Adaptive Optics Prototype System at 1-m NVST[J]. IEEE Photonics Journal, 2017, PP(2):1-1.

[10] Bargshady N, Garza G, Pahlavan K. Precise Tracking of Things via Hybrid 3-D Fingerprint Database and Kernel Method Particle Filter[J]. ieee sensors journal, 2017, 16(24):8963-8971.

[11] Wang T. Unsupervised video multi-target tracking based on fast resampling particle filter[J]. Journal of supercomputing, 2020, 76(2):1293-1304.

[12] Liu Z, J Shang, Hua X. Smart City Moving Target Tracking Algorithm Based on Quantum Genetic and Particle Filter[J]. Wireless Communications and Mobile Computing, 2020, 2020(8):1-9.

Journal of Physics: Conference Series **2209** (2022) 012011 doi:10.1088/1742-6596/2209/1/012011

Analysis and Research on the Interference of Civil Aviation Radio Navigation Equipment

Xianhao Zhang[1*], Zhongmei Luo[1], Guo Kang[1]

[1]Civi Aviation Flight University of China,Guanghan,Sichuan,618307,China

[*]Corresponding author's e-mail: xianhaozhang@cafuc.edu.cn

Abstract—Aiming at the problem that civil aviation radio navigation equipment is susceptible to interference more and more frequently, through the study of aviation radio interference theory, the categories of civil aviation radio interference are classified, and then the source and harm of radio interference are analyzed, and specific measures to solve the interference problem are proposed. This is of great help to solve the difficult problem of radio interference signals, and is conducive to reducing the probability of civil aviation radio navigation equipment being interfered, thereby ensuring aviation flight safety.

1. Introduction

After decades of development, my country's civil aviation industry has continued to grow and has now become the world's leading civil aviation country. Its passenger volume, cargo volume and flight traffic are all showing a rapid growth trend. In 2019, the entire industry completed a total transportation turnover of 129.325 billion ton-kilometers, an increase of 7.2% over the previous year. The total transportation turnover of domestic routes was 82.951 billion ton-kilometers, an increase of 7.5% over the previous year. Among them, the Hong Kong, Macao and Taiwan routes completed 1.690 billion ton-kilometers, down 3.5% from the previous year; the total transportation turnover of international routes was 46.374 billion ton-kilometers, an increase of 6.6 %[1].

With the continuous development of radio technology, broadband China, the "Internet +" action plan, "Made in China 2025" and other major national strategies have been implemented one after another. The application of radio technology in broadband communications, Internet of Things, intelligent manufacturing and other fields is expanding. The research and development of the fifth generation mobile communication (5G) system has been carried out in full swing around the world. Various new technologies, services, and equipment are emerging in an endless stream and rapidly popularized, bringing people a convenient and intelligent life experience. The electromagnetic environment is becoming increasingly complex. In addition, some market supervision is not in place, and some "three noes" products and radio equipment assembled and used by criminals have also entered the society, resulting in frequent other radio interferences under the condition of crowded spectrum resources. Interference to the dedicated radio frequency of civil aviation may lead to flight flow control, and aircraft cannot take off and arrive normally; at worst, it may cause the failure of civil aviation ground-to-air communication equipment and navigation equipment, causing the pilot to fail to hear the controller and execute wrong instructions and flight. Deviations in the trajectory, etc[2][3].

2. Civil Aviation Radio Interference Theory

2.1 Civil aviation radio navigation equipment

Civil aviation radio navigation equipment mainly includes non-directional beacons, VHF omnidirectional beacons, rangefinders and instrument landing systems. The main frequencies used by civil aviation are: 190kHz-1750 kHz, 108MHz-136.975 MHz, 328.6MHz-335.4 MHz, 960MHz-1215 MHz, 1030MHz-1090 MHz, 1250MHz-1350 MHz, 2700MHz-2900 MHz, 9340MHz-9400 MHz, etc. The frequency dedicated to civil aviation is mainly line-of-sight propagation, and direct wave propagation is the main propagation process. When affected by terrain conditions, it will also have the characteristics of reflection, scattering, and diffraction.

2.2 Definition of radio interference

Radio interference refers to the effect of useless energy generated by a certain emission, radiation, induction or their combination on the reception of radio communication systems, which reduces the performance of the reception effect or fails to receive signals. This effect is called radio interference.

According to the nature of the interference source, it is divided into active interference and passive interference. Passive interference refers to the radio waves formed by the reflection of the ionosphere, the diffraction of obstacles, the reflection of reflectors, and the re-radiation (secondary radiation) of the re-radiator (secondary radiation) during the propagation of electromagnetic waves emitted by the ground station transmitter. Wave, the interference caused by receiving the useful radio signal of this transmitter. Active interference refers to the amount of radio frequency emitted, radiated, induced or leaked by the radiation source of non-aviation navigation equipment or facilities through antennas, feeders, power lines, control lines, housings, output terminals, etc., through conduction or radiation. Interference caused by receiving useful radio signals.

2.3 Types of radio interference

2.3.1 Co-channel interference

Any interference that is sent by other signal sources at the same frequency as the useful signal and enters the receiver's intermediate frequency passband in the same way is called co-channel interference [4] [5]. In the mobile communication system, in order to improve the frequency utilization, the same frequency channel must be used repeatedly after a certain distance. This method is often called co-channel reuse. The problem caused by co-channel reuse is co-channel interference. The closer the reuse distance, the greater the co-channel interference, and the higher the frequency utilization; the farther the reuse distance, the smaller the co-channel interference and the lower the frequency utilization. Since the same frequency interference signal and the useful signal are also amplified and detected, when the two signals have carrier frequency difference, it will cause beat interference; when the modulation degree of the two signals is different, it will cause distortion interference; when two signals exist The phase difference can also cause distortion interference. The larger the interference signal, the smaller the output signal-to-noise ratio of the receiver. When the interference signal is large enough, it can cause blocking interference of the receiver[6][7].

2.3.2 Adjacent channel interference

The so-called adjacent channel interference is the interference caused by signals in or near the radio frequency passband of the receiver that fall into the intermediate frequency passband after frequency conversion. Adjacent channel interference will reduce the signal-to-noise ratio and sensitivity of the receiver; strong interference signals can cause blocking interference in the receiver. When a communication network is composed of many radio stations, the problem of adjacent channel interference is often prone to occur. This kind of interference is mostly caused by the technical indicators of radio equipment that do not meet national standards. In terms of transmitters, if the frequency stability is too poor or the modulation is too large, the transmission spectrum is too wide, which can cause adjacent channel interference to other radio stations. If the factors that affect the bandwidth of the transmitter are not strictly controlled, it is easy to generate unnecessary out-of-band radiation; in the receiver, when the IF filter is poorly selected, it is easy to cause interference or make the interference serious.

2.3.3 Intermodulation interference

When two or more signals of different frequencies are added to a non-linear device, the non-linear transformation of the non-linear device can generate many combined frequency signals, and some of the combined frequency signals may fall into the receiver. In-band, this forms intermodulation interference. According to the intermodulation location, intermodulation interference can be divided into the following types: transmitter intermodulation interference, receiver intermodulation interference, and intermodulation interference between transmission circuits and communication systems. There are three conditions for the generation of intermodulation interference: first, a fixed frequency combination relationship; second, the interference is sufficiently large; third, there are related stations working at the same time. Technically speaking, it is quite difficult to analyze and find intermodulation interference, and the impact it causes is relatively large, and the damage it poses is also great.

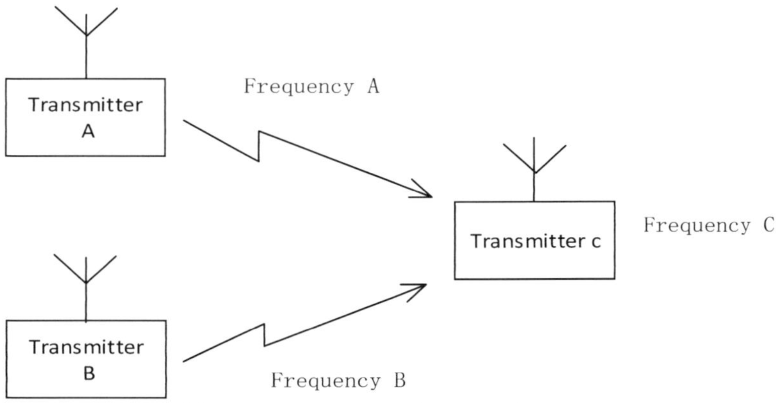

Figure 1 Intermodulation interference model

2.3.4 Blocking interference

When there is a strong interference signal in the outside world, although the frequency does not cause intermodulation or co-frequency or adjacent frequency interference, it can still cause useful signal gain due to the nonlinearity of the receiver after it acts on the front-end circuit of the receiver. Decrease (suppressed) or increase the noise, so that the sensitivity of the receiver decreases. This phenomenon is the blocking of the receiver. This interference is called blocking interference.

2.3.5 Bandwidth interference

When the useless power signal sent by the transmitter outside the adjacent frequency band is received by the receiver in the normal working frequency band, the interference that causes the decrease of the system signal-to-noise ratio is called out-of-band interference. Usually can be divided into transmitter spurious frequency interference and receiver spurious frequency interference. A general VHF transmitter uses a crystal oscillator to generate a basic frequency, and after multiple frequency multiplications, a carrier frequency is generated. Due to the non-linear characteristics of frequency multipliers and high-frequency amplifiers, various harmonic components will be generated. Intermediate frequency signals are formed through the combined frequency action. These harmonic components are amplified and emitted to form useless interference signals. The receiver caused interference.

3. Civil Aviation Radio Interference Analysis

3.1 Sources of interference to radio navigation equipment

3.1.1 Illegal installation of radio stations

With the continuous development of social economy, in order to pursue convenience, a large number of cordless phones have been used in the society. In order to improve the efficiency of use, some manufacturers or users will increase their rated power. If the vertical phone is near the airport during the application process , It will cause telephone voice interference to terrestrial stations or airplanes. Especially in the economically underdeveloped areas, especially in the mountainous areas or islands with inconvenient traffic, the telecommunications department may complete the transmission of voice signals through wireless differential transfer in order to ensure communication safety, which may cause interference.

3.1.2 Interference from high-power radio stations

High-power radio stations mainly refer to the broadcasting stations set up by the radio and television departments. Broadcasting stations include voice broadcasting stations (mainly FM broadcasting) and TV broadcasting stations. The transmission power is very large, easily tens of kilowatts, or even several kilowatts; the antenna height is high, and it is mostly set in the "three high" locations (high mountains, high buildings, off towers) , And most of the antennas share a tower, which is very easy to cause intermodulation interference; and the transmission frequency band of the FM broadcasting service is 87MHz-108MHz, which is adjacent to the communication and navigation frequency band 108MHz-137MHz, which is a dedicated frequency band for civil aviation radio. Due to the limited frequency resources of radio stations, coupled with the vigorous development of television stations and radio stations in various regions for the needs of publicity work, the radio and television frequencies in various regions have been expanding upwards, and the rapid development of the aviation industry has also led to the downward expansion of civil aviation frequencies. As a result, too many stations are set up in a limited frequency band. During use, similar frequencies interfere with each other. In particular, the fast-developing and large-scale broadcasting and television services can easily interfere with civil aviation services [8][9].

3.1.3 Interference caused by cable TV signal leakage

The carrier of the cable TV programs we usually watch is transmitted through the cable TV system, and some carriers have already occupied the frequency band of civil aviation communication and navigation during use, as shown in Table 1. During the transmission process, the leakage of radio frequency energy will cause interference, and there will be broadcast voice in the frequency band of civil aviation communication and navigation. Through the receiver, the interference signal and the useful signal are equally amplified, detected and output. When the strength of the interference signal reaches a certain limit, it will cause the blocking interference of the receiver.

Table 1. Comparison of frequency bands used by cable TV and aviation

Cable Supplementary Channel	Frequency Range	aviation
Z1	111-118.75MHz	VOR,LOC
Z2	119-126.75MHz	Civil aviation distress, airport scene communication
Z3	127-134.75MHz	Civil Aviation Mobile Business
Z4	134-142.75MHz	Civil Aviation Mobile Business

3.1.4 High-voltage power transmission line interference

High-voltage power transmission line interference is also one of the main interference factors. The corona effect of the power transmission system and the radio noise caused by gap discharge are affected by electromagnetic waves. In practical applications, high-voltage transmission lines, as tall metal objects, will reflect and re-radiate radio navigation signals during the transmission process. Some harmful and wide-band electromagnetic waves will appear during the corona process, and the generated wavebands will change the navigation. The air field pattern of the signal causes passive interference to the aviation system.

3.1.5 Interference caused by internal equipment of civil aviation

In order to ensure the safety of airport flight transportation, the airport and various communication stations set up at the airport need to be equipped with a large number of electronic equipment, and these equipment will generate electromagnetic radiation in the application. The interior of the airport is a complex electromagnetic radiation system, so it must be completely It is impossible to eliminate electromagnetic radiation in the airport, but the interference caused by reasonable application is limited. If it cannot be applied reasonably and scientifically, the electromagnetic compatibility problems generated by it are not handled well, and the use of equipment is improperly managed, which may cause damage to civil aviation. The dedicated radio frequency causes harmful interference and endangers flight safety.

3.2 Harm of interference

Radio interference can affect flight safety through many links, causing direct or indirect harm. The adverse effects caused by this interference can be roughly divided into three situations:

(1) The impact on the work of the ground navigation station. This interference may cause the flying group hovering in the air to receive ground error instructions, which may lead to the possibility of a crash due to an incorrect judgment. Airport navigation systems mostly work in the 117MHz-140MHz range, which is the most easily used frequency band for other communication systems. If other communication systems interfere with the navigation in the terrestrial space, the flight crew may get the wrong indication signal, which will cause the flight trajectory to deviate, which is extremely harmful.

(2) Impact on ground-to-air communications and air traffic control. At present, the communication system used in civil aviation ground-to-air communication makes it susceptible to interference from radio signals. It generally works on VHF and short-band ordinary AM signals and single-sideband signals. Its communication methods are analog signals and voice. Communication, coupled with high flight altitude, long receiving communication distance, and high sensitivity of the aircraft. Such interference signals can easily interfere with ground-to-air communications and affect normal controlled calls. In severe cases, the quality of ground navigation and aircraft communication is extremely poor, and normal instructions cannot be obtained, which can lead to serious consequences. Under normal circumstances, pilots who are not disturbed can correctly talk to the navigation station to obtain information. When there is serious interference, the noise in the channel is too large, and the pilot can't hear the controller's commands and misjudges the descending altitude of his own aircraft. Because of the trouble of air traffic control, the serious ones are in danger of crashing, especially when the plane takes off and landing.

(3) The impact on the airborne radio instrument and signal transmission system. Such interference can cause the pilot to operate incorrectly and thus make the flight dangerous. Many airborne equipment is for automatic navigation and to improve the aircraft's take-off and landing capabilities, such as: automatic directional receivers, ranging receivers, VHF omnidirectional beacon and instrument landing receivers. Their working frequency range is medium wave to very high. Frequency bands, some are in the UHF and microwave range. After receiving the interference, the pilot of the aircraft makes wrong judgments and operations, which puts the flight in danger.

4. Conclusion

Through the above research, we can summarize some specific measures to avoid interference:

(1) Antenna isolation

In order to avoid interference between different systems affecting the communication quality, it is generally required that the coupling loss between the transmitting and receiving antennas of different systems is greater than the minimum threshold that will cause inter-system interference, and the coupling loss is the isolation [10][11].

(2) Add filter and anti-jammer

The anti-interference approach is generally to add a filter in the equipment, which is used to suppress high-frequency interference and suppress equipment interference. The former suppresses the influence on the equipment in the alternating current system, and the latter suppresses the interference of the equipment such as high-frequency switches on the alternating current system.

(3) Policy measures

The government and relevant radio management departments formulate relevant laws and regulations to regulate and restrict people's disturbance and illegal use of frequencies. Taking into account the electromagnetic environment, radio stations at all levels operate normally, requiring all units and individuals that use radio stations to consciously Restrict one's own abuse of radio, and formulate some tough rewards and punishments.

Acknowledgment

The main work of this paper is a part of the project "electromagnetic environment compatibility assessment system of civil aviation airport and aviation station" (2019YFG0301), which is a Key Research and Development Project of Sichuan Provincial Science and Technology Department.

References

[1] Civil Aviation Administration of China.(2020)Civil Aviation Industry Development Statistical Bulletin.Beijing.
[2] Peng,C.(2016)Research on the Interference of Civil Aviation VHF Ground-to-Air Communication.J. Sci. Commun.,000(011):113-114.
[3] Xie,Y.X.(2016)Analysis of civil aviation VHF ground-to-air communication interference problem.J. Sci. Commun.,4:219-220.
[4] Yang,T.(2018)Interference Types of Civil Aviation VHF Radio.J. Sci. Commun.,000(012):77-78.
[5] Ma,W.H.(2018)Analysis and research on anti-interference of VHF radio. Dalian University of Technology.
[6] Li,Y.G.(2014)Types of Civil Aviation Radio Interference Sources and Analysis of Investigation and Treatment.J. Sci. Commun.,13:101-102.
[7] Zhang,M.H.(2019)Research on Aerial Monitoring and Analysis of Civil Aviation Radio Interference Based on UAV.Civil Aviation Flight University of China.
[8] Li,Z.R.(2019)Analysis and testing of radio interference in civil aviation ground-to-air communications.J. Sci. Commun., 21(13): 36.
[9] Hao,T.Y.,Hu,S.C.(2019)The influence of radio interference on aircraft and ground navigation equipment and its cause analysis.J. Sci. Commun.,4(15):100-101.
[10] Shuai,J.L.(2020)Discussion on the characteristics and countermeasures of civil aviation radio interference.J. Sci. Commun.,6:55-56.
[11] Xue,C.J.(2016)Calculation Analysis and Management Countermeasures of Civil Aviation Radio Interference.Mongolian university.

Lightweight IT Operation and Maintenance Integrated Monitoring Method for APP System

Xuyong Wang [1*], Rui Chen[1]

[1] Shenzhen Power Supply Co., Ltd., Shenzhen, China

[*]Corresponding author's e-mail: wangxuyong@sz.csg.cn

Abstract. Through the operation and maintenance system, the centralized control of the operation state of enterprise equipment can be realized. In order to solve the problem of unstable monitoring effect of traditional methods, untimely risk early warning and high error rate of early warning information, a lightweight it operation and maintenance integrated monitoring method for APP system is proposed. Analyze the actual needs of the monitoring function of the lightweight IT operation and maintenance system, design a targeted IT operation and maintenance monitoring function module according to the analysis results, and give the lightweight IT operation and maintenance process. A lightweight IT operation and maintenance database is established based on fuzzy theory, and data with similar characteristics is simulated to minimize the dispersion of sample data. Finally, in order to achieve integrated monitoring of lightweight IT operation and maintenance, and to improve monitoring efficiency and monitoring accuracy, it is necessary to reduce the dimensionality of multi-dimensional data, which is mainly achieved through principal component analysis. The experimental results show that compared with the traditional method, the method in this paper can obtain accurate early warning information in time in monitoring, indicating that its monitoring effect is better, and it has a certain positive effect on ensuring the safe operation of the IT operation and maintenance system.

1. Introduction

The operation and maintenance system can effectively control related equipment in various industries, such as network equipment, power equipment, monitoring equipment and other IT equipment for maintenance and control [1], through the operation and maintenance system, centralized control of the operation status of enterprise equipment can be realized, to achieve unified management, once a problem occurs in a certain link, it can be dealt with in time [2,3]. The traditional operation and maintenance monitoring means mainly rely on manual implementation, and the fault identification and location are realized through manual inspection. However, due to the reasons of manual technology, the operation and maintenance monitoring effect is poor, and it will consume a lot of human, material and financial resources, so the manual monitoring does not have a certain ductility [4]. Under the above background, it is necessary to study an intelligent method to effectively monitor it operation and maintenance problems and improve the effectiveness of operation and maintenance [5].

In view of the shortcomings of manual monitoring, experts in related research fields have conducted in-depth research on IT operation and maintenance management issues, and have obtained certain research results, which have played a certain reference role for operation and maintenance work. Reference [6] designed a software operation and maintenance monitoring system. First, the characteristics of the monitoring system were analyzed, and the system architecture was designed according to the analysis results. Analyze the specific implementation plan of the system according to the function of each module in the system, and under the interaction of each module, improve the safety of the system, and then improve the level of system monitoring. Experimental verification shows that the system can ensure the safe operation of software operation and maintenance, but when there are certain interference conditions, the monitoring effect is unstable. Reference [7] designed an IT operation and maintenance management system based on Fiori. From the aspects of improving the level of refined management, improving user service experience and timely warning capabilities, the IT operation and maintenance management system was designed, which not only can improve operation maintenance management efficiency can also meet user experience requirements, and is suitable for application in the field of refined IT operation and maintenance management. However, the system also has some problems, such as untimely risk warning and poor stability, which affect the management effect. In addition, some relevant researchers have proposed a cloud network integrated operation and maintenance monitoring system for operators. The system focuses on the system architecture design, function module deployment, system capacity estimation and other aspects, and designs a monitoring system integrating scene monitoring, data analysis and business processing to provide guarantee for cloud network integrated operation and maintenance. The experimental analysis shows that the system has the advantage of large capacity and can monitor multiple objects and targets at the same time. However, when there is a problem in a certain link, there is a certain error in the early warning information of the system, which will affect the emergency treatment effect.

Although the existing methods have made some contributions in operation and maintenance monitoring, there are still some problems, such as unstable monitoring effect, untimely risk early warning and high error rate of early warning information. Aiming at the above problems, a lightweight IT operation and maintenance integrated monitoring method for app system is proposed.

2. Analysis of lightweight IT operation and maintenance system

2.1. Monitoring function requirements of lightweight IT operation and maintenance system

With the continuous improvement of Internet technology and computer performance, the applicability of IT systems has been improved, which has attracted a certain degree of attention. In this context, the integrated monitoring of IT operation and maintenance systems is currently urgently needed. problem. Through operation and maintenance monitoring, the operation effect of the IT system is improved, the probability of risks in the operation of the system is reduced, and the stable operation of the system is guaranteed [8]. Lightweight IT operation and maintenance systems can be oriented to a variety of objects, such as substations, communication networks, power systems, etc. This article is mainly for APP systems to study lightweight IT operation and maintenance monitoring issues, so as to improve the performance and operational effects of IT systems, and build a Lightweight, safe and efficient operation and maintenance system provides reference [9]. First, analyze the monitoring function requirements of the lightweight IT system:

The requirements for the integrated monitoring function of lightweight IT operation and maintenance for the APP system can be mainly divided into two categories: on the one hand, the monitoring of the equipment belongs to the underlying monitoring function; on the other hand, the user can log in to the various functional modules of the system on the page, viewing and configuration functions belong to user application requirements [10]. Under the above two levels of functional requirements, this article combines the monitoring requirements of relevant operation and maintenance personnel, and specifically divides the requirements of the lightweight IT operation and maintenance integrated monitoring function for the APP system into the following categories:

(1) First, starting from the basic requirements of APP system operation and maintenance monitoring, real-time monitoring of memory capacity, traffic monitoring, resource utilization and other content;

(2) Since the APP system will contain various APPs, a large amount of data will be generated under the conditions of joint operation. Therefore, a database for the APP system must be established to

transmit, store, and collect the data generated during the operation. In response to this problem, it is necessary to ensure the space utilization rate and data integrity of the database during the operation and maintenance process;

(3) For business monitoring in the APP system, including data exchange, administrator authority setting and IT operation and maintenance personnel work monitoring, etc.

2.2. IT operation and maintenance monitoring function module design

According to the lightweight it operation and maintenance integrated monitoring function requirements for APP system analyzed in Section 2.1, the IT operation and maintenance monitoring function module is designed. Since the monitoring function mainly targets two types of objects, namely administrators and operation and maintenance personnel [11], the targeted monitoring function module is designed by analyzing the role characteristics and permissions of different objects, it is shown in Figure 1.

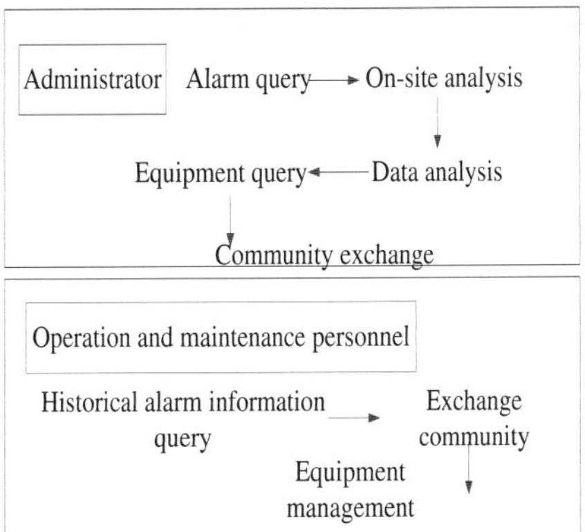

Figure 1 IT operation and maintenance monitoring function module

According to Figure 1, in the APP system operation and maintenance process, the administrator can not only query the current alarm information of the system, but also query historical information, and grasp the system operation status through comprehensive analysis of the alarm information. In addition, the administrator can also monitor the system business by obtaining monitoring data, and obtain the system evaluation report through data analysis and statistics, and send it to the operation and maintenance center to facilitate all operation and maintenance personnel to obtain relevant information [12]. In order to improve the system operation and maintenance effect, the administrator can also perform user operations in the system, establish a communication community, and improve the operation and maintenance effect.

In addition to administrators, operation and maintenance personnel are also an effective guarantee for the normal operation of the IT system. The main responsibility of operation and maintenance personnel is to maintain all devices and APPs in the APP system, obtain monitoring views through the monitoring equipment in the system, and obtain alarms. Based on the information, configure the monitoring indicators. In addition, operation and maintenance personnel can also communicate with other staff in the communication community to ensure the circulation of information.

3. APP system-oriented lightweight IT operation and maintenance integrated monitoring method

3.1. Lightweight IT operation and maintenance process

With the increase and popularization of mobile clients, the scale of malware is gradually growing in the upgrading of computer technology. The escalating attack and defense war in cyberspace continues to strengthen the concealment of malware attack means, which poses a great challenge to the current malware identification methods [13]. Therefore, it is necessary to conduct in-depth research on lightweight it operation and maintenance threatened by network attacks. The virtualization of APP system makes the application of hardware resources more flexible, and can transform hardware resources requiring special operations into virtual resources that can take into account several operating systems. The APP system itself has a fixed operating environment, which can be used to manage physical hosts to ensure the parallel reliability of lightweight IT operation and maintenance. The following describes the specific process of lightweight IT operation and maintenance under the APP system, as shown in Figure 2.

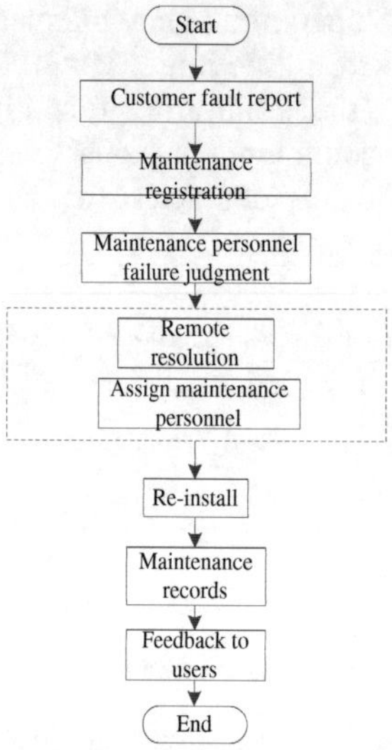

Figure 2 Lightweight IT operation and maintenance flow chart

In the process of lightweight it operation and maintenance, users, administrators and operation and maintenance personnel in the operation and maintenance link need to be verified to prevent bad personnel from attacking the system, stealing user information in the system and causing user losses. This paper uses identity authentication to verify the personnel entering the operation and maintenance system. As a means of confirming the user's identity information on the Internet, identity authentication can verify whether the physical identity of the operator is consistent with the digital identity retained in the database, and it is also the key threshold for users to use the equipment [14]. As the first line of defense of network security, identity authentication provides a security guarantee to prevent illegal users from invading and protect the security of users' personal information.

After the server receives the user's operation request, the traditional operation and maintenance system sends the request information back to the browser for processing, and finally displays the processing results to the user, which usually makes the waiting time of the operation user longer and consumes more network communication resources [15]. To solve the above problems, this paper proposes a new identity authentication method, which creatively establishes all operations on intelligent terminals on a unified Ajax engine. The specific identity authentication process is introduced below:

(1) Add an intermediate layer between the IT operation and maintenance system server. When the two communicate for the first time, the intermediate layer downloads user operation request information on the smart APP terminal to help the server process some user requests that do not need to load the entire page;

(2) Use the Ajax engine to share the user request information in the server, thereby alleviating the server request queuing situation, speeding up the processing and response to user requests, and improving the timeliness of IT operation and maintenance monitoring;

(3) On the basis of the above links, test the network environment to ensure the suppression effect of interference factors in the operation and maintenance process.

3.2. Construction of lightweight IT operation and maintenance database

The growing network scale and the rapid development of communication technology have played a certain role in promoting the frequent occurrence of network attacks such as extortion and secret theft. Since such attacks are mostly carried on malware, efficient and accurate detection of malware is a key point of lightweight it operation maintenance monitoring. Lightweight IT operation and maintenance data is a description of the different state attributes of the equipment under attack during the operation and maintenance process. The database summary contains a large amount of data representing the attributes and content of the attack, and most of the data contains attribute information. The establishment of lightweight it operation and maintenance database is conducive to the operation and maintenance personnel to quickly and effectively extract relevant information when operating and maintaining the equipment in the APP system, help them quickly locate the fault interval and fault type, and then make rapid judgment, make decisions in the shortest time and reduce the losses caused by faults.

In the construction of lightweight IT operation and maintenance database, fuzzy theory is mainly used to simulate data with similar characteristics in a large number of user samples, operation and maintenance data and other data types to minimize the dispersion of sample data. The specific methods are as follows:

In the process of building a lightweight IT operation and maintenance database, assume that D represents the normal vector of the data projection space, which can be expressed by formula (1):

$$D_k = A^S G_k \quad (1)$$

In the formula, A^S represents the projection of the data sample; k represents the data dimension; G_k represents the high-dimensional data in the data projection space. Then the data cluster center in the data projection space is expressed as:

$$R_n = \sum_{n=1}^{N} I'_n(x,y)\sin\left(\frac{2\pi n}{N}\right) \quad (2)$$

In the formula, I'_n represents the total error value of the data sample; R_n represents the cluster center constructed on the basis of the sample data; n represents the total amount of data; N represents the data type; x and y both represent a constant.

On the basis of formula (1) and formula (2), the divergence problem of various data in the database is transformed into a maximization objective function, and the expression is:

$$F(x) = \left[\frac{2\pi x}{d} + \phi(x,y)\right] \quad (3)$$

In the formula, d represents the fuzzy set of data; $\phi(x,y)$ represents the consistency between the data.

The optimal projection direction of the data is obtained by solving formula (3), and its expression is:

$$W(a,b) = \frac{(k-1)N}{2M} \times f(k) \quad (4)$$

In the formula, a and b represent the coordinates of a certain data in the database; $f(k)$ represents the intra-class divergence of the data, which can accurately measure the different goals of the maximum and minimized divergence, and the minimum value is the minimization of the intra-class divergence of the sample data. According to the obtained optimal projection direction, a complete data space can be obtained, and all the data in the space are jointly formed into a lightweight IT operation and maintenance database.

3.3. Integrated monitoring of lightweight IT operation and maintenance

At this stage, the business of an enterprise is gradually expanding and involves more data. For this reason, it needs to be managed through an IT system. Therefore, ensuring the normal operation of the IT system is the key to enterprise development. In the operation and maintenance of enterprise IT systems, it is necessary to conduct integrated monitoring of its status to avoid the impact of a certain link failure on the overall operation of the system. Normally, most companies mainly monitor the operation and maintenance conditions through manual methods, but there is a certain degree of subjectivity in the form of manual monitoring, and the judgments made in the monitoring will waste a certain amount of time. Therefore, it is necessary to study a fast and effective method to assist system operation and maintenance. A lightweight IT operation and maintenance database is constructed through Section 3.2. Since there is a large amount of data in the database, it will affect the monitoring effect. Therefore, in the integrated monitoring of lightweight IT operation and maintenance, the data is reduced in dimensionality.

In order to achieve integrated monitoring of lightweight IT operation and maintenance, improve monitoring efficiency and monitoring accuracy, it is necessary to reduce the dimensionality of multi-dimensional data. This paper mainly implements the principal component analysis method. Construct a data observation matrix Y in the lightweight IT operation and maintenance database [16]:

$$Y = \begin{bmatrix} y_{11} & y_{12} & y_{1m} \\ y_{21} & y_{22} & y_{2m} \\ y_{m1} & y_{m2} & y_{mM} \end{bmatrix} \quad (5)$$

In the formula, M represents the observation factor, which can display the status of IT operation and maintenance; y_i represents the data sampling frequency in the monitoring process.

Standardize the observation factor to obtain the safety factor of the IT system under different operating conditions:

$$R_0 = \left(R_{0x}R_{0y}\right) = \left[\sum_{x=0}^{n}\sum_{y=0}^{n}(x,y)\sum_{x=0}^{n}\sum_{y=0}^{n}(a_0,b_0)\right] \quad (6)$$

In the formula, R_{0x} and R_{0y} both represent IT system operation progress monitoring; a_0 and b_0 both represent IT system maintenance progress monitoring.

In order to facilitate the calculation, R' is used to describe the standardized IT system operation and maintenance monitoring data [17], the standardized IT system operation and maintenance correlation coefficient is r_{xy}, which can be calculated by the following formula:

$$r_{xy} = \sum_{x,y=1}^{n} R_{0n}(x,y) \quad (7)$$

On this basis, the eigenvector of matrix Y is obtained as:

$$Y_r = (y_{r1}, y_{r2}, \cdots, y_{rk}) \quad (8)$$

Finally, use the principal component analysis method to obtain z principal components in the IT system operation and maintenance monitoring, namely:

$$z^p = \sum_{i=1}^{n}\left|z_{t,i} - 1\right| \quad (9)$$

In the formula, z^p represents the characteristic value of IT system operation and maintenance;

$z_{t,i}$ represents the principal component contribution rate of IT system operation and maintenance.

On this basis, the main component expression of the monitoring data is carried out, namely:

$$Z_{n\times m}=\begin{pmatrix} z_{11} & z_{12} & \cdots & z_{1n} \\ z_{21} & z_{22} & \cdots & z_{2n} \\ \cdots & \cdots & \cdots & \cdots \\ z_{n1} & z_{n2} & \cdots & z_{nm} \end{pmatrix} \quad (10)$$

Keeping the principal components in the matrix can complete the integrated monitoring of the lightweight IT operation and maintenance for the APP system.

4. Simulation experiment analysis

In order to verify the effectiveness and universality of the lightweight IT operation and maintenance integrated monitoring method for the APP system proposed in this paper, simulation experiments are carried out. The method of reference [6] and the method of reference [7] are used as comparison methods, and experimental conclusions are drawn through the comparison between different methods.

4.1. Experimental environment and parameter settings

Before the start of the simulation experiment, in order to ensure the validity of the experimental results and the consistency of the experimental conditions, it is first necessary to set the operating parameters of the lightweight IT operation and maintenance system. The specific parameters are shown in Table 1.

Table 1 Operating parameters of lightweight IT operation and maintenance system

Parameter	Specification
Smart Meter	APMB10
4G router	ZLAN8303-7
Video camera	CS-C5C-3B1WFR
Routing Protocol	OSPF/EIGRP
Monitoring tools	VMware ESXi

Under the operating parameters of the lightweight IT operation and maintenance system shown in Table 1, the conventional IT operation and maintenance system database is selected as the experimental data source. The selected database contains a variety of configuration item data, including user information, passwords, alarm values, Information backup port, device type, network IP address and serial number and other parameters. With the support of this database, the operation and maintenance monitoring effect is evaluated and relevant conclusions are drawn.

4.2. Results and analysis

On the basis of the above experimental environment and parameter settings, the method of this article, the method of reference [6] and the method of reference [7] are compared, and the following are specifically carried out from three aspects: monitoring stability, risk warning timeliness and warning information error rate For comparison, the results are analyzed as follows.

4.2.1. Monitoring stability verification

First, the stability of the IT operation and maintenance system is used as an experimental indicator. This indicator mainly refers to whether the monitoring method can effectively suppress the influence of interference factors during the monitoring process, so as to obtain a more stable monitoring result. In this paper, the stability coefficient is used as a measurement index, and its specific value is 0.1-1.

The larger the value, the more stable the monitoring result. On the contrary, the monitoring result fluctuates more. The specific experimental results are shown in Figure 3.

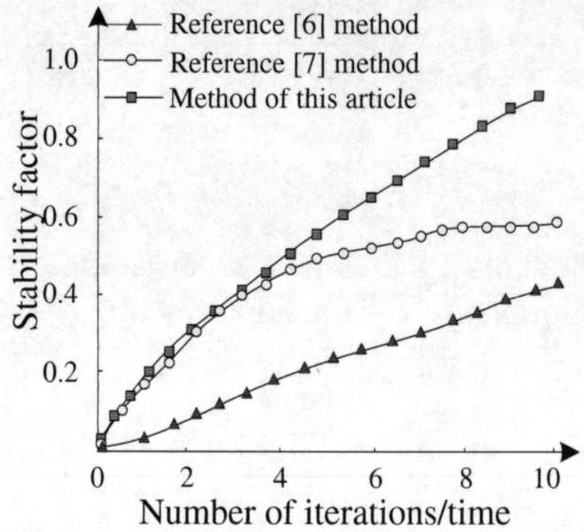

Figure 3 Comparison results of monitoring stability

By analyzing Figure 3, it can be seen that the stability of it operation and maintenance monitoring of the three methods shows a continuous growth trend under different iteration times. Among them, the growth trend of reference [6] method is consistent with that of this method. Although the change trend of reference [7] method also shows a growth trend on the whole, when the number of iterations reaches 4, the growth trend gradually slowed down. By comparing the stability coefficient values of different methods, it can be seen that the maximum stability coefficient of the method in this paper reaches 0.9, while the maximum stability coefficient of the method in reference [6] does not reach 0.5, and the maximum stability coefficient of the method in reference [7] does not reach 0.6. Through the above comparison, it can be seen that in the process of monitoring the lightweight it operation and maintenance system by using this method, it can always maintain a stable trend and play a good supporting role in obtaining accurate monitoring results.

4.2.2. Timeliness verification of risk early warning

In the operation of the IT operation and maintenance system, various APPs have a certain probability of being attacked by abnormalities, which will cause certain failures. Therefore, it is necessary to promptly warn the APP risks in the IT operation and maintenance system to achieve the purpose of timely warning. The comparison results of the risk warning time of different methods are shown in Table 2.

Table 2 Comparison results of risk warning time

Number of iterations/time	Method of this article	Reference [6] method	Reference [7] method
20	0.5	0.9	1.0
40	0.8	1.3	1.3
60	0.9	1.7	1.8
80	1.1	2.0	2.1
100	1.3	2.5	2.6

Analyzing the data in Table 2 shows that as the number of iterations increases, the risk warning time of the three methods has shown a continuous increase trend, but through comparison, it can be seen that the growth interval of the method in this paper is small, the span is small, and the risk warning time The lowest value is 0.5s, the highest value is 1.3s, and the method of reference [6] and the method of reference [7] have a longer risk warning time. It shows that the risk early warning of the method in this paper is timely and can realize effective early warning of APP risk in the IT operation and maintenance system in the shortest time.

4.2.3. Verification of error rate of early warning information

The error rate of early warning information can directly affect the effect of system operation and maintenance. If the error rate is too large, it will mislead the operation and maintenance personnel and affect the processing effect. Therefore, the error rate of early warning information is used as an experimental indicator to compare the monitoring effects of different methods. Select 5 representative APPs in the IT operation and maintenance system to monitor for a period of 3 months, obtain the fault warning information of the APP during the monitoring period, and calculate the average value. The final result is shown in Figure 4.

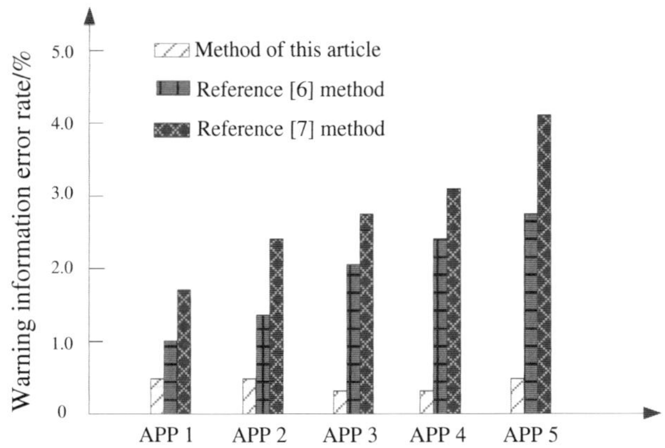

Figure 4 Error rate comparison results of early warning information

According to the analysis of Figure 4, there is a large gap between the early warning information error rate of the method in this paper and the two traditional methods. The early warning information error rate of the five apps is less than 1%, while the early warning information error rate of the method in reference [6] and the method in reference [7] is large, especially the gap between the early warning information error rate of the method in reference [7] and the other two methods is obvious. It can be seen that the early warning information of this method is reliable.

In conclusion, through the comparison of multiple indicators, compared with the traditional methods, this method can obtain accurate early warning information in time, which shows that its monitoring effect is good, and plays a positive role in ensuring the safe operation of it operation and maintenance system.

5. Conclusion

With the wide application of cloud computing, virtualization and other technologies, the network structure becomes more and more complex, which affects the use effect of APP system, and brings great challenges to the operation and maintenance staff. It is necessary to study the operation and maintenance of APP system. Therefore, this paper proposes a lightweight it operation and maintenance integrated monitoring method for APP system. The method performance verification stage mainly studies the three aspects of monitoring stability, risk warning timeliness and warning information error rate. It is concluded that the method in this paper has a good application effect, which can effectively guarantee the safe operation of IT operation and maintenance. The next stage will focus on improving the anti-interference issues in the operation and maintenance process to meet the current performance design and operation and maintenance requirements of diversified APPs.

References

[1] Chen X , Eder M A , Shihavuddin A , Zheng D . (2021) A Human-Cyber-Physical System toward Intelligent Wind Turbine Operation and Maintenance[J]. Sustainability, 13(2):561.
[2] Hernandez-Callejo L , Gallardo-Saavedra S , Alonso-Gomez V . (2019) A review of photovoltaic systems: Design, operation and maintenance[J]. Solar Energy, 188(8):426-440.
[3] Seino N . (2019) SAP global LCM services to realize global operation and maintenance[J]. Fujitsu Scientific & Technical Journal, 55(1):59-63.
[4] Moens P , Bracke V , Soete C , Hautte S V , Hoecke S V . (2020) Scalable Fleet Monitoring and Visualization for Smart Machine Maintenance and Industrial IoT Applications[J]. Sensors, 20(15):4308.
[5] Ciani L . (2019) Future trends in I&M: Diagnostics, maintenance and condition monitoring for cyber-physical systems[J]. IEEE Instrumentation and Measurement Magazine, 22(4):48-49.
[6] Shan J F . (2019) Design and implementation of operation and maintenance monitoring system for software[J]. Heilongjiang Science, 10(10):116-117.
[7] Xing B W . (2020) Design and Implementation of IT Operation and Maintenance Management System Based on Fiori[J]. Energy Science and Technology, 126(11):94-98.
[8] Sikora M , Szczyrba K , U Wróbel, Michalak M . (2019) Monitoring and maintenance of a gantry based on a wireless system for measurement and analysis of the vibration level[J]. Eksploatacja i Niezawodnosc - Maintenance and Reliability, 21(2):341-350.
[9] Ma J , Yan L . (2019) Design and Implementation of Hospital IT Operation Management System[J]. China Digital Medicine, 14(1):111-113.
[10] Zhao L , Zhou Y , Matsuo I , Korkua S K , Lee W J . (2020) The Design of a Remote Online Holistic Monitoring System for a Wind Turbine[J]. IEEE Transactions on Industry Applications, 56(1):14-21.
[11] Qu Y , Qian B L , Zhang C Y , Long Y , Jia J . (2020) Design and implementation of intelligent maintenance system based on blockchain technology[J]. Telecommunications Science, 36(5):152-158.
[12] Chen H Y , Qi H W , Yin H . (2019) The intelligent operation and maintenance integration solution for cloud data center in the aerospace test system[J]. Microelectronics & Computer, 36(5):33-37.
[13] Zhang G Z , Zhao W G , Zhang H . (2019) Design and Development of Digital Operation and Maintenance System for Hutong Yangtze River Bridge[J]. Journal of the China Railway Society, 41(5):16-26.
[14] Peng Z Q , Zhang Q B , Su D W , Huo X S , Zhang X Y . (2019) Remote operation and maintenance technology of substation supervisory control system based on GPS[J]. Electric Power Automation Equipment, 39(4):210-216.
[15] Han C H , Yang Y . (2019) Research on Automatic Inspection of IT Operation and Maintenance Technology[J]. New Technology & New Process, (10):76-80.
[16] Celen M , Djurdjanovic D . (2020) Integrated maintenance and operations decision making with imperfect degradation state observations[J]. Journal of Manufacturing Systems, 55(2):302-316.
[17] Tong C L , Xie S J . (2019) Implementation of Automatic Operation and Maintenance Based on Zabbix and ITIL System[J]. Journal of Beijing Electronic Science and Technology Institute, 27(1):1-6.

AFESSD: Object Detector for Natural Gas Pipeline Construction Scene

Zhen Yang[1,a], Qinhao Zheng[1,b*], Jie Zhou[1,c], Zhijian Yin[1,d]

[1]School of Communication & Electronics, Jiangxi Science & Technology Normal University, Nanchang, Jiangxi, 330000, China

[a]e-mail: yangzhenphd@aliyun.com, [b*]e-mail: 1020170701@jxstnu.com, [c]e-mail: foryu0923@aliyun.com, [d]e-mail: zhijianyin@aliyun.com

Abstract. Object detection is a challenging task in computer vision, which aims to classify and locate objects. SSD (Single Shot MultiBox Detector) is one of the classical algorithms in the current object detection field, which uses multiple feature maps with different scales to detect all objects in an image. Based on SSD, in this paper, we propose an improved AFESSD for object detection in natural gas pipeline construction scene: Firstly, the attention mechanism module is introduced; Secondly, Feature Fusion Block and Feature Enhancement Block are designed to achieve high-to-low level feature information fusion and enhancement. Finally, we verify AFESSD on two object detection datasets of natural gas pipeline construction scene. Experimental results on the first dataset demonstrate that SSD and AFESSD achieve 81.35% and 83.13% mean average precision (mAP) respectively. On the second dataset, SSD and AFESSD achieve 50.55% and 50.99% mAP respectively. Therefore, AFESSD improves SSD's object detection accuracy in natural gas pipeline construction scene.

1. Introduction

Recently, with the development of low-carbon energy structure, natural gas will play an increasingly important role in the energy consumption structure. Pipeline is the main way of natural gas resource distribution. Whether the behavior is standardized during the construction process of natural gas pipelines is related to the life security of relevant personnel. Therefore, it is more and more significant to inspect irregularities of relevant construction personnel in monitoring video. Since manual processing is too time-consuming and laborious, it is very important to use object detection system. Currently, object detection methods can be divided into two types: 1) traditional object detection method based on hand-crafted features [1-4]; 2) object detection method based on deep learning. The traditional machine learning method firstly preprocess the image, then determine candidate regions through the sliding-window paradigm and extract features of these candidate regions, finally, the classifiers are used to judge the category of extracted features. As hand-crafted features cannot meet the needs of multiple features in complex scene, their generalization ability is poor, which brings many difficulties to the detection of related object in construction scene. In addition, the process of feature extraction is too complex, resulting in slow detection speed. Eventually, the object detection effect is poor. In recent years, with the rapid development of deep learning, image processing technology has been continuously improved. Meanwhile, it also has promoted the development of object detection to a large extent. Subsequently, various object detection methods based on Convolutional Neural Networks (CNN) have been proposed. Compared with traditional methods, these methods have stronger feature extraction capability, better generalization ability and higher accuracy.

2. Related Works

2.1. Two-Stage Method

At present, the CNN-based object detection methods can be divided into two embranchments: two-stage method and single-stage method. The object detection of two-stage method is divided into two steps: firstly, a large number of candidate regions are generated on the image; secondly, the object category and regression information are output through CNN. Typical two-stage methods include R-CNN [5], SPP-Net [6], Fast R-CNN [7], Faster R-CNN [8] and R-FCN [9]. Early methods such as R-CNN and Fast R-CNN use selective search algorithm to generate regional proposal candidates, which results in time-consuming running of the framework. Later, Faster R-CNN innovatively proposed the idea of Anchor, pre-setting prior boxes of different scale and aspect ratio in feature maps. Then the Region Proposal Network (RPN) is used to perform binary classification and rough position adjustment on these prior boxes. Finally, final classification and secondary position refinement of these prior boxes are carried out through a sub-network, realizing the end-to-end training.

2.2. One-Stage Method

The two-stage method generally has high detection accuracy, but due to complex network structure and too many model parameters, its speed is too slow to meet the requirements of real-time detection. Based on this, the one-stage method was subsequently proposed. Different from the two-stage method, the one-stage method does not need to generate candidate regions, instead directly classifying and regressing the object in one step to achieve trade-off between speed and accuracy. The representative methods are YOLO [10] and SSD [11]. YOLO divides an image into $S \times S$ grids, abandoning regional proposal candidates, and each grid is responsible for predicting the object on which the center point falls. Finally, YOLO directly outputs the object category and location regression at the end of the network, which greatly improves the detection speed. However, YOLO does not consider objects with different sizes in the image and uses full connection layer, which will make spatial location information lost and lead to miss detection of intensive objects. Consequently, YOLO's Recall is underperforming. SSD combines the idea of Anchor and Regression and simultaneously adopts multiple convolution layers with different scales to detect object, giving consideration to both detection accuracy and speed. However, SSD does not take into account the relationships between different layers and take full advantage of the semantic information from different layers.

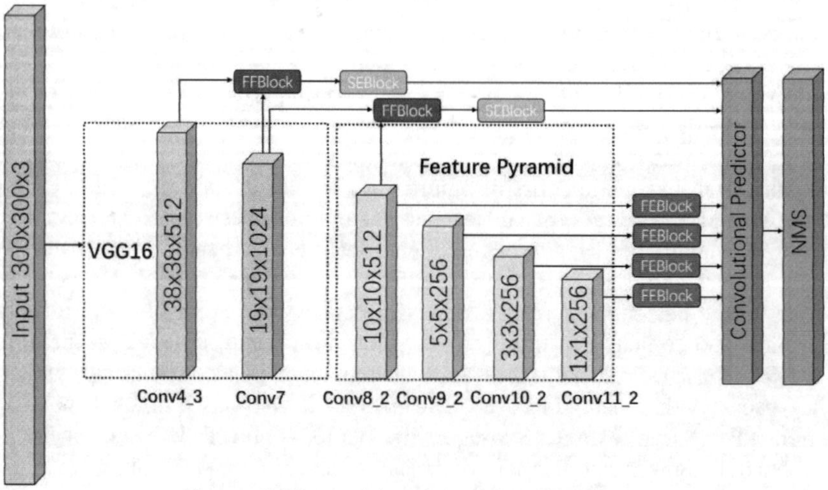

Figure 1. The architecture of our AFESSD.

3. Methods

The AFESSD architecture we proposed is illustrated in Figure 1. Since AFESSD is based on the SSD, in this section, we first briefly review the structure of SSD and then introduce our improved module.

3.1. SSD Framework

The Single Shot MultiBox Detector (SSD) uses the VGG-16 [12] as based network and adds extra convolution layers to the end of it. Then selecting some layers of different sizes from basic and extra convolution layers and to predict scores and offsets for some prior boxes on each scale. The shallow and deep feature maps are responsible for detecting small and large object respectively.

3.2. FFBlock

Although, the SSD's shallow feature map contains more detailed information, it has limited semantic information and insufficient feature representation ability, resulting in the detection performance for small objects on shallow layer is far weaker than that for large objects on deep layer. Based on this, we proposed Feature Fusion Block (FFBlock) to fuse different information from high levels to low levels and apply it to three detection layers of Conv4_3, Conv7, and Conv8_2. The specific structure of FFBlock is shown in Figure 2. In order to fuse the two adjacent feature maps in the original SSD, it is necessary to double spatial resolution of the latter layer to unify the size of feature maps. To avoid increasing the number of parameters, bilinear interpolation is used instead of deconvolution for up-sampling. Finally, the output results of two adjacent detection layers are multiplied by elements to obtain the output results of FFBlock.

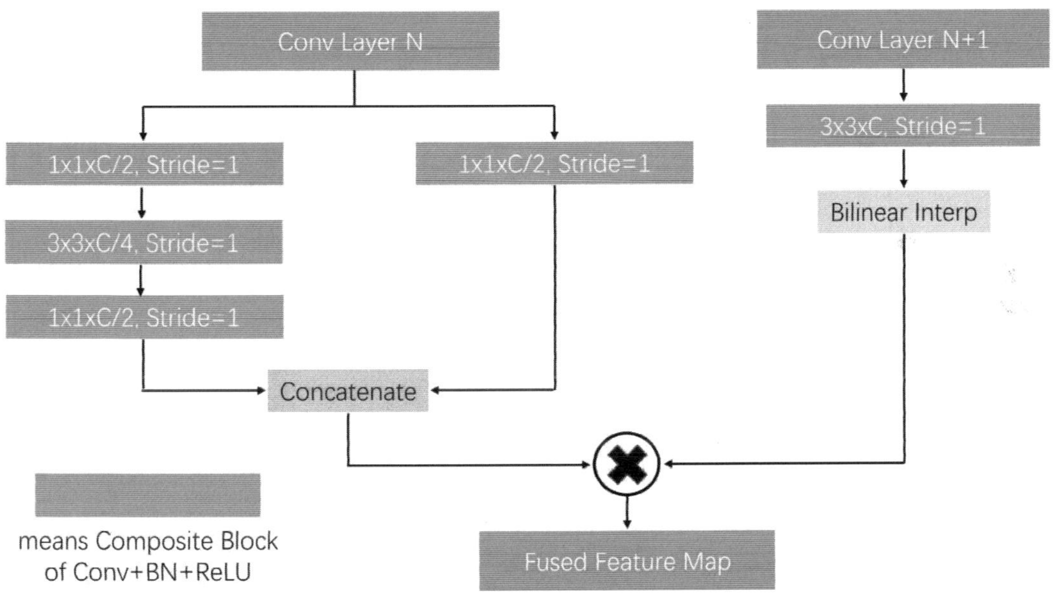

Figure 2. Illustration of the FFBlock.

3.3. SEBlock

To enrich the attention features of FFBlock output feature maps, SEBlock [13] is connected after each FFBlock, which can improve the relationship between feature channels, facilitate the extraction of global information and improve the global receptive field of feature maps. Referring to SENet, more detailed introduction about SEBlock can be comprehended.

3.4. FEBlock
In order to fully enhance feature extraction and regression capability to object location of deep network, Feature Enhancement Block (FEBlock) was designed and applied in three detection layers Conv8_2, Conv9_2, Conv10_2 and Conv11_2, which is shown in Figure 3. Firstly, according to the Partial Transition [14] idea, input features maps are divided into two branches using two 1x1 convolution, one of which is connected to the residual module. Then the output features maps of two branches are connected in series. Finally, 1x1 convolution is used for feature fusion. Unlike CSPNet, which directly divides the features map into two parts according to the channel, we believe that using two 1x1 convolution can further improve the reusability of features, reducing the amount of computation concurrently. Combining residual module [15] and CSP strategy not only enables the network to learn more nonlinear relationships, but also different layers avoid to learn repeated gradient information, which can effectively strengthen the feature learning ability of detection layer and enrich the feature information of deep network.

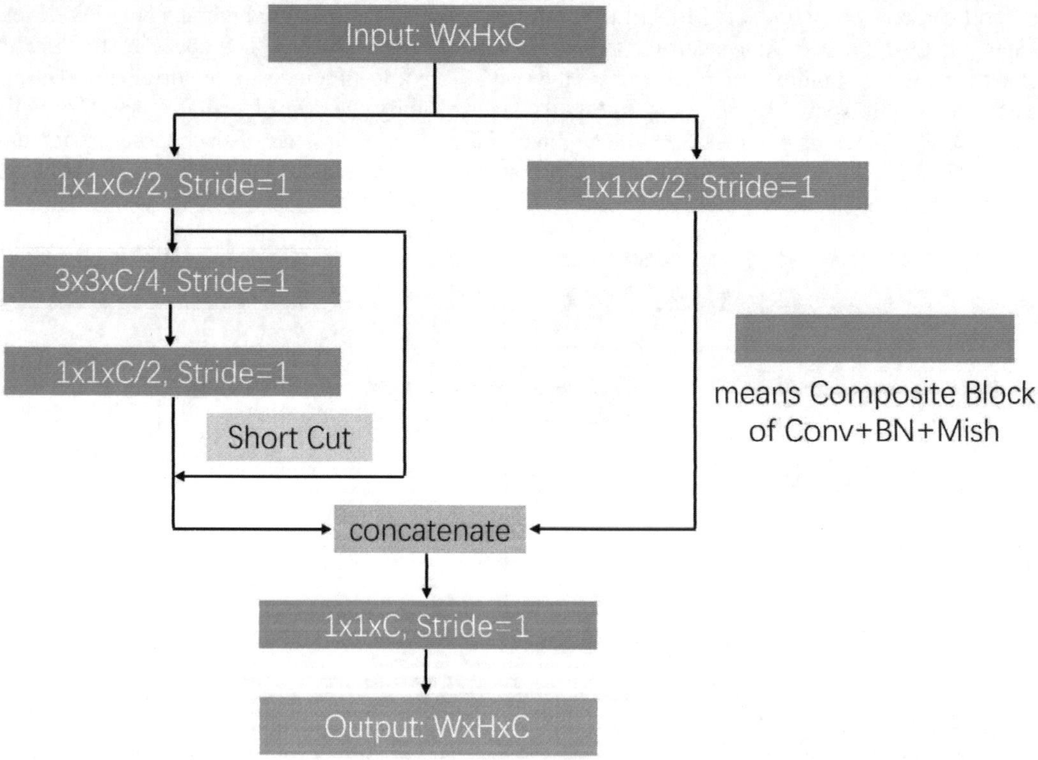

Figure 3. Illustration of the FEBlock.

4. Experiments

4.1. Dataset Establishment
Since there is no public dataset for object detection of natural gas pipeline construction scene, we had made the natural gas pipeline construction scene object detection dataset by ourselves. Most of pictures come from the photographing of dozens of natural gas pipeline construction scene by UAV, and a small part of pictures come from the web crawler. We screen for these pictures in the first place, removing ones which only contain background or very small number of objects. Then we mark images by labelImg. With this marking tool, when manually operating, we only need to mark various user-defined objects in the image, and the tool can automatically generate corresponding configuration file. Finally, they are saved in the format of Pascal VOC detection dataset [16], and the relevant files are

stored in three folders, Annotations, ImageSets and JPEGImages. Data annotation was completed in two stages, forming two datasets consequently, named First-Phase Dataset and Second-Phase Dataset. The labeled objects of datasets in the two periods are different. The specific object categories examples are shown in Figure 4, where 'yes' and 'no' indicate whether the pipeline is sealed. The number of pictures in First-Phase Dataset is 4267, with 8 positive sample categories. The number of pictures in Second-Phase Dataset is 3374, with 7 positive sample categories. Then the two datasets were divided into train set and test set in the ratio of 9:1.

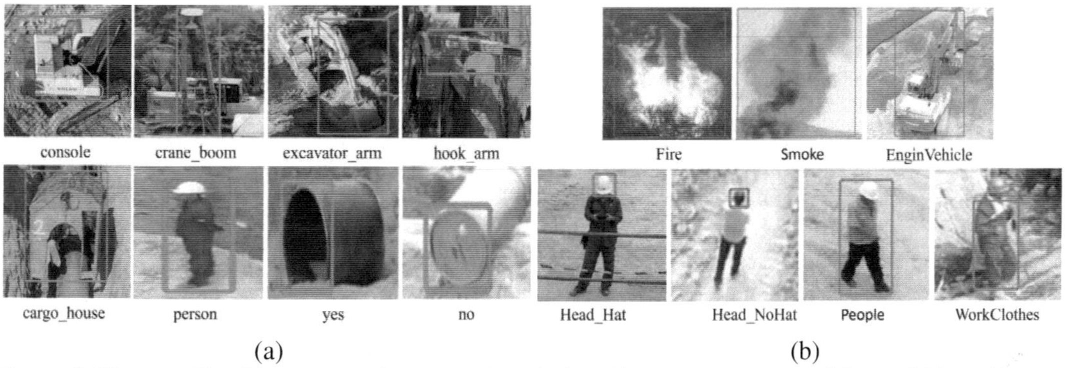

Figure 4. The specific object categories examples of First-Phase Dataset (a) and Second-Phase Dataset (b)

4.2. Training Details

Our experiment was carried out on a computer whose GPU is NVIDIA Geforce GTX 1080Ti, and implementation code of the method is based on the Pytorch framework. In order to obtain a more robust and better performance model, it is necessary to make data enhancement for training images, including random clipping and flipping, which is consistent with the original SSD. Furthermore, we use Adam to optimize the objective functions of AFESSD, which are composed of the Cross Entropy Loss for classification and the Smooth L1 Loss for location. Due to the memory limitation of GPU, when the input image size is 300, we set the batch size to 16 and initial learning rate to 0.0005, training 100 epochs.

4.3. Experimental Results

Based on the divided two datasets, SSD and AFESSD are compared on mAP. As shown in Figure 5 and Figure 6, under our experimental conditions, in First-Phase Dataset, the mAP of SSD is 81.35% and that of AFESSD is 83.13%. In Second-Phase Dataset, the mAP of SSD is 50.55% and that of AFESSD is 50.99%. These experimental results show that AFESSD method we proposed increases SSD's mAP by 1.78% and 0.44% respectively, achieving better detection performance.

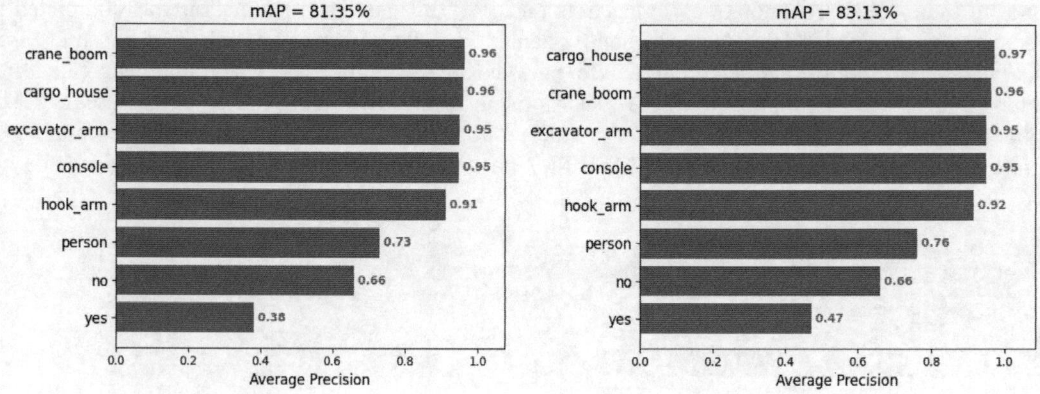

(a) (b)

Figure 5. First-Phase Dataset experimental results of SSD (a) and AFESSD (b)

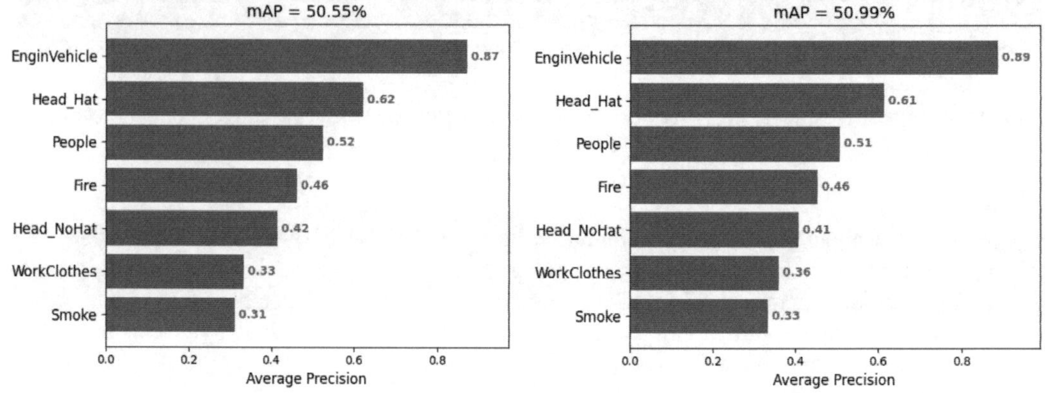

(a) (b)

Figure 6. Second-Phase Dataset experimental results of SSD (a) and AFESSD (b)

4.4. Detection Samples

In order to intuitively compare the performance of AFESSD and SSD, we randomly selected a certain number of images from the test set and visualized the detection results. As shown in Figure 7, AFESSD detects more objects than SSD in the same image. In addition, AFESSD uses FFBlock to combine each layer from Feature Pyramid, thereupon gaining more image context information to improve the detection accuracy. FEBlock that we put forward can effectively enhance the semantic information for feature maps. Therefore, we can see clearly that AFESSD's object confidence score is better than SSD.

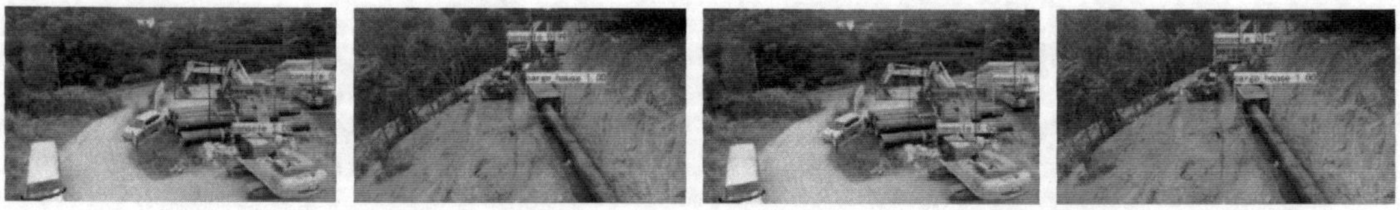

(a) (b)

Figure 7. Detection samples of SSD (a) and AFESSD (b)

5. Conclusions

We use attention mechanism module and design Feature Fusion Block and Feature Enhancement Block to improve SSD. Experiments are carried out on two object detection datasets of natural gas pipeline construction scene to verify our approach. Experimental results show that the accuracy of AFESSD method proposed in this paper is better than SSD, achieving better detection effect. AFESSD can be applied to detect violations of relevant construction personnel in surveillance video, protecting the life security of relevant personnel. In future work, we hope to enhance AFESSD with stronger backbone networks, in addition to applying the proposed fusion module and enhancement module to other computer vision tasks such as semantic segmentation.

Acknowledgments

This work was supported by the National Natural Science Foundation of China (61866016, 62061019), the General Project of Jiangxi Natural Science Foundation(20202BABL202014), the Key Project of Jiangxi Education Department(GJJ201107), and the Youth Top Talent Foundation of Jiangxi Science and Technology Normal University (2018QNBJRC002).

References

[1] Ahonen, T., Hadid, A., Pietikäinen, M. (2006) Face Recognition with Local Binary Patterns. IEEE Trans. Pattern Anal.Mach. Intell. 28: 2037–2041.

[2] Lowe, D.G. (2004) Distinctive Image Features from Scale-Invariant Key points.Int. J. Comput. Vis. 60: 91–110.

[3] Dalal, N., Triggs, B. (2005) Histograms of oriented gradients for human detection. In: Proceedings of the IEEE Conference on Computer Vision and Pattern Recognition, San Diego, CA, USA. pp. 886–893.

[4] Lienhart, R., Maydt, J. (2002) An extended set of Haar-like features for rapid object detection. In: Proceedings of the International Conference on Image Processing, Rochester, NY, USA. pp. 900–903.

[5] Girshick, R., Donahue, J., Darrell, T., Malik, J. (2014) Rich feature hierarchies for accurate object detection and semantic segmentation. In: Proceedings of the IEEE Conference on Computer Vision and Pattern Recognition, Columbus, OH, USA. pp. 580–587.

[6] He, K., Zhang, X., Ren, S., Sun, J. (2014) Spatial pyramid pooling in deep convolutional networks for visual recognition. In: Proceedings of the European Conference on Computer Vision, Zurich, Switzerland. pp. 346–361.

[7] Girshick, R. (2015) Fast r-cnn. In: Proceedings of the IEEE International Conference on Computer Vision, Santiago, Chile. pp. 1440–1448.

[8] Ren, S., He, K., Girshick, R., Sun, J. (2015) Faster r-cnn: Towards real-time object detection with region proposal networks. In: Proceedings of the Advances in Neural Information Processing Systems, Montreal, QC, Canada. pp. 91–99.

[9] Dai, J., Li, Y., He, K., et al. (2016) R-fcn: object detection via region-based fully convolutional networks. Advances in neural information processing systems, 29: 379-387.

[10] Redmon, J., Divvala, S., Girshick, R., Farhadi, A. (2016) You only look once: Unified, real-time object detection. In: Proceedings of the International Conference on Computer Vision & Pattern Recognition, Las Vegas, NV, USA. pp. 779–788.

[11] Liu, W., Anguelov, D., Erhan, D., Szegedy, C., Reed, S., Fu, C.-Y., Berg, A.C. (2016) Ssd: Single shot multibox detector. In: Proceedings of the European Conference on Computer Vision, Amsterdam, The Netherlands. pp. 21–37.

[12] Simonyan, K., Zisserman, A. (2014) Very deep convolutional networks for large-scale image recognition. arXiv preprint, arXiv:1409.1556.

[13] Hu, J., Shen, L., Sun, G. (2018) Squeeze-and-excitation networks. In: Proceedings of the IEEE Conference on Computer Vision and Pattern Recognition, Seattle, USA. pp. 2011-2023.

[14] Wang, C-Y., Liao, H-Y. M., Wu, Y-H., et al. (2020) CSPNet: A new backbone that can enhance learning capability of CNN. In: Proceedings of the IEEE/CVF conference on computer vision and pattern recognition workshops, pp. 390-391.

[15] He, K., Zhang, X., Ren, S., Sun, J. (2016) Deep residual learning for image recognition. In: Proceedings of the IEEE Conference on Computer Vision and Pattern Recognition, Las Vegas, NV, USA. pp. 770–778.

[16] Everingham, M., Eslami, S.M.A., Gool, L.V., Williams, C.K.I., Winn, J., Zisserman, A. (2005) The Pascal Visual Object Classes Challenge. A Retrospective.Int. J. Comput. Vis, 111 (1) :98-136.

Journal of Physics: Conference Series **2209** (2022) 012014 doi:10.1088/1742-6596/2209/1/012014

Research on Application of Artificial Intelligence in Communication Network

Zijun Bai

High school attached to Northeast Normal University, Changchun, 130000, China

robbbertbai@163.com

Abstract——Artificial intelligence has been widely utilized in people's lives, especially in communication. The article provides a review of the recent research status of the application of artificial intelligence in communication, the usefulness of big data, machine learning, and cloud computation in this field. Several examples of how artificial intelligence is used to communicate are illustrated to discuss better and form a framework of modern artificial intelligence applications. The aspects of concern in ethics of using artificial intelligence are also mentioned, as ethics are the principles judging whether the technologies are suitable for humans to use or not. In order to predict the future development of communication, an introduction of previous progress in the field of communication is provided and sorted into categories of wireless communication, ethical concerns, and network monitor and control. This study confirms that the future development of artificial intelligence is mainly in two aspects: SDN (Software-defined network) and NFV (Network function visualization). AI will thus be used in enhancing communication efficiency.

1. Introduction

In recent years, the development of big data, cloud computation, and machine learning inevitably incite the utilization of artificial intelligence. AI has been increasingly used in individuals 'everyday lives. After making several signs of progress, such as driverless cars, artificial intelligence is also being increasingly brought into focus by the public. People are interested in how this advanced technology can be utilized to benefit society.

People nowadays are increasingly paying attention to the safety of their private information when using the internet, as they are terrified by the abuse of private information. To solve the intriguing problem, network security should be improved. Additionally, the moral principles that should be concerned during the designing of communication networks with the help of artificial intelligence are also important. The authority should ensure that individuals 'rights and interests are inviolable while benefiting from the technology, so it is crucial to investigate which aspects of ethical concerns should be applied.

The purpose of the study will firstly focus on reviewing the history of development in artificial intelligence and communication, which summarizes the previous efforts made by the masters in the field. Communications is a sector with heavy ICT use, dealing with various consumer demands on individualization requirements, multimedia services, and precision management, which has made network security more and more critical.[1] Thus, the article will subsequently summarize the modern research of how artificial intelligence is applied in communication networks. To enable the required technological developments and responses, AI researchers and practitioners will need to take moral, societal, and legal value into account in the design of AI systems.[2] Therefore, the third part of the article mentions ethical concerns in the present-day development of communication. In the fourth sector, there will be a prediction of the future application of AI in communication networks and future developments.

2. Artificial intelligence and communication network

2.1. *Concept of artificial intelligence*

AI, known as artificial intelligence, has been defined differently during various periods. However, the modern definition of AI can be understood as intelligence demonstrated by machines instead of the natural intelligence displayed by humans or animals. According to what is defined in the book A Collection of Definitions of Intelligence (Technical report), AI is the field as the study of "intelligent agents": any system that perceives its environment and takes actions that maximize its chance of achieving its goals.[3]

The main advantage of utilizing artificial intelligence in the modern world is that AI can simulate or implement human learning behaviors, which is realized by the technique called machine learning. The technologies in machine learning of artificial intelligence significantly contribute to AI's advanced application in modern industry.

Using the technologies in artificial intelligence, the communication network can benefit from the strengths of AI, especially in two aspects------learning and reasoning. With the strengths of learning and reasoning, AI is suitable for enhancing communication efficiency due to advanced learning and reasoning strategies.

2.2. *Concept of communication network*

The communication network is not a tangible network that enables communications between people. As communication generally refers to the conversations between individuals through a specific medium, the communication network's definition can be considered the link connecting people through remote communicating devices, which ensures the share of information between people, people, computers, and computers. The communication field at present is mainly telecommunication, consisting of various nodes connected by telecommunication links, enabling the transmission of information and messages.

2.3. *Communication*

Communication is directly linked with people's living conditions, as it is essential for individuals to connect and chat with others to send and receive messages. The main characteristic of modern communication is mobile, and mobile communication technology is also constantly updated. The communication technology when the first cell phone came out was 1G, which adopted an analog cellular network. The second version of mobile communication technology was 2G, using digital voice transmission technology as the core. The 2G technology is only valuable for enabling calls and messages but is too outdated to serve mobile phones' advanced functions in recent years. The emergence of 3G ensured the delivery of images, though the data transmission rate was relatively slow. 4G was the more effective version of 3G, allowing better quality of the images conveyed and the rate of delivering information. 5G is the recent development in telecommunication with the advantages of high speed, large capacity, low delay.

3. Analysis of the application status of artificial intelligence in communication networks

The recent study of communication mainly converges in improving communication networks. There are three core sectors which are the application of AI in wireless communication, the ethical concerns of using AI in communication and network monitor and control.

3.1. *Wireless communication*

Communication, which used to be constrained by geographical factors, would take much time to transfer. However, the efficiency of communication has been significantly prompted after the invention of wireless communication. The recent research of wireless communication is increasingly paying attention to the detection of abnormal communication flow.

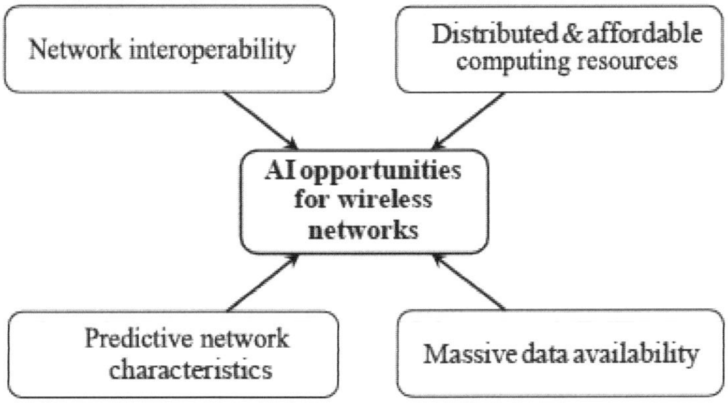

Fig. 1. Favorable conditions for the adoption of machine intelligence techniques in the next generation wireless networks [4]

Due to the advantages demonstrated in the graph, the usage of AI significantly contributes to wireless communication. As people utilize wireless communication systems, asking for network services, certain network flow will be generated.[5] The management and inspection for communication flow is undoubtedly an essential part of the telecommunication field—the most common way to monitor and manage networks in detection technology. People can detect abnormal flow through established and matched rules and models. As the attacking technologies are growing more and more advanced simultaneously, the old way to detect the flow seemed outdated. The computation for a high-speed network is massive nowadays, so the key to enhancing flow detection methods is to utilize big data and machine learning strengths. Based on the previous attack experience, the database for vulnerability can be improved, thus protecting the network from more attacks.

As a result, the modern method for detecting abnormal communication flow takes advantage of big data, though learning from the previous attacks to enhance the database record to prevent attacks.

3.2. Ethical concerns

The author of [6] maintains that people must meet the moral requirements that enhance human well-being when developing AI. As the extent of autonomy in the systems based on AI has been improved, more responsibility should be taken into account. For example, the decision-making process of AI should be more transparent, and the resource of the data should be proposed.

The artificial intelligence system designed for communication networks should consider responsibility, accountability, and transparency. Responsibility, in this case, means the AI systems' role of making decisions and take part in recognizing errors. Accountability refers to explaining and showing the exact process of individuals' actions to other network users to consider the decisions. Transparency means the ability to analyze, reproduce, and inspect the mechanisms for making decisions from AI during communication.

In a word, the ethical concerns of applying artificial intelligence in communication networks are recently focused on accountability, responsibility, and transparency.

3.3. Network monitor and control

Monitoring and controlling the network, which significantly enhances the network security, is in charge of network supervision, regulation, and inspection. Routing has a significant impact on the network's performance, and it is a wild-studied topic in communication.[7] Machine learning techniques have been applied to deal with routing problems such as shortest past routing, adaptive routing, and multicasting routing. Researchers utilized several algorithms with core knowledge in machine learning, such as the genetic algorithm applied to solve multicasting problems.

Research in monitoring the network, primarily focusing on traffic, using AI to analyze and predict traffic demand. According to the author of [8]'s opinion: basically, traffic tendencies can be divided into two types: short-term tendencies, such as temporary traffic increases during the events, and long-term tendencies, from which anomalous tendencies such as temporary traffic increase during the events have been removed. AI is applied to learn the characteristic of the two trends in order to make appropriate predictions. AI can analyze the network information to distinguish whether the network quality is at any normal state.

AI is generally significant in improving network security with an advanced machine learning basis. The security of the Internet is also the key for individuals to use the network for communicating safely.

4. Future application of AI in communication

With the rapid development of communication technologies and the increasing demand for high communication efficiency, the IP network seems outdated with huge network flow problems highly enhance the amount of work required. Therefore, the performance of the network should be enhanced.

4.1. SDN

SDN (software-defined network) is a network architecture approach that enables the network to be intelligently and centrally controlled, or 'programmed,' using software applications.[9] There are three aspects that SDN benefits communication. Firstly, SDN allows network operators to adjust the network behavior and avoid being constrained by the proprietary platforms. Furthermore, SDN is of centralized intelligence to control logically, enabling the better control and management of resources on the internet. The openness of SDN is also advantageous, as the network supports are compatible with applications in the same software environment.

4.2. NFV

The increase in communication efficiency can be realized by NFV (Network Function visualization) as well. The article [10] mentions that NFV is a way to reduce cost and accelerate service deployment for network operators by decoupling functions like a firewall or encryption from dedicated hardware and moving them to virtual servers. The advantages of NFV are mainly to the operators that it brings a new way for network maintenance so that the cost and time for maintaining the network can be reduced. Additionally, NFV decreases network power consumption, as NFV techniques make networks' functions, especially in hardware to the software, thus reducing network power consumption.

5. Conclusion

The article reviews the characteristics of artificial intelligence and the development of communication technology, explaining what artificial intelligence is and how applying AI technology benefits communication. The study emphasizes how AI is utilized in the field of communication. The modern study of applying AI to a communication network mainly concerns three aspects: wireless communication, ethical concerns, and the monitor and control of the network. Wireless communication promotes the importance of detecting abnormal traffic flow, while ethical concerns emphasize communications accountability, responsibility, and transparency. The monitor and control of the network increase the security of individuals' privacy during communication.

The future research focus of applying AI in communication networks should pay attention to SDN and NFV technologies. Although SDN and NFV are applied in modern society in some instances, the spread of the technologies has to be reinforced to provide a more effective way of communicating. However, the new systems for communicating face problems such as lack of experience and being replaced by updated methods. As people are generally less familiar with the rationales of SDN and NFV, the potential problems are not clearly shown unless more time for application. It is also uncertain whether SDN and NFV will soon be replaced, as the developing rate in AI is amazingly rapid. The future is virtually unpredictable—accountability, responsibility, and transparency.

Reference

[1] XU, G. B. & Mu, Y. B. & Liu, J. L., (2018). Inclusion of artificial intelligence in communication networks and services. ITU Journal ICT Discoveries, Vol 1, No.1, pp. 33-135.

[2] Dignum, V., (2018). Responsible artificial intelligence: Designing AI for human values. ITU Journal ICT Discoveries, Vol 1, No.1, pp. 1-135.

[3] Legg, S. & Hutter, M., (2007). A collection of definitions of intelligence. Frontiers in Artificial Intelligence and Applications, Vol.157 pp. 17-24

[4] Bogale, T. E. & Wang, X. B. & Le, L. B., (2018). Machine intelligence techniques for next-generation context-aware wireless networks. ITU Journal ICT Discoveries, Vol 1, No.1, pp. 109-135.

[5]Otani, T. & Toube, H. & Kimura, T. & Furntani, M., (2018). Application of AI to mobile network operation. ITU Journal ICT Discoveries, Vol 1, No.1, pp. 59-135.

[6] Dignum, V., (2018). Responsible artificial intelligence: Designing AI for human values. ITU Journal ICT Discoveries, Vol 1, No.1, pp. 1-135.

[7] Samek, W. & Stanczak, S. & Wiegand, T., (2018). The convergence of machine learning and communication. ITU Journal ICT Discoveries, Vol 1, No.1, pp. 49-135.

[8] Otani, T. & Toube, H. & Kimura, T. & Furntani, M., (2018). Application of AI to mobile network operation. ITU Journal ICT Discoveries, Vol 1, No.1, pp. 59-135.

[9]What is SDN.[online]Available at<https://www.ciena.com/insights/what-is/What-Is-SDN.html>(02/09/2021 23:07)

[10]What is Network Function Visualization(NFV).[online]Available at< https://www.blueplanet.com/resources/What-is-NFV-prx.html>(02/09/2021 23:58)

Handover Scheme of LEO Satellite Network Based on Geographical Division Scenario

Feng Liu[1], Yurong Zhai[2]

[1] School of Electronic and Engineering, Beihang University, Beijing, China, liuf@buaa.edu.cn

[2] School of Electronic and Engineering, Beihang University, Beijing, China, 610087168@qq.com

Abstract. On the basis of completing the division of geographical regions, for the problem of satellite mobile handover, this paper proposes a satellite selection algorithm for selecting the satellite with the longest coverage time to reduce the handover rate. Also based on the idea of "centralized location management", this paper proposes the satellite handover scheme based on the geographical division scenario to update the satellite routing table entries and ensure the correct routing of data packets. Regarding the user handover problem, based on the idea of "decoupling data transmission and address update", a user handover scheme based on geographical division scenario is proposed, which reduces handover delay and signaling overhead compared to traditional satellite network handover schemes.

1. Introduction

With the development of space technology, the need for real-time and reliable communication between mobile hosts and ground users has become more obvious. The future trend of designing global communication systems is to combine satellite communications with terrestrial communications. Mobile handover is a key technology of mobile satellite communication networks. When a mobile host hands over between different access points, a mobile handover scheme is required to maintain the continuity of data transmission between the two parties.

LEO satellites revolve around the earth and hand over between ground stations when passing through different areas on the earth's surface; spacecraft or ground mobile nodes communicate with other ground nodes through relay satellites, and the satellites act as routers to forward data. However, in this scenario, the satellite moving speed is high, and users within the satellite coverage area need to hand over every few minutes on average [1]; and the satellite needs to maintain a routing table entry for each user, resulting in the satellite's on-board routing table too long, and the routing table needs to be updated every time the user changes the IP address, so it is difficult to establish an address aggregation relationship between the satellite and the ground user terminal.

In the geographical division scenario, the satellite and geographical area "one-to-one" access not only maintains the address convergence relationship, but also reduces the number of on-board routing table entries, speeds up routing convergence, and avoids ground users' frequent and passive handover caused by high satellite moving speeds. We stipulate that when the midpoint of the division is within the coverage of the satellite, the access is successful; once the midpoint of the division leaves the coverage of the dominant satellite, handover will occur.

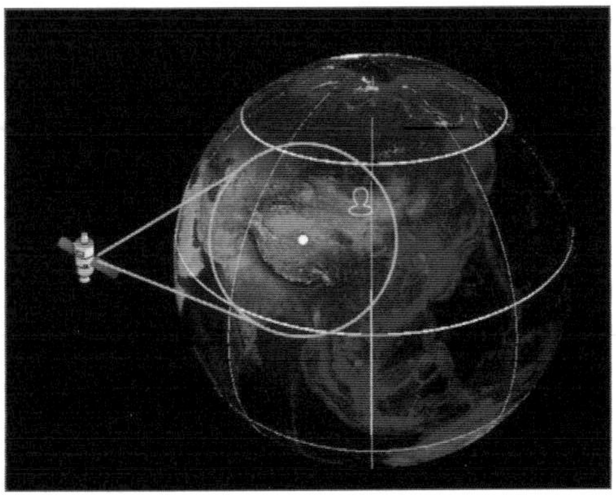

Figure 1. Two handover roles in geographical division scenario.

In this paper, we do not pay attention to how the IP address is allocated during the handover process, nor do we pay attention to the specific routing mechanism. Related work has been implemented in [2] [3]. Based on the assumption that the geographical division has been completed, the specific handover process has been studied. Compared with the traditional satellite network scenario, there are two handover roles in the geographical division scenario, including satellite handover and user handover, as shown in figure 1. This paper focuses on the handover execution process, and proposes a corresponding handover scheme, which reduces the handover rate, reduces the time of user communication interruption, and realizes the correct routing of data packets. And through theoretical analysis and simulation, explains the advantages of the proposed algorithm or scheme, and provides certain ideas and references for the handover problem of low-orbit satellite networks.

2. Related work

2.1. Satellite selection algorithm

When the satellite hands over between different geographic regions, there are 2 to 3 satellites available for the current region. The greater the latitude, the more candidate satellites. [4] proposed a satellite-to-ground link handover algorithm, which selects the shortest distance satellite. Although this method is simple to implement, there is no guarantee that the satellite will still be able to cover this subarea at the next moment, resulting in a higher handover rate. [5] proposed a satellite selection algorithm based on signal strength. This method needs to calculate the neighbor cell list of the terminal, and needs to update the handover weights at all times. There is a certain algorithm time-consuming, which will increase the handover delay.

In recent years, there have also been many methods that combine these satellite selection criteria with graph theory [6], fuzzy neural networks, etc., but these methods have two main problems. One is that the design of input parameters is very complicated. For example, in [7], one of the input parameters is the reference signal received power (RSRP). Although this parameter is a key parameter in the terrestrial LTE network, it is difficult to measure this parameter in real time in the actual satellite network. Secondly, these methods require a large amount of training data sets, and we lack data support and environmental support for these algorithms.

2.2. Handover scheme

As for the satellite handover process, there is currently no relevant research work. For the handover of ground mobile terminals, it is mainly divided into hard handover based on mobile IP and soft handover based on SCTP protocol [8]. The handover delay of the mobile IP handover scheme is relatively large;

and if the satellite uses soft handover when hands over between different geographic regions, the satellite needs to have two data antennas, and the midpoint of the geographic region must be in the overlapping area covered by the old and new satellites. In actual scenario, the new access handover process will be initiated when the old satellite no longer covers the geographic area; if the user terminal adopts soft handover when hands over between different geographic areas, the user terminal needs to be in the overlapped area of the geographic area. However, there is no overlapping area of coverage between geographic regions, which is inconsistent with the actual scenario. Therefore, only hard handover can be used for both satellite handover and user handover.

3. Scheme design

3.1. Satellite selection algorithm

In view of the geographic division scenario as described above, handover occurs when the satellite no longer covers the midpoint of the geographic region. How to choose the dominant satellite in the current region can make the coverage time the longest and the handover rate the lowest. To solve this problem, a prediction model of coverage time based on satellite ephemeris is proposed in this section.

Figure 2 shows the relative geometric relationship between the satellite and the earth, S represents the position of the satellite, U represents the location of the user terminal, H represents the location of the sub-satellite point, R_e represents the radius of the earth, h represents the height of the satellite, and t represents the elapsed time after the terminal enters the satellite coverage, $\theta(t)$ is the current elevation angle. When the terminal is at the boundary of the satellite coverage area, it has the minimum elevation angle θ_{min}. In the Earth-centered Earth-Fixed Coordinate System (ECEF), the angular velocity of low-orbit satellites changes very little (less than 3%) in most circular orbits [9], so we assume that the satellite's angular velocity ω in the Earth-centered Earth-fixed coordinate system is a fixed value, at this time the trajectory of the sub-satellite point is a large arc.

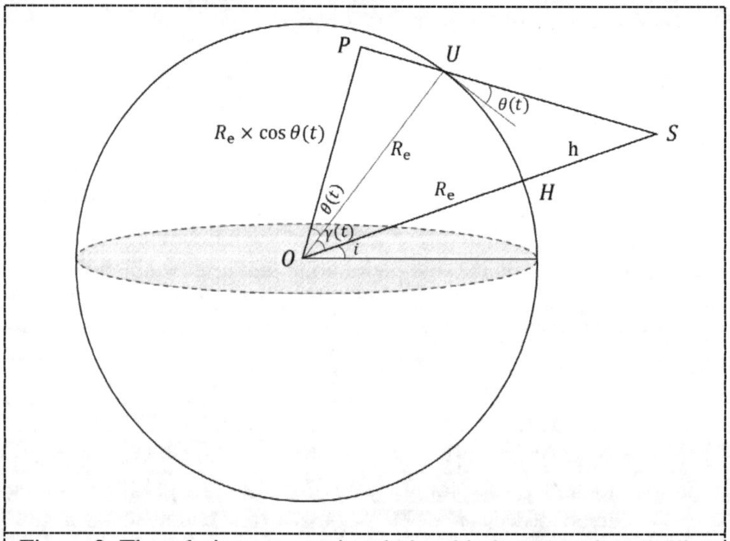

Figure 2. The relative geometric relationship between the satellite and the earth.

The satellite has a fixed trajectory, so we can use the effective information of the satellite ephemeris to realize the satellite selection algorithm. From [10], the real-time elevation angle prediction model of the satellite can be obtained, as shown in the formula (1).

$$\theta(t) = \tan^{-1}\left\{\frac{\cos\left\{(\omega_e \cos i - \omega_s)t + \cos^{-1}\left(\frac{\cos\left(\cos^{-1}\left(\frac{R_e}{R}\cos\theta_{min}\right) - \theta_{min}\right)}{\cos\gamma_m}\right)\right\} \times \cos\gamma_m - \frac{R_e}{R}}{\sin\left\{\cos^{-1}\left\{\cos\left((\omega_e \cos i - \omega_s)t + \cos^{-1}\left(\frac{\cos\left(\cos^{-1}\left(\frac{R_e}{R}\cos\theta_{min}\right) - \theta_{min}\right)}{\cos\gamma_m}\right)\right) \times \cos\gamma_m\right\}\right\}}\right\} \tag{1}$$

and

$$R = R_e + h \tag{2}$$

ω_s is the angular velocity of the satellite in the Earth's Inertial Coordinate System (ECI), which is a fixed value, ω_e is the angular velocity of the earth's rotation, i is the orbital inclination of the satellite constellation, and γ_m is the trace angle, which represents the minimum angular distance between the sub-satellite point position of the satellite and the user terminal.

When handover occurs, first calculate the elevation angle of the midpoint of the current zone relative to each satellite, and then substitute the current elevation angle according to the real-time elevation angle prediction model proposed in [10] to obtain two times t_{p1} and t_{p2}. According to the timestamp in the broadcast message, judge whether the satellite is far away or close to the partition. By choosing between the two times, the elapsed time t_p from the minimum elevation angle to the handover time can be obtained.

In a low-orbit satellite constellation, the satellite coverage time is exactly twice the elapsed time for the satellite elevation angle to change from θ_{min} to θ_{max}, so the total satellite coverage time t_{cov} is shown in equation (3):

$$t_{cov} = 2\frac{\cos^{-1}\left(\frac{\cos\gamma_0}{\cos\gamma_m}\right)}{\omega_s - \omega_e \cos i} \tag{3}$$

and

$$\gamma_0 = \cos^{-1}\left(\frac{R_e}{R}\cos\theta_{min}\right) - \theta_{min} \tag{4}$$

Finally, the difference between the two times t_{cov} and t_p can be used to obtain the remaining coverage time of each satellite, and finally the satellite with the longest coverage time is selected among all the candidate satellites.

3.2. Handover scheme

Compared with the traditional satellite network scenario, the handover in the geographical division scenario is a dual handover, including the handover of satellites between different geographical divisions and the handover of users between different geographical divisions.

The topology of the satellite network has been changing dynamically. The handover of satellites between different geographical regions will cause the change of the satellite IP address. If the OSPF dynamic routing protocol is adopted, these two factors will cause frequent updates of the satellite routing tables. So when the route has not yet converged, it will cause the wrong route of the data packet, resulting in a lot of packet loss. Because the satellite has a constant motion pattern, we can use static routing to transmit data packets. During initialization, the routing table is generated at each time according to the changes in the satellite network topology. When each satellite transmits data packets, it only needs to find the next hop according to the destination address on the routing table at the current time, and the route convergence time is 0.

However, the static routing algorithm only solves the problem of changes in the satellite network topology. When the geographic area changes the dominant satellite, that is, when the IP address of the satellite changes, the corresponding IP address in the routing table entry needs to be updated in time to send the data packet to the correct destination. The specific satellite network topology is shown in figure 3, and the connection relationship between the nodes is given. We have introduced a ground control center to achieve centralized location management. We use high-speed laser links between satellites, and use low-speed links between satellites and ground stations, and all ground stations are connected with the control center and connected with the corresponding satellites.

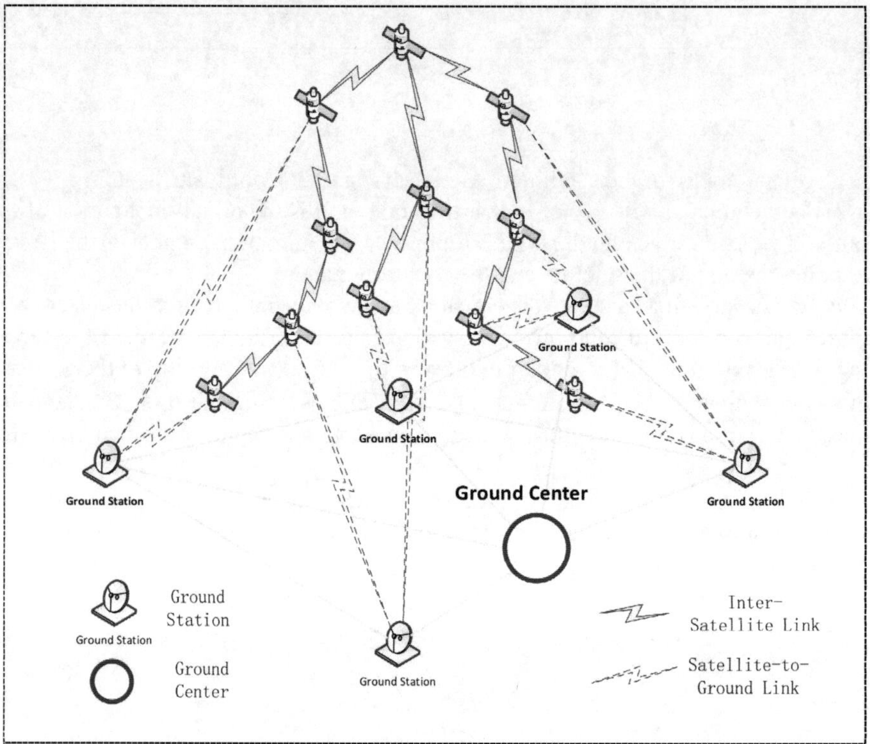

Figure 3. Satellite network topology.

During initialization, the data packet routing table at different times is generated according to the topological connection of the satellite network at each time, and this table is updated regularly for data packet transmission. When the midpoint of the geographic area is not within the coverage of the current dominant satellite, the satellite with the longest coverage time is selected according to the satellite selection algorithm, and then the specific handover process is performed. During the handover, the satellite will send the updated geographic region and IP address of the dominant satellite mapping relationship to the ground control center, and then the ground control center will directly broadcast the changed geographic region and IP address of the dominant satellite mapping relationship to each satellite.

After each satellite receives it, it will change the IP address of the corresponding satellite in the routing table entry, indicating that the routing path of the data packet to this satellite has not changed, but the IP address has changed. The signaling interaction sequence diagram of the entire satellite handover process is shown in figure 4.

Another type of handover in the geographical division scenario is user handover. Since there is no overlapping coverage area in the geographical area, user handover in the geographical division scenario can only be a hard handover. The reason why the hard handover based on mobile IP has a large handover delay and low efficiency is because it has a restriction that a local agent entity must be set for each network segment, and in the handover process, the local agent can forward the data packet from the communication node after completing the registration process.

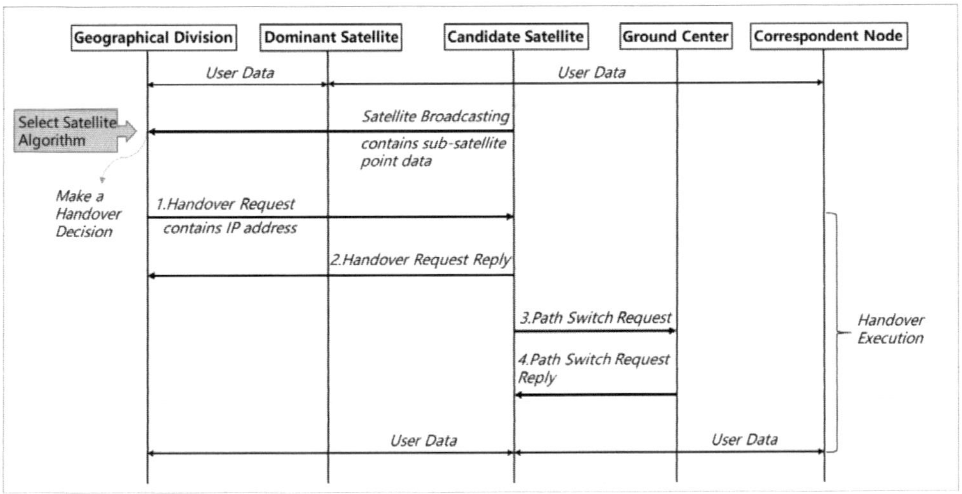

Figure 4. Signaling interaction sequence diagram of satellite handover process.

In the user handover solution proposed in this paper, we can decouple the data transmission process and the address update process, adopt centralized location management, set up a ground control center, and lift the restriction of one local agent for each network segment. When the user hands over, the newly-accessed satellite sends the address mapping relationship to the ground control center for maintenance, and at the same time informs the correspondent node to route the data packet directly to the new address. So the data transmission process and the address update process are implemented in parallel. This not only reduces the handover delay, but also improves the transmission efficiency of data packets. The signaling interaction sequence diagram of the entire handover process is shown in figure 5.

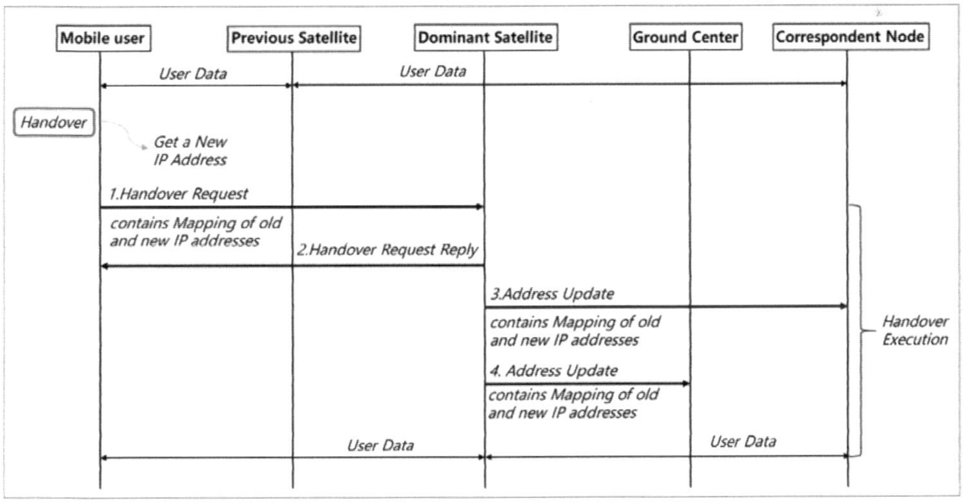

Figure 5. Signaling interaction sequence diagram of user handover process.

4. Simulation analysis and discussion

4.1. Simulation scenario

The simulation software uses STK to build an Iridium constellation. Each orbital surface is circular, the orbital inclination and the orbital radius are fixed, the orbital height is 780km, and the minimum elevation angle is 8.2°. This paper adopts a uniform zoning scheme, that is, a geographical division scheme in which a circle inscribes a trapezoid. Based on the simulation results of this division scheme, it can be concluded that when handover occurs in the middle and low latitudes, there are 2~3 candidate satellites available for the current division (all are idle satellites, without access to any geographical division). The greater the latitude, the more candidate satellites.

The simulation duration of the scenario is set to 24 hours, which is 86400s, and the static routing table is generated with a time step of 1s, the topology update cycle is also 1s, the link delay is set to 3ms, and the processing delay of each signaling is set to 3ms. The inter-satellite link bandwidth is set to 1Gbps, and the satellite-to-ground link bandwidth is set to 100Mbps. We only consider the one-way data transmission process from the communication node to the user in the geographical division. The communication node uses the Poisson process to generate data flow. The number of data packets generated by the communication node per unit time is a random variable and obeys the Poisson distribution. The average value is λ, the value of λ is set to 100, and the time interval for sending data packets is 1ms.

4.2. Satellite selection algorithm

The sampling interval of sub-satellite data samples in the satellite coverage area selected in the simulation process is 1s. For the Iridium constellation, there are 675 sampling points. The accuracy of the elevation prediction model has been verified in [10]. In this paper, we have selected multiple coverage handover scenarios in the same geographical division at different times, and compared the coverage duration of the proposed satellite selection algorithm and the nearest satellite selection method at each handover time. The result obtained is shown in figure 6.

According to the result of figure 6, we can see that the coverage duration of the satellite selection algorithm proposed in this paper is significantly increased, which ensures that the dominant satellite selected during handover has the longest coverage time. At the same time, we also noticed that at some handover moments, the coverage durations obtained by the two satellite selection algorithms are the same, that is, the satellite with the longest coverage time is the nearest satellite, indicating that at this time, the midpoint of the division is near the sub-satellite track of the access satellite orbit.

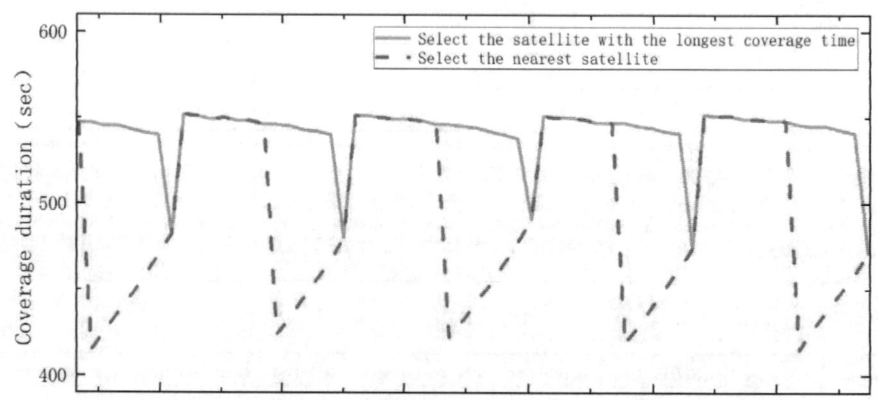

Figure 6. Comparison of coverage duration.

We also found in the simulation process that when the nearest satellite is selected as the dominant satellite in one handover, the next handover will hand over to the last satellite with the longest coverage time, as shown in table 1. For example, at 3901 seconds, the access satellite (608) selected by the nearest satellite algorithm is actually the access satellite (608) selected by the satellite selection algorithm in this paper during the last handover. That is to say, the handover times of the satellite selection algorithm using the nearest satellite in the same time is twice the handover times of the satellite selection algorithm proposed in this paper. Therefore, the satellite selection algorithm proposed in this paper can greatly reduce the number of handovers and avoid frequent handovers.

Table 1. Access satellites selected by two satellite selection algorithms at different handover moments.

Handover time/sec	3604	4138	4668	5196	5720	6141	6697	7252	7806
The access satellite selected by the satellite selection algorithm in this paper	608	607	606	605	102	102	101	111	110
Handover time/sec	3604	3901	4138	4464	4668	5025	5196	5584	5720
The access satellite selected by the nearest satellite algorithm	107	608	106	607	105	606	104	605	103

In summary, we can conclude that the closest satellite is often not the best choice when handover is performed in a geographical division scenario. The coverage time prediction model based on the satellite ephemeris proposed in this paper has high accuracy, can obtain the satellite with the longest coverage time, and reduces the handover rate.

4.3. Handover scheme

For the handover of satellites between different geographical division, first assign IP addresses to the satellite-to-ground ports according to the different network segments where the geographical division is located, then set the link connection relationship, and realize the interconnection and intercommunication between the various protocol models. When handover occurs, the geographical division first selects the next satellite to access according to the result of the satellite selection algorithm, and then executes the satellite handover process to update the address, ensuring the correctness of the data packet routing.

For the user handover between different geographical division, the data transmission process of users in the geographical division and communication nodes is first simulated. Then three different handover schemes, namely MIPv4, MIPv6, and user handover scheme based on geographical division scenario, are used for simulation verification.

We compared the signaling overhead and handover delay of the three handover schemes, as shown in figure 7. The handover scheme proposed in this paper reduces the signaling overhead of the MIPv6 handover scheme by 60%; the handover delay is 59% lower than the handover delay of the two traditional satellite network handover schemes.

In the MIPv4 handover scheme, the original satellite will forward the data packets sent by the communication node to the newly-accessed satellite through the inter-satellite link, which is relatively inefficient in routing, that is, "triangular routing." When the MIPv6 handover scheme performs handover, data packets must be received through the "triangular routing" process of the previous satellite. But when the user terminal receives the data packet forwarded by the previously connected satellite, it will realize that the correspondent node does not know the current new IP address of its own, so it will initiate the registration process to the correspondent node. Therefore, although the MIPv6 handover solution solves the "triangular routing" problem of subsequent data packet transmission, the signaling overhead during the handover process is relatively large. The user handover scheme we proposed does not need to register with the previously accessed satellite, but requests the current access satellite to notify the communication peer node of the new IP address, so it will get lower handover delay and smaller signaling overhead.

Then we compared the throughput changes of the three handover schemes, as shown in figure 8. We can see that when no handover occurs, the ground user terminal can normally receive the data packet from the communication node. After the start of the handover process, the throughput of the three schemes will drop to 0, because the three schemes are based on hard handover. However, we can see that the user handover scheme proposed in this paper has the shortest throughput time of 0. The reason why the throughput of the MIPv6 handover scheme will increase after the handover is completed instead of returning to the original normal level is that the MIPv6 scheme solves the "triangular routing" problem through the registration process with the CN and speeds up the subsequent data packet transmission. Superimposed with data packets that have not been sent before, resulting in a temporary increase in throughput.

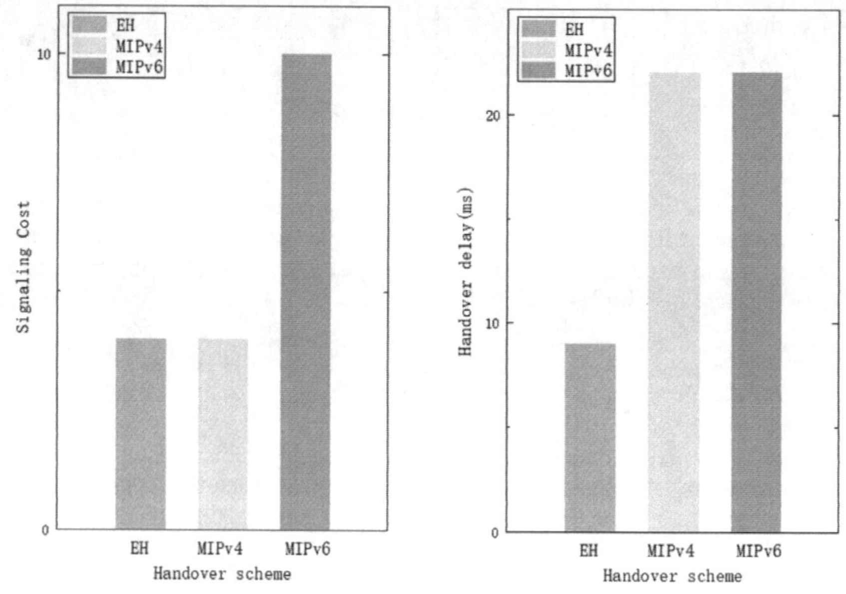

Figure 7. Comparison of signaling overhead and handover delay of handover schemes.

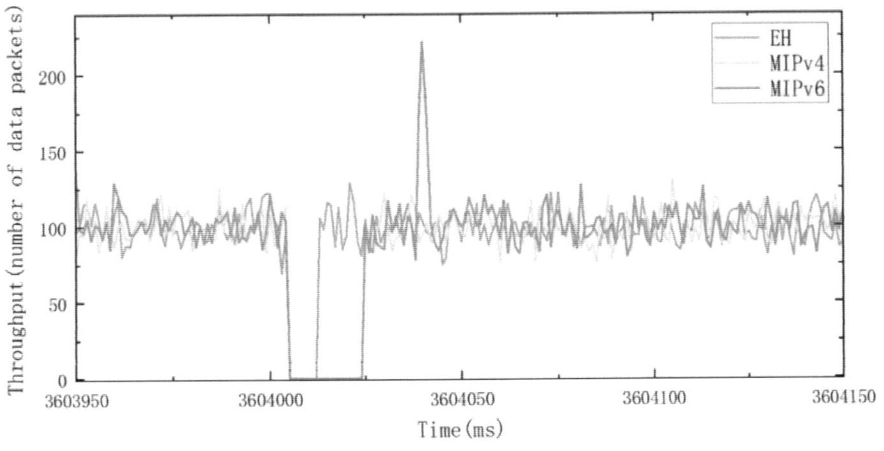

Figure 8. Comparison of throughput of handover schemes.

To sum up, the satellite handover scheme based on geographical division scenario proposed in this paper realizes the dynamic update of addresses and solves the problem of incorrect routing of data packets. The user handover scheme based on geographical division scenario achieves the least number of interactions, the least handover delay, and the least packet loss on the basis of hard handoff, thereby ensuring the continuity of communication and the quality of service.

5. Conclusion

In the geographical division scenario, for the satellite mobile handover problem, this paper proposes a satellite selection algorithm based on the geographical division scenario, selects the satellite with the longest coverage time, and reduces the handover rate. And based on the idea of "centralized location management", proposes a satellite handover scheme based on geographical division scenario, updating the satellite routing table entries to ensure the correct routing of data packets. For user handover issues, based on the idea of "decoupling data transmission and address update", a user handover scheme based on geographical division scenario is proposed. Compared with the traditional satellite network handover schemes, it reduces signaling overhead and handover delay. The above-mentioned scheme is simulated and verified in the Iridium constellation scenario. The simulation results show the advantages of the scheme proposed in this paper, which can better guarantee the continuity of communication and the quality of service.

References

[1] S. L. Kota, K. Pahlavan, P. Leppanen, "Broadband Satellite Communications for Internet Access," Kluwer Academic Publishers,2004.

[2] Xiao Y, Zhang T, Liu L. Addressing Subnet Division Based on Geographical Information for Satellite-ground Integrated Network[J]. IEEE Access, 2018, PP:1-1.

[3] Fu W, Xu Z, Gao Z. IP Addressing and Address Management of Space-Based Network Based on Geographical Division[M].2021.

[4] Ye X, Xiao F, Sun L, et al. Research on Handover Algorithm of satellite mobile communication network [J]. Computer science, 2009, 36 (9): 4.

[5] Shen J, Chen X, Zhao X. A beam cell handover algorithm in LEO satellite communication system [J]. Telecommunications technology, 2020, 060 (004): 439-447.

[6] Wu Z, Jin F, Luo J, et al. A Graph-Based Satellite Handover Framework for LEO Satellite Communication Networks[J]. IEEE Communications Letters, 2016, 20(8):1547-1550.

[7] Li N, Gong B, Deng Z. A Handoff Algorithm Based on Parallel Fuzzy Neural Network in Mobile Satellite Networks[J]. Journal of Communications, 2017, 12(7):395-404.

[8] Chowdhury P K, Atiquzzaman M, Iva Nc Ic W. Handover Schemes in Space Networks: Classification and Performance Comparison[C]// Space Mission Challenges for Information Technology, 2006. SMC-IT 2006. Second IEEE International Conference on. IEEE, 2006.

[9] I. Ali, N. Al-Dhahir and J. E. Hershey, "Doppler characterization for LEO satellites," in IEEE Transactions on Communications, vol. 46, no. 3, pp. 309-313, March 1998, doi: 10.1109/26.662636.

[10] Seyedi Y, Rahimi F. A Trace-Time Framework for Prediction of Elevation Angle Over Land Mobile LEO Satellites Networks[J]. Wireless Personal Communications, 2012, 62(4):793-804.

The Electronic and Ferroelectric Properties in Strontium and Zirconium Doped $BaTiO_3$ from First-Principles Calculations

Haolei Shen[1], Peng Wang[1], Rongri Tan[1a*], Yousheng Yu[2b*]

[1] Jiangxi Science and Technology Normal University, Nanchang 330013, China

[2] School of Science, East China University of Technology, Nanchang 330013, China

Corresponding authors's e-mail: [a]rogertanr@hotmail.com, [b*]ysyu@ecut.edu.cn

Abstract—The structure, electronic and ferroelectric properties of BSZT were investigated based on first-principles calculations. The result of relaxation structure and two-dimensional charge density indicates that the tetragonal crystalline structure is distorted after Sr and Zr doping. The double-well curve fits well with the phenomenological Landau-Devonshire theory. Moreover, the smaller Zr ion displacement leads to a lower spontaneous polarization of BSZT than the original $BaTiO_3$. The hybridization between the 3d states of Ti and the 2p states of O, and the hybridization between the 4d states of Zr and the 2p states of O are observed in the density of states, which is the reason for the ferroelectricity of BSZT materials.

1. Introduction

Barium titanate with general ABO_3 formula have been described as a promising material category in recent decades. Barium titanate materials have excellent properties such as piezoelectricity, ferroelectricity, high dielectric constant and thermoelectricity[1–3].

Pure $BaTiO_3$ shows a smaller piezoelectric response, the largest leakage current, a lower Curie temperature (T_c), and a smaller breakdown strength. In order to develop the dielectric, ferroelectric and piezoelectric properties of $BaTiO_3$ ceramics, efforts have been made in various directions, including substitution of hetero valent ions by solid solution formation and field engineering, but with limited success [4,5]. The current analysis strategy is realized by simulating a new system and replacing Ba^{2+} or Ti^{4+} with alternative ions with comparable ion sizes. For example, the Nb shows a significant effect on the dielectric properties of $BaTiO_3$ [6]. Benlahrache et al. [7,8] found that the microstructure of barium titanate changed significantly and the dielectric constant increased by adding $NaNbO_3$. With the increase of Sr content, the Curie temperature of (Ba,Sr) TiO_3 (BST) ceramics decreases linearly [9]. Ciomaga et al. found that Zr with different substitution concentrations has a strong impact on the properties of $BaZr_xTi_{1-x}O_3$ (BZT) ceramics. [10]. Alshoaibi et al. discussed the influence of Y atom doping Ba and Ti of $BaTiO_3$ on the electronic structure and optical properties of the system. The results show that Y doping can significantly improve the dielectric properties of $BaTiO_3$ [11]. However, as far as we know, there is no theoretical study on the ferroelectric properties of BSZT materials.

In this work, we have studied the properties of BSZT ferroelectric materials using first principles calculations. The substitution effect of Ba and Ti by strontium and zirconium into $BaTiO_3$ on the structure, electronic and ferroelectric properties were discussed.

2. Computation details

We performed the first-principles calculations based on the density functional theory (DFT) with generalized gradient approximation (GGA), as implemented Vienna Ab-initio Simulation Package (VASP) [12,13]. Select a 2 × 2 × 2 super cell, which contains 40 atoms, belongs to the P4*mm* space group, and the initial lattice constant a = b = 3.9905, c = 4.023. Cut-off energy was set to 600 eV. The pseudopotentials were constructed according to the electronic configuration of Ba $5s^25p^66s^2$ states, Sr $4s^24p^6$ states, Ti $3s^23p^63d^24s^2$ states, Zr $4s^24p^64d^25s^2$ states and O $2s^22p^4$ states. Monkhorst-Pack k-point 5 × 5 × 5 was used for geometry optimization and electronic structure calculations. Using GGA + U methods, the Hubbard energy exchange correlation potentials of the 3d orbitals of Ti atoms and the 4d orbitals of Zr atoms were corrected by U = 10 eV and U = 6 eV, respectively. The convergence standard of the interaction force between atoms is 0.01 eV/Å, and the convergence standard of the system energy is 1×10^{-8} eV/Å. The spontaneous polarization calculation adopts the standard Berry-phase method.

3. Results and discussion

3.1. Electron charge density

Figure 1 shows the crystal structures of the ferroelectric and paraelectric phases of BSZT after relaxation and the two-dimensional charge density. One-eighth of Ba^{2+} and Ti^{4+} were replaced by Sr^{2+} and Zr^{4+}. The positive relaxation of Ti and Zr along the c axis in the ferroelectric phase deviates from the octahedral center compared to the paraelectric phase. From fig.1(c), it can be found that the electronic charge density of Ti ion overlaps with upper oxygen ions (O3) and the right oxygen ions (O1) in the same plane to form a covalent bond. The absence of hybridization of Ti1 and O4 exhibits ionic bonding. Zr and the surrounding O ions have similar chemical bond characteristics.

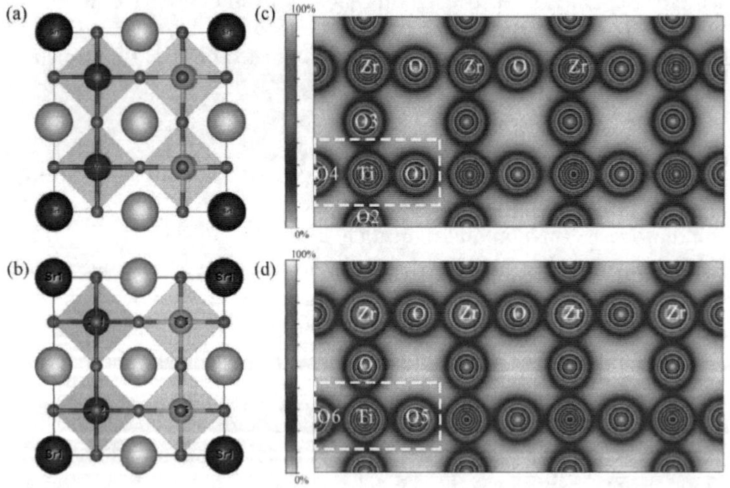

Fig.1. Optimized structure of (a) ferroelectric and (b) paraelectric BSZT. Two-dimensional charge density of (c) ferroelectric and (d) paraelectric BSZT.

3.2. Double-well potential

In order to further explore the influence of Ti and Zr ion shifts on the ferroelectric properties of BSZT, we used the modern polarization theory to calculate the spontaneous polarization of BSZT. The spontaneous polarization of ferroelectrics is defined as the polarization change when the paraelectric phase transforms into the ferroelectric phase.

Figure 2 shows the variation of energy difference with polarization function of BSZT. The phenomenological Landau-Ginsburg-Devonshire theory is used to fit the energy polarization curve as follows,

$$\Delta G = \frac{1}{2}\alpha P^2 + \frac{1}{4}\beta P^2 + \frac{1}{6}\gamma P^6$$

Where α, β and γ are coefficient constants, and P is the polarization of each unit cell.

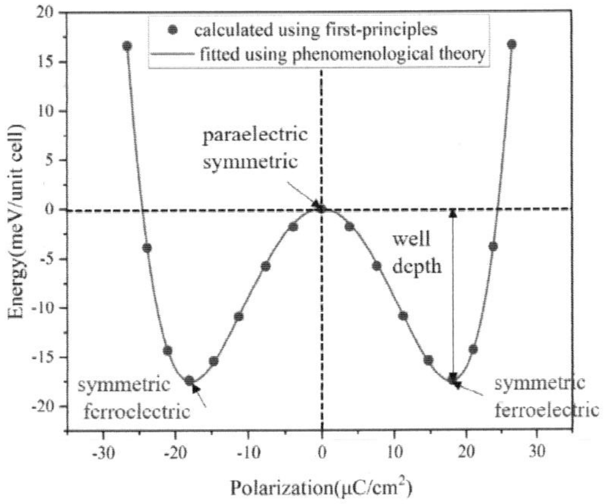

Fig.2. Double-well potential of BSZT versus polarization

For BSZT, the potential curve fitted well with the LGD model. In Fig.2, the symmetry point of the minimum on both sides represents two stable polarization states of equal magnitude and opposite directions. Paraelectric corresponds to P=0 in the figure, which has no Ti displacement as shown in Fig. 1(b, d). We calculated the polarization value of BSZT is 18.07 μC/cm^2, which is a decrease compared to undoped $BaTiO_3$. The reduction of spontaneous polarization indicates that the ferroelectric performance of BSZT is weaker than that of undoped $BaTiO_3$.

3.3. Band structure and density of states

Fig.3(a, b) illustrates the band structure of BSZT, which is calculated in the Brillouin region with and without Hubbard energy adjustment in the direction of high symmetry. The calculated BSZT band gap without Hubbard correction is 1.92 eV, which is much lower than the original $BaTiO_3$ bandgap value which is attributed to the limitations of GGA functional. Taking into account the Hubbard potential adjustment, the band gap is increased to 3.14 eV, but it is still slightly smaller than the 3.2 eV band gap of pristine $BaTiO_3$.

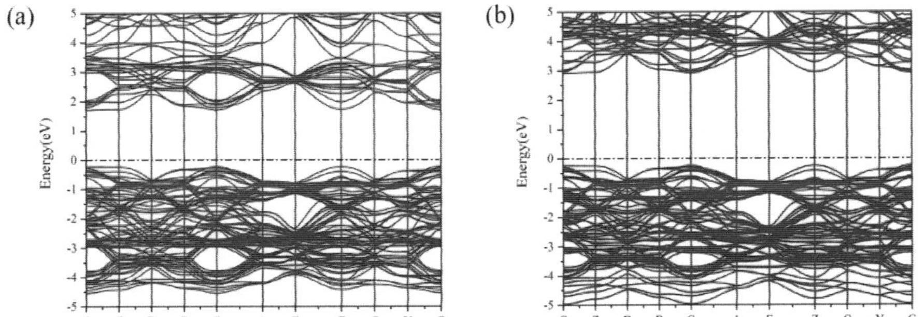

Fig.3. Calculated band structure of BSZT (a) without and (b) considering Hubbard adjustment.

Fig. 4 shows the density of states (DOS) and partial density of states (PDOS) of BSZT ferroelectrics without and with Hubbard adjustment. In the fig. 4(a), the energy band of ~34 eV is composed of 3p state of Ti atom. The energy band at ~26 eV is composed of 6s state of Ba atom. The energy band from ~17 eV to ~15 eV is mainly composed of the 2s state of O atom. From ~11 eV to ~10 eV, it is mainly composed of 5p state of Ba atom. In the Fig.4 (c, d), the 3d state of Ti atom at the conduction band is improved by Hubbard potential correction. The valence band of -5 to 0 eV is mainly composed of 2p states of O atom and 3p and 3d states of Ti atom. The conduction band of 2 eV to 8 eV mainly comes from the 3d state of Ti atom and the 2p state of O atom. It is proved that the strong hybridization between 3d orbital of Ti and 2p orbital of O, Zr 4p and O 2p states is the source of ferroelectricity of BSZT material.

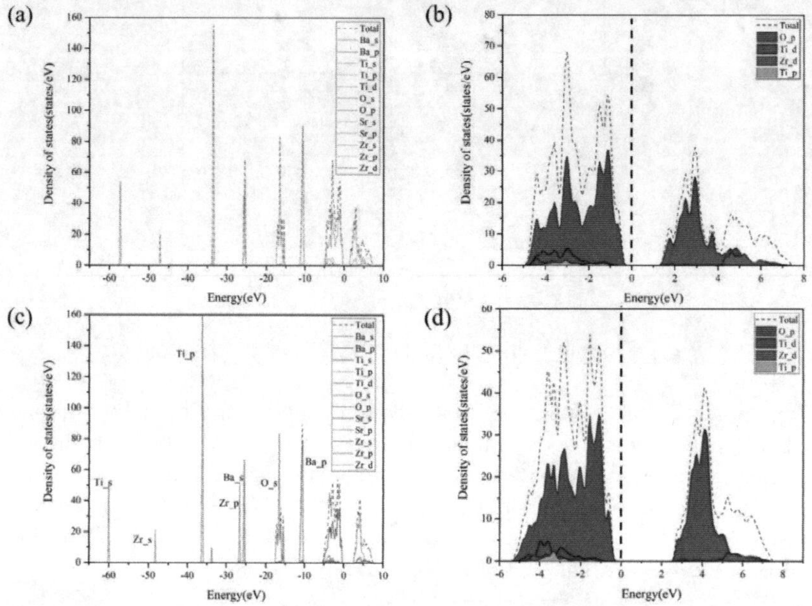

Fig.4. Calculated density of states of BSZT without Hubbard (a, b) and considering Hubbard energy correction (c, d).

4. Conclusion

According to first-principles calculations, we have investigated the structural, electronic and ferroelectric properties of the (BaSr)(ZrTi) material. The double-well curves fit well with the LGD model. The squeezing of Sr ions on the oxygen octahedra and the smaller displacement of Zr ions lead to the reduced spontaneous polarization, revealing the weaker ferroelectricity of BSZT compared to pristine $BaTiO_3$. After adding Hubbard energy adjustment, the band gap of BSZT is 3.14 eV, slightly lower than 3.2 eV of original $BaTiO_3$. The result of density of states demonstrates that the ferroelectricity of BSZT material originates from the hybridization between the 3d states of Ti and the 2p states of O, and the hybridization between the 4d states of Zr and the 2p states of O.

Acknowledgments

This work was financially supported by the National Natural Science Foundation of China (Grant No. 11564015, 12164002), the Fund for Distinguished Young Scholars of Jiangxi Science & Technology Normal University (Grant No.2015QN-BJRC002) and the Educational Commission of Jiangxi Province of China.

References

[1] Ianculescu, A., & Mitoseriu, L. (2010). Ba (Ti, Zr) O (-3)-Functional Materials: From Nanopowders to Bulk Ceramics (K). Nova Science Publishers, Incorporated.

[2] Bai, Y., Han, X., Zheng, X. C., & Qiao, L. (2013). Both high reliability and giant electrocaloric strength in BaTiO 3 ceramics. Scientific reports, 3(1), 1-5.

[3] Liu, X., & Tan, X. (2016). Giant strains in non-textured (Bi1/2Na1/2) TiO3-based lead-free ceramics. Advanced materials, 28(3), 574-578.

[4] Yu, Z., Ang, C., Guo, R., & Bhalla, A. S. (2002). Piezoelectric and strain properties of Ba (Ti 1− x

Zr x) O 3 ceramics. Journal of applied physics, 92(3), 1489-1493.

[5] Takahashi, H., Numamoto, Y., Tani, J., & Tsurekawa, S. (2006). Piezoelectric properties of BaTiO3 ceramics with high performance fabricated by microwave sintering. Japanese journal of applied physics, 45(9S), 7405.

[6] Benlahrache, M. T., Benhamla, N., & Achour, S. (2004). Dielectric properties of BaTiO3–NaNbO3 composites. Journal of the European Ceramic Society, 24(6), 1493-1496.

[7] Cao, W. Q., Xu, L. F., Ismail, M. M., & Huang, L. (2016). Colossal dielectric constant of NaNbO3 doped BaTiO3 ceramics. Mater. Sci.-Poland, 34(2), 322-329.

[8] Qi, H., Zuo, R., Fu, J., & Dou, M. (2017). Thermally stable electrostrains of morphotropic 0.875 NaNbO3-0.1 BaTiO3-0.025 CaZrO3 lead-free piezoelectric ceramics. Applied Physics Letters, 110(11), 112903.

[9] Luo, B., Wang, X., Tian, E., Qu, H., Zhao, Q., Cai, Z., ... & Li, L. (2018). Chemical composition and temperature dependence of the energy storage properties of Ba1-xS rxTiO3 ferroelectrics. Journal of the American Ceramic Society, 101(7), 2976-2986.

[10] Ciomaga, C., Viviani, M., Buscaglia, M. T., Buscaglia, V., Mitoseriu, L., Stancu, A., & Nanni, P. (2007). Preparation and characterisation of the Ba (Zr, Ti) O3 ceramics with relaxor properties. Journal of the European Ceramic Society, 27(13-15), 4061-4064.

[11] Alshoaibi, A., Kanoun, M. B., Ul Haq, B., AlFaify, S., & Goumri-Said, S. (2020). Insights into the Impact of Yttrium Doping at the Ba and Ti Sites of BaTiO3 on the Electronic Structures and Optical Properties: A First-Principles Study. ACS omega, 5(25), 15502-15509.

[12] Kresse, G., & Joubert, D. (1999). From ultrasoft pseudopotentials to the projector augmented-wave method. Physical review b, 59(3), 1758.

[13] Kresse, G., & Furthmüller, J. (1996). Efficient iterative schemes for ab initio total-energy calculations using a plane-wave basis set. Physical review B, 54(16), 11169.

Dynamic Power Allocation in D2D Communications Using Deep Reinforcement Learning

Jun Zhou[1]

[1] School of Electronics and Information Engineering, Shenzhen University, Shenzhen, Guangdong,518000, China

*Corresponding author's e-mail: 17369423604@163.com

Abstract. Due to the development of communication technology, mobile devices continue to increase. As one of the critical technologies of 5G, D2D communication is an up-and-coming technology. In this paper, multiple D2D pairs usually multiplex the same channel, which will cause serious channel interference. To solve this problem, a distributed deep reinforcement learning framework is proposed, which can well adapt to the power allocation in a dynamic environment. The simulation results show that compared with other benchmark methods, the proposed scheme can improve the overall D2D rate.

1. Introduction

In the future, wireless network data transmission has experienced explosive growth, the density of users is increasing, and the interference between users is also increasing, how to improve the capacity of the system, which is very important. In this paper, the problem of maximizing system capacity is investigated.

Different from data-driven methods, many articles have studied the issue of power control. In traditional schemes, Game theory is used for the performance of device-to-device (D2D) communication links through carrier sense multiple [1]. Weighted MMSE (WMMSE) is used as a linear beamformer to maximize system throughput [2]. In [3], a fractional programming algorithm (FP) is used to optimize discrete variables. The above algorithm requires a lot of timely channel state information (CSI).

Machine learning is a new method of resource allocation in the field of wireless communication [4], especially reinforcement learning. The authors in [5] propose centralized Q-learning and distributed Q-learning methods, and shows the advantages of distributed Q-learning. In [6], authors propose a two-stage resource allocation strategy, which guarantee the quality-of-service (QoS) of D2D users. The author proposes a Q-learning-based cognitive radio network transmission scheduling mechanism, the proposed algorithm can transmit data packets better, with lower power consumption and packet loss rate [7]. The authors propose a fully autonomous multi-agent Q-learning algorithm for heterogeneous cellular networks with multiple base stations. The algorithm can show faster convergence after a few iterations [8].

In this paper, a distributed deep reinforcement learning method, called Deep-Q Network (DQN), is proposed, which can be well adapted for power control in dynamic environments [9]. The main contributions are summarized as follows:

1) The proposed method uses centralized training and distributed execution methods, which fully utilizes the computing power of the base station.

2) The proposed algorithm relies on the data generated by the environment to train the network, and is well adapted to power control in a dynamic environment.

2. System Model and Problem Formulation

2.1 System Model

As shown in Fig. 1, in this paper, considers a full-power multiplexed full-synchronization dedicated D2D network. The network contains N pairs of D2D links, it is represented as $N = \{1,2,...,N\}$. The base station covers the entire cell and is located in its center, the D2D links are evenly distributed within the coverage of the base station. Each user's transmitter and receiver are equipped with an antenna.

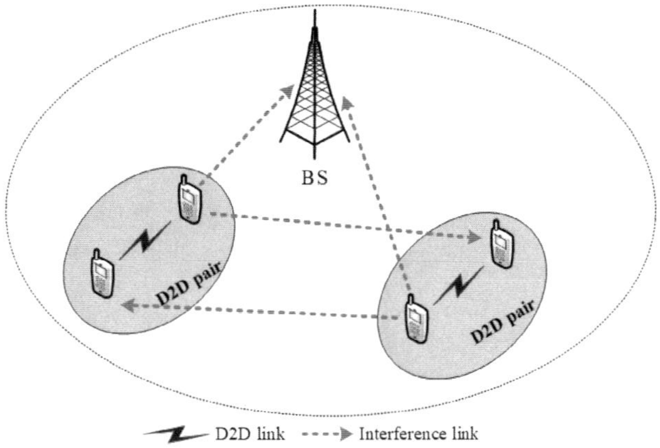

Fig.1 System Model

At slot t, $g_{x,y}^{(t)}$ represents the channel gain in time from the D2D transmitter x to the receiver y, which can be written as,

$$g_{x,y}^{(t)} = | h_{x,y}^{(t)} |^2 \beta_{x,y} \quad (1)$$

where $h_{x,y}^{(t)}$ is the small-scale fading, small scale fading is described using Rayleigh fading, $\beta_{x,y}$ is the large-scale fraction containing path loss and shadow fading, which includes path loss and shadow fading. According to Jake's model, Rayleigh fading is modeled as the first sequence of complex Gaussian Markov processes:

$$h^{(t)} = \varepsilon h^{(t-1)} + \sqrt{1 - \varepsilon^2} e \quad (2)$$

The correlation $\varepsilon = J_0(2\pi f_d T)$, f_d is maximum Doppler frequency, J_0 is the zeroth-order Bessel function of the first kind. $h(0)$ and $e(0)$ follow a complex normal distribution.

The received signal-to-interference-plus-noise ratio (SINR) of D2D link n, which is written as follows:

$$\gamma_{d,n}^{(t)} = \frac{p_{n,k}^{(t)} g_{n,n}^{(t)}}{\sum_{j \neq n}^{N} p_{j,n}^{(t)} g_{j,n}^{(t)} + \sigma^2} \quad (3)$$

At the slot t, $g_{n,n,k}^{(t)}$ represents the channel gain of the D2D link n in time. $g_{j,n}^{(t)}$represents the channel gain from the transmitter of D2D link k to the receiver end of D2D link n, σ^2 represents Gaussian white noise.

The rate obtained by D2D link n, which is expressed as follows:

$$U_n^{(t)} = \log_2(1 + \gamma_{d,n}^{(t)}) \quad (4)$$

2.2 Problem Formulation

This paper investigates the D2D total rate maximization problem, users can select a reasonable transmission power to minimize the interference between users. The problem can be planned as follows:

$$\max_{p} \sum_{n=1}^{N} U_n^{(t)}$$
$$s.t.\ p_n \le p_{\max}, \forall n \in N \tag{5}$$

(5) restricts the maximum transmission power of D2D pair not to exceed $p_{\max}$.

3 Deep Reinforcement Learning Based Power Control

3.1. Overview of reinforcement learning

At time slot t, each agent takes action a^t according to the current state s^t, The environment will feedback to the agent a reward r^t, and agent from the current state s^t to the next state s^{t+1}. The agent achieves dynamic balance by constantly interacting with the environment. In the reinforcement learning, the goal of reinforcement learning is to maximize the cumulative discount reward, which is written as follows,

$$R^{(t)} = \sum_{k=0}^{\infty} \mu^k r^{t+k+1} \tag{6}$$

where $\mu \in [0,1]$, which is called the cumulative discount factor.

The most classic reinforcement learning algorithm is Q-learning, it maintains a Q-table, and the agent keeps interacting with the environment to keep up with the new Q-table. In the current state, the Q-value is expressed as follows:

$$Q^{\pi}(s, a) = E[R^t \mid s^t = s, a^t = a] \tag{7}$$

In order to find the optimal Q value, the update of the Q function obeys Bellman's equation, which is given as follows:

$$q(s^t, a^t) = q(s^t, a^t) + \alpha[\gamma^{t+1} + \mu \max_{a'} q(s^{t+1}, a^{t+1}) - q(s^t, a^t)] \tag{8}$$

3.2 Deep Multiagent Reinforcement Learning

Reinforcement learning contains several basic elements, the action space A, the state space S,,and the reward function R , basic elements represented by a tuple $< S, A, R >$, which is defined as follows:

State space: To maximize the objective function (6), and enable the agent to better adapted to the environment, consider measuring the local state of the agent in two time slots, and the interfering neighbor users and the interfered neighbor users consider the information of the previous one time slot. In this article, the state of each agent is composed of three different state information, local observation information, information about interfering neighbor users, and information about interfered neighbor users.

Local Information: The user's local information takes into account the information of the first two time slots, which allows the agent to better perceive changes in the environment. the transmit power of agent n, $p_n^{(t-1)}$,the spectrum efficiency achieved by agent n, $U_n^{(t-1)}$, the instant channel gain of the agent n, $g_{n,n,k}^{(t)}$, the channel gain of the agent n, $g_{n,n}^{(t-1)}$ 与 $g_{n,n}^{(t-2)}$, Interference received by agent n, $\sum_{j\neq n}^{N} p_{j,n}^{(t-1)} g_{j,n}^{(t-1)}$ and $\sum_{j\neq n}^{N} p_{j,n}^{(t-2)} g_{j,n}^{(t-2)}$.

Interfering Neighbors State: Interfering with the spectrum efficiency of neighbor agent j, $U_j^{(t-1)}$, interference to agent n, $p_j^{(t-1)} g_{j,n}^{(t-1)}$, channel gain of interfering neighbor user j, $g_{j,j}^{(t-1)}$, interfering with the channel gain from neighbor user j to agent n, $g_{j,n}^{(t-1)}$.

Interfered Neighbors State: Spectrum efficiency obtained by interfered neighbor i, $U_i^{(t-1)}$, interference of agent n to interfered neighbor i, $p_j^{(t-1)} g_{j,n}^{(t-1)}$, the direct channel gain of interfered neighbor user i, $g_{i,i}^{(t-1)}$, channel gain from agent n to interfered neighbor i, $g_{i,n}^{(t-1)}$.

Action space: The agent needs to choose the appropriate power, which can reduce the interference between users, $p_{\max}$ is maximum transmit power of D2D link, which is discretized into L levels, the action space A is given as follows:

$$A = \{p \mid p \in \{0, \frac{p_{\max}}{L}, \frac{2p_{\max}}{L}, \ldots, p_{\max}\}\} \tag{9}$$

Reward function: Reinforcement learning is driven by a reward function, which is related to the rate of acquisition of each agent, which can better optimize the objective function (5), the reward function of the agent consists of two parts. The first part is the rate obtained by agent n.

$$U_n^{(t)} = \log_2(1 + \frac{p_n^{(t)} g_{n,n}^{(t)}}{\sum_{j \neq n}^{N} p_{j,n}^{(t)} g_{j,n}^{(t)} + \delta^2}) \tag{10}$$

When the interfered neighbor is interfered, the rate obtained will be reduced, and the second part is the reduced rate. When the interference of agent n is not considered, the rate obtained by agent i is denoted by $U_{i\backslash n}^{(t)}$,which can be given as:

$$U_{i\backslash n}^{(t)} = \log_2(1 + \frac{p_i^{(t)} g_{i,i}^{(t)}}{\sum_{i \neq n}^{N} p_j^{(t)} g_{j,j}^{(t)} + \delta^2}) \tag{11}$$

Total spectral efficiency reduction value of the interfered neighbor. which is given as follows:

$$\overline{U}_n^{(t)} = \sum_{i \in M_R^n} (U_{i\backslash n}^{(t)} - U_i^{(t)}) \tag{12}$$

The set representing the disturbed neighbor is denoted by M_R^n

Finally, the reward function obtained by agent n, which is given as:

$$r_n^{(t)} = U_n^{(t)} - \overline{U}_n^{(t)} \tag{13}$$

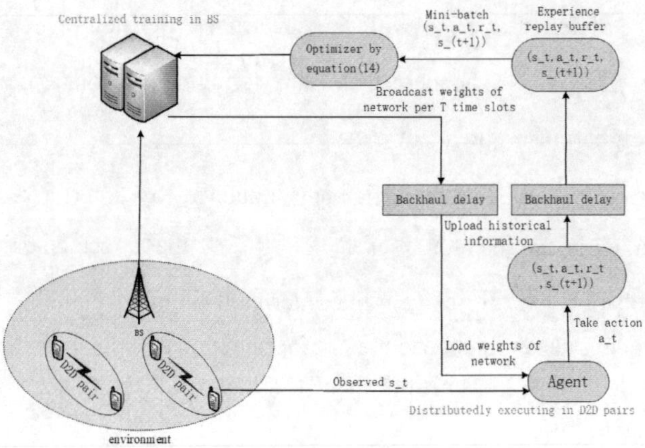

Fig.2 DRL algorithm.

In order to make full use of the computing resources of the base station, the neural network adopts centralized training, distributed execution at each D2D transmitter, and centralized training to simplify implementation and improve stability. Each agent n shares the same set of DQN parameters. As shown in the Figure 2, in the training phase, the centralized network trainer fetches small batches of data from the experience replay buffer, and the it is a queue, which is minimized by optimizing the equation (14). The network training is carried out at the base station, which can make full use of the powerful computing power of the base station, and use the information of all agents to train the same network, which allows the agents to learn faster and accelerates the convergence of the network. After a while, the base station sends the trained network weights to each D2D pair through the backhaul link, and then the D2D transmitter loads the trained neural network parameters, and the D2D user's transmitter selects the appropriate one according to the current state. behavior.

There are two reasons why reinforcement learning can maintain intelligence. First, the way of experience replay makes good use of past experience, and secondly, the correlation of the data is broken by the method of target network. In the time slot t, the agent notices the current state s, takes actions a, the agent transitions to the next state, and obtains a reward, and then the experience samples are stored in the experience replay buffer. After collecting enough experience samples, one B samples are taken from the experience replay buffer to calculate the loss function, which is written as follows:

$$Loss(\theta) = \frac{1}{B}\sum_{i=1}^{B}(y_n - Q(s, a) \mid \theta) \tag{14}$$

y_n is the output Q-value of the target DQN, which is given as follows:

$$y_n = r_n^{t+1} + \mu q(s_n^{t+1}, a_n^{t+1} \mid \theta') \tag{15}$$

4 Simulation Results

Consider a single-cell uplink communication scenario, users are randomly distributed within the coverage of the base station. The maximum distance from the D2D transmitter to the receiver is 60m, the maximum transmission power of the user is 5.42781W, and the number of D2D links is $N \in [5, 17]$. Due to the actual digital signal processing, the user's maximum SINR is 30dB, and Gaussian white noise is -114dBm/Hz. The pathloss model for D2D links is 128.1 + 37.6log(dkm) The Shadowing standard deviation is 10dB.

This paper considers a five-layer neural network, one input layer, one output layer, and three hidden layers. The number of neurons in the hidden layer is 100, 50, and 50. In order to ensure the convergence of the network, the hidden layer uses "Relu" as the activation function, and the size of the experience replay buffer set to 2000, the minibatch size is set to 128, Adam optimizer is used to minimize the loss function, which can Speed up network convergence and find the optimal solution quickly, the learning rate is set to 0.001.

The paper uses four benchmark algorithms to show the superiority of the proposed algorithm. The Q-learning algorithm, Random power allocation, Fractional Programming (Fp), power allocation, and P-max power allocation schemes.

In Fig 3, the performance of the same learning rate and different batch size algorithms is compared. It can be seen from the figure that as the training time increases, the algorithm learning under different batch size conditions reaches a stable state, the learning performance of the algorithm is not sensitive under the conditions of different batch sizes.

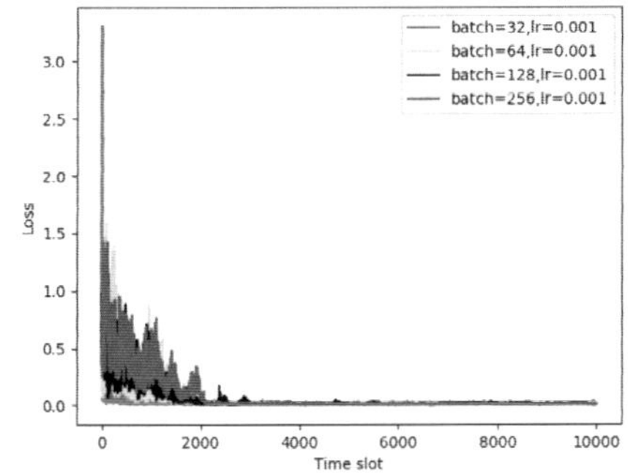

Fig.3 Algorithm learning performance with same learning rate and different batch sizes.

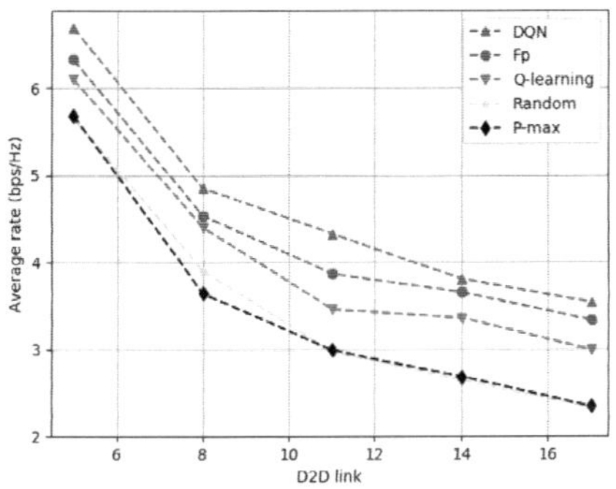

Fig.4 The performance of the average rate obtained by the DQN algorithm different links

Fig. 4 When the D2D link in the system increases, the interference between users becomes more and more serious, and the average rate obtained by the D2D link will decrease. Random power allocation and maximum power allocation are relatively simple to allocate power, and the interference between links is the largest. When the number of links increases, the average rate of the links drops the fastest. It can be seen from the figure that the proposed algorithm is compared with the other four algorithms. With the increase of D2D links, users can obtain a higher average rate and can better select the appropriate transmit power in a dynamic environment.

5 Conclusion

This article studies the problem of power allocation in a dynamic environment. The power allocation problem is formulated as a D2D total rate maximization problem. In this article, a distributed power control method is proposed. Each agent can take appropriate actions according to the current state, and can adapt to power control problems in a dynamic environment. In the future, the power control problem will be applied to the scenario of millimeter-wave communication, and the joint optimization of channel and power will be considered, the current environment will be closer to the natural environment.

References

[1] Lyu,J., Chew,H,Y.,Wong,W,C., (2016)A Stackelberg Game Model for Overlay D2D Transmission With Heterogeneous Rate Requirements. in IEEE Transactions on Vehicular Technology.,65: 8461-8475.
[2] Shi, Q., Razaviyayn, M., Luo, Z.Q, He, C., (2011) An Iteratively Weighted MMSE Approach to Distributed Sum-Utility Maximization for a MIMO Interfering Broadcast Channel. in IEEE Transactions on Signal Processing., 59: 4331-434.
[3] Shen, K., Yu, W., (2018) Fractional Programming for Communication Systems—Part II: Uplink Scheduling via Matching, in IEEE Transactions on Signal Processing., 66: 2631-2644.
[4] Lee, W., Kim, M., Cho, D., (2019)Deep Learning Based Transmit Power Control in Underlaid Device-to-Device Communication. in IEEE Systems Journal., 13: 2551-2554.
[5] Nie, S., Fan, Z., Zhao, M., Gu, X., Zhang, L. (2016) Q-learning based power control algorithm for D2D communication. In 2016 IEEE 27thAnnual International Symposium on Personal. Valencia., pp: 1-6.
[6] Maghsudi, S., Sta´nczak, S., (2016) Hybrid Centralized–Distributed Resource Allocation for Device-to-Device Communication Underlaying Cellular Networks. IEEE Transactions on Vehicular Technology.,65: 2481-2495.
[7] Zhu, J., Song, Y., Jiang, D., Song, H., (2018) A New Deep-Q-Learning-Based Transmission Scheduling Mechanism for the Cognitive Internet of Things. in IEEE Internet of Things Journal., 5: 2375-2385.
[8] Asheralieva, A., Miyanaga, Y., (2016) An Autonomous Learning-Based Algorithm for Joint Channel and Power Level Selection by D2D Pairs in Heterogeneous Cellular Networks. in IEEE Transactions on Communications., 64: 3996-4012.
[9] Nasir, Y.S., Guo, D., (2019) Multi-Agent Deep Reinforcement Learning for Dynamic Power Allocation in Wireless Networks. in IEEE Journal on Selected Areas in Communications., 37: 2239-2250.

A Design of Wideband Microstrip-to-Microstrip Vertical Transition for RF Circuits

Shangbin Sun[1], Yuanxiang Chen[1*], Jia Fu[1], Shangjing Lin[1], Hu Zhu[2], Xiaobo Zhao[2] and Jianguo Yu[1]

[1] School of Electronic Engineering, Beijing University of Posts and Telecommunications, Beijing, 100876, China

[2] Accelink Technologies Co, Wuhan, Hubei, 430205, China

[*]Corresponding author's e-mail: chenyxmail@bupt.edu.cn

Abstract. With the miniaturization of RF circuits, the transmission of signals in different layers becomes important. How to realize the vertical transition structure with wide transmission band and low loss has become the focus of research. In this paper, we proposed a broadband vertical transition structure design based on cavity coupling. This model achieves broadband coupling between the upper and lower microstrips by adding parabolic patches, which improves the model transmission bandwidth. The simulation results show that the optimized model has ultra-wideband transmission characteristic, the 1dB bandwidth is 3.32 GHz-10.42 GHz and the fractional bandwidth is more than 103%. What' more, the thick ground layer of the model improves the heat dissipation performance. Therefore, this structure is very suitable for high-speed and high-density RF circuits.

1. Introduction

With the improvement of signal transmission rate and the miniaturization of devices, multilayer integrated circuits are widely used. And in multilayer integrated circuits, microstrip-microstrip vertical transition structures are essential. One of the methods to achieve vertical transition is based on via [1]. However, this method will introduce parasitic effect at high frequencies, which will damage the signal transmission. Besides, the fabrication can be difficult. Slot-coupled is the second way to achieve this goal [2-4]. This kind of vertical transition structure based on this method has a wide bandwidth but poor thermal performance. At the same time, the thickness of the ground plane of their model was assumed to be infinitesimally thin, which cannot be realized in reality. Another way is based on cavity-coupled. Cavity-coupled has lots of advantages, such as low loss, wide transmission band and good heat dissipation performance [5]. Most of previous studies focus on the design of cavity [5-6], however, the model performance can also improve by adding patches.

In this paper, a novel microstrip-microstrip vertical transition structure based on cavity-coupled is proposed. The coupling between two different layers is enhanced by adding parabolic patches. The optimized model not only has the advantages of traditional cavity coupling structure, but also further reduces the insertion loss and improves the transmission bandwidth. Through the simulation, we get that the 1dB bandwidth of the model is 3.32-10.42GHz. The minimum insertion loss in the transmission band is 0.229dB and the fractional bandwidth is more than 103%.

2. Transition Circuit Design

2.1. Design of the transition structure

The layout of the proposed transition model based on cavity-coupled is shown in Figure 1. The model consists of upper and lower microstrips, two parabolic patches, a thick ground plane and a metallic cavity. The material of the two dielectric substrates is Rogers RO4350, the thickness is 0.5 mm, the relative permittivity is 3.66. The width of the microstrips is 1.1 mm, so the characteristic impedance of microstrips is 50 Ω. In addition, two parabolic patches are opposite to each other through a metal cavity at the ends of the two microstrips in different layers respectively. In this way, the coupling of upper and lower signal is enhanced.

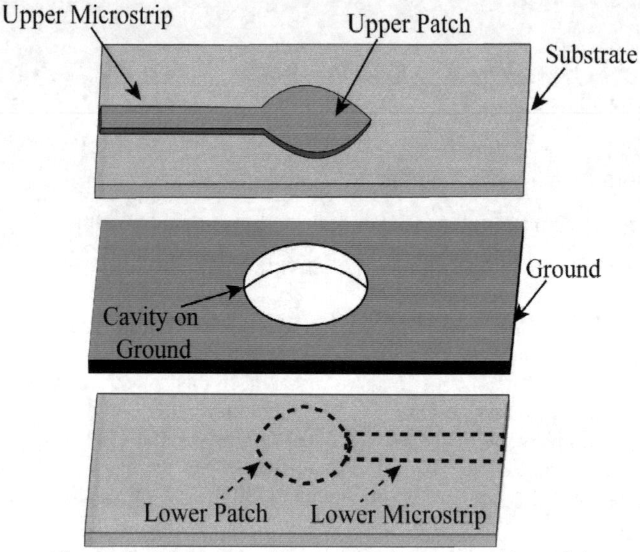

Figure 1. Schematic diagram of the designed model

In this design, we consider the thickness of the common ground plane. An electrically thick ground plane can be used as a radiator for on-board active components. It can also provide isolation between different circuit layers or modules [5]. If the ground plane thickness is too large, it will deteriorate the coupling of upper and lower signal. In contrast, if the thickness is too small, it will make the heat dissipation performance of the model worse, and make the fabrication difficult as well. In consideration, we set the thickness of the common ground plane as 0.1 mm. The radius of the cylindrical metal cavity **Rcav** is set to 3.5 mm, and the resonant frequency of the cavity is 39.2 GHz. The purpose of this design is to make the resonant frequency of the cavity much higher than the transmission frequency, so that it can work in non-resonant mode, to achieve a wider transmission bandwidth. The metallic cavity and the common ground plane are shown in figure 2.

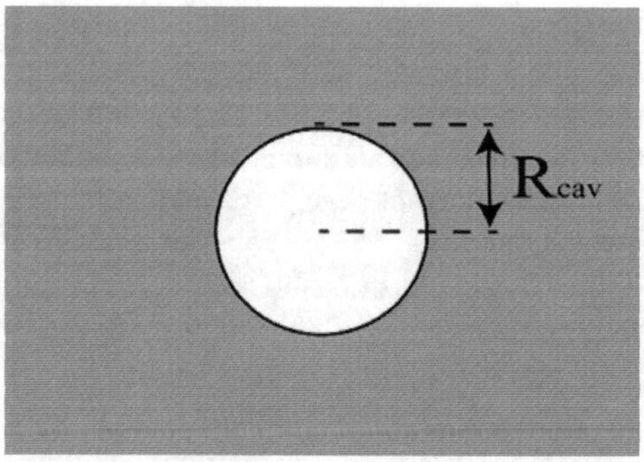

Figure 2. Layout of the metallic cavity and the common ground plane

2.2. Patches design based on parabolic shape

We add two patches at the end of the microstrips to achieve the strong coupling of the upper and lower signal. In size selection, according to the transmission line theory, the length of the patches should be 1/4 central wavelength of the operation to ensure the best transmission performance. In this paper, the patches are designed to be 7 mm in length. In addition, when the shape of the patch resembles a tapered coupled structure, an almost constant coupling value can be obtained in wider frequency band, this feature is helpful to realize the broadband transmission of the model. In this proposed model, we set the patch as a combination of two parabolas to achieve this feature, as shown in figure 3.

The quadratic coefficient of curve 1 is set as **a**, and the quadratic coefficient of curve 2 is set as **-a**. Adjusting the value of **a** is to adjust the width of the patches. If the width is too small, the strong coupling between the upper and lower signal cannot be realized. If the width is too large, the characteristic impedance of microstrips change sharply, which will increase the return loss of the model, and damage the transmission performance. Therefore, the value of quadratic coefficient of parabola is a critical parameter for the proposed model.

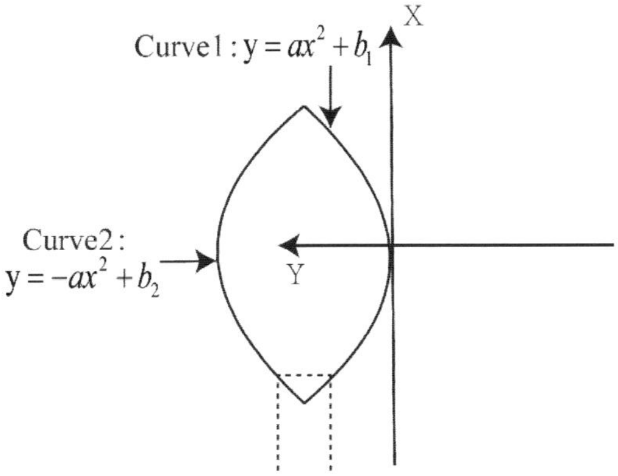

Figure 3. Layout of the parabolic patch

In order to select the best parameters, we obtain the S parameters of this model at the frequency of 1-16 GHz. By changing the absolute value of the quadratic coefficient of the curves, the best performance of this model is obtained. The results are shown in table I.

Table 1. Different values of quadratic coefficients correspond to the transmission performance of the model

Value of the quadratic coefficient	Minimum insertion loss (dB)	1dB bandwidth (GHz)
0	0.264	3.5-5.1
0.12	0.224	3.4-10.1
0.15	0.223	3.3-10.4
0.20	0.283	3.2-9.5
0.25	0.278	3.8-7.5

3. Result and Discussion

The results make known that under this current transmission scenario, when the parabolic quadratic coefficient **a** is 0.15, the model transmission performance is best. On this premise, the minimum insertion loss of this model is 0.223 dB at 4.33 GHz, the 1dB bandwidth is 3.32-10.42 GHz. In the transmission band, the return loss of this model is better than 10dB. It shows that the model we designed has the advantages of low loss and wide transmission band. The S-parameter simulation results are shown in figure 4.

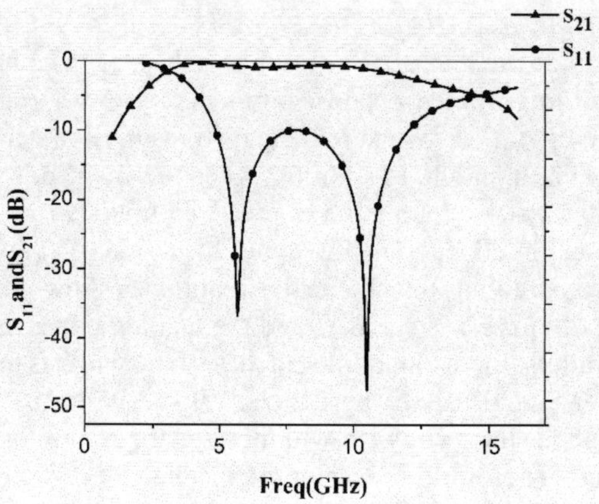

Figure 4. S_{11} and S_{21} of the proposed model

In order to verify the performance of the model by adding patches, we compare the transmission performance of the optimized model with that of the model without patches. Besides, to judge whether the model has ultra-wideband characteristic, the fractional bandwidth is also a critical factor. In this work, the transmission bandwidth of the proposed model is 3.32 GHz-10.42 GHz. So, the fractional bandwidth of the proposed model is 103.3%. The simulation results show that the proposed model has ultra-wideband transmission performance.

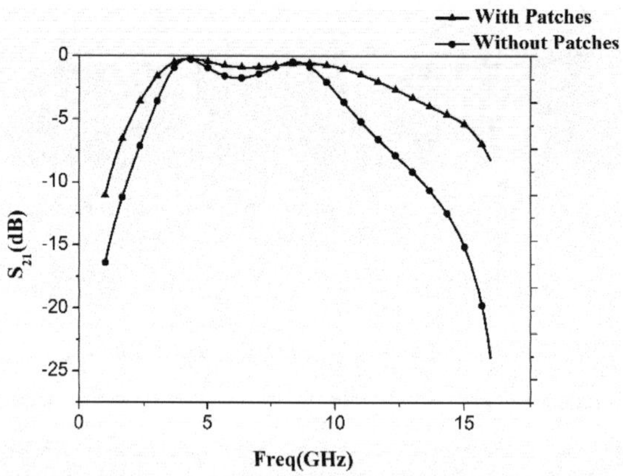

Figure 5. Comparison of model transmission coefficients

4. Conclusion

In this paper, a microstrip-microstrip vertical transition structure based on cavity-coupled is designed. This vertical signal transition structure with high bandwidth and low loss provides a solution for signal transition at different layers of RF circuits. This shows that this work is meaningful and necessary.

By adding parabolic patches at the end of the microstrips, the coupling of the upper and lower signals is enhanced. When the absolute value of parabolic quadratic coefficient a is 0.15, the model can achieve the best performance. The return loss of this optimized model is better than 10dB. A minimum insertion loss of 0.223 dB is obtained at 4.33 GHz. The 1 dB bandwidth of the proposed model is 3.32-10.42 GHz, and the fractional bandwidth of the proposed model is 103.3%. At the same time, the thick ground plane provides good heat dissipation performance, which can be well used for high frequency microwave circuits.

In the future work, the sensitivity of the proposed structure will be studied to prove that the structure is robust and its performance will not be excessively damaged by the dimensional error during the fabrication processes.

Acknowledgments

This work was supported by the National Key Research and Development Program of China (2018YFB2200903), the National Natural Science Foundation of China (NSFC, NO. 61875239, 61701034, 61531007 and 61875016) and BUPT Excellent Ph.D. Students Foundation (CX2019124).

References

[1] Tsai C. C., Cheng Y. S., Huang T. Y., et al. (2011) Design of microstrip-to microstrip via transition in multilayer LTCC for frequencies up to 67 GHz. IEEE T COMP PACK MAN., vol. 1, no. 4, 595–601.

[2] L. Zhu and K. Wu. (1999) Ultra broadband vertical transition for multilayer integrated circuits. IEEE Microw. Guided Wave Lett., vol. 9, no. 11, 453–455.

[3] X. B. Huang and K.-L. Wu. (2012) A broadband and vialess vertical microstrip-to-microstrip transition. IEEE Trans. Microw. Theory Techn., vol. 60, no. 4, 938–944.

[4] Ziwen Tao, Jianpeng Wang, Yan Dou. (2016) Broadband Microstrip-to-Microstrip Vertical Transition Design, IEEE MICROW WIREL CO., vol. 26, no. 9, 1-3.

[5] Li, J. Cheng, and C. Lai. (2006) Designs for broadband microstrip vertical transitions using cavity couplers, IEEE Trans. Microw. Theory Tech., vol. 54, no. 1, 464–472.

[6] D. M. Pozar, et al. (2003) Analysis and design of cavity-coupled microstrip couplers and transitions, IEEE Trans. Microw. Theory Tech., vol. 51, no. 3, 1034–1044.

Design of Multi-band Compatible Trigeminal Monopole Array Antenna for Unmanned Aerial Vehicle

Zhiwei Zhang, Bin Lin*, Yuting Huang, Ye He, Haoyuan Chen, Meijie Zheng, Shuyao Wen

Xiamen University Tan Kah Kee College, Fujian Zhangzhou 363105, China

*181318339@qq.com

Abstract—Compared with the performance index of UAV system antenna, a compatible trigeminal monopole array antenna is proposed. This antenna uses photosensitive resin as the substrate material, and can be made with UAV shell using 3D printing technology, realizing the integration of the antenna system and UAV. The antenna uses a trigeminal monopole antenna as the basic radiation unit, which is arranged in accordance with the Sierpinski fractal structure to form fractal radiation array. While enhancing the radiation through the array, it ensures that the antenna has a larger working bandwidth. In radiation performance test, the antenna is operating over a frequency spectrum of 1.038-11.916 GHz, and the minimum S_{11} is -37.92 dB. The proposed antenna has multi-band compatibility, better radiation performance, and greater performance redundancy.

1. Introduction

UAV is one of the greatest and most practical inventions in the 21st century. It is a revolutionary improvement on manned aircraft. It is the crystallization of a variety of advanced technologies such as aircraft power technology, automatic control technology, wireless transmission technology and so on. It has the characteristics of small size, low cost, vertical or short-range take-off and landing, simple structure, flexible operation, etc, and at the same time has important military and civilian value. Antenna, a communication tool between UAV and its controller, is an important component of UAV. The flight control, wireless data transmission, and function command transmission of UAV all depend on antenna. The performance of antenna directly determines the performance of UAV. Using the UAV shell material as the antenna dielectric plate, and fusing the antenna with the UAV shell, it is possible to design a UAV conformal antenna with excellent performance and good working stability.

At present, the main frequency bands used for UAV communication are the Bluetooth system 2.400～2.4835 GHz[1-4], the second to fifth generation mobile communication frequency bands 0.905-0.960 GHz, 1.710-1.980 GHz, 2.010-2.400 GHz, 2.570-2.620 GHz, 3.300-4.990 GHz[5-7], ultra-wideband system 3.100～10.600 GHz[8-10], GPS system 1.164～1.577 GHz, Beidou system 1.204～1.564 GHz. The UAV antenna must be able to work in the above frequency bands at the same time, have multi-band compatibility, good radiation performance, and greater performance redundancy.

2. Introduction to Trigeminal Monopole Antenna

Figure 1 show the schematic diagram of trigeminal monopole antenna structure.It consists of a feeding area, a monopole radiating arm and two folded monopole radiating arms. The monopole radiating arm mainly covers the frequency bands of Beidou system, Bluetooth system, and the second to fifth generation mobile communication. The folding monopole radiating arms mainly cover the fifth-generation mobile communication and ultra-wideband system frequency bands. Monopole radiation structure has high antenna gain, which can make the antenna achieve strong radiation intensity. The radiating arms are composed of multiple corner-cut fractal radiation patches, and the self-similarity of the fractal structure can effectively expand the working bandwidth of the antenna.

Fig.1 Schematic diagram of trigeminal monopole antenna structure

3. Introduction to Fractal Array Structure

Multiple antennas form an array according to a certain arrangement rule, which can effectively enhance radiation. The Fractal Array Structure is an antenna array method arranged according to the fractal law, which can give full play to the characteristics of the fractal structure in broadband work. The arrangement method ensures that the antenna has a larger working bandwidth while enhancing the radiation through the array. Sierpinski Fractal Structure is a common surface fractal structure, which is widely used in the design of radio frequency antennas. Figure 2 show its iterative process. Arranging the array element antennas according to the Scherpinski fractal law can enhance the radiation intensity and ensure the broadband performance.

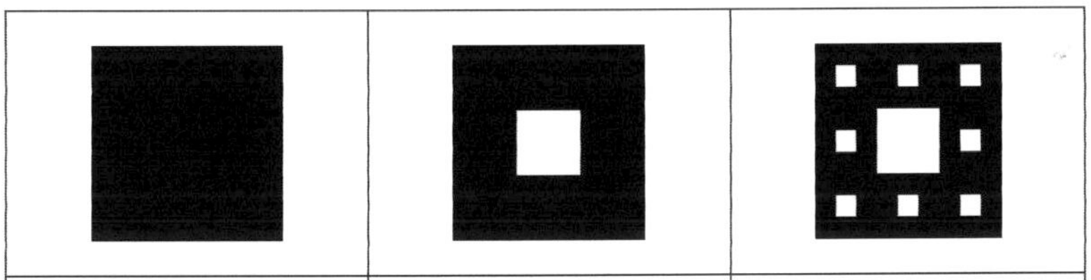

(a) Original graphics	(b) 1st iteration	(c) 2nd iteration

Fig.2 Schematic diagram of Sierpinski's fractal iteration process

4. Antenna Structure Design

The antenna uses photosensitive resin as the substrate material, and its εr=6.5. The overall size of the antenna is 30mm×30mm×1mm. Figure 3 show the integral structure of the antenna radiator. The radiator is composed of 8 trigeminal monopole antennas according to the first-order Sierpinski fractal structure to form a fractal antenna array. In this way, the antenna radiation intensity can be enhanced while maintaining the broadband performance of the trigeminal monopole antenna to the maximum extent, and the antenna can completely coverage the frequency bands of Beidou system, Bluetooth system, the second to fifth generation mobile communication and ultra-wideband system.

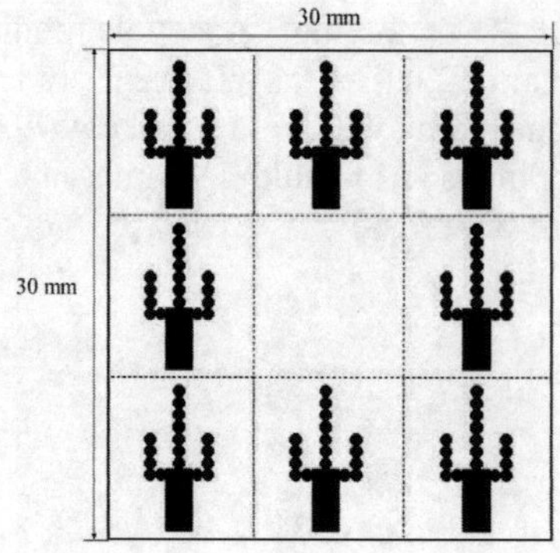

Fig.3 Schematic diagram of antenna radiation patch structure

5. Antenna radiation performance test

We have made antenna samples and conducted actual tests on their radiation performance and pattern performance. The results are shown in Figure 4 and Figure 5.

According to the actual test results, the working frequency range of the antenna is 1.038～11.916 GHz and the minimum S_{11} is -37.92dB. The antenna has multi-band compatibility, better radiation performance, greater performance redundancy. It can be seen that the electric and magnetic planes of the antenna pattern cover all radiation angles and have excellent omnidirectional radiation capabilities.

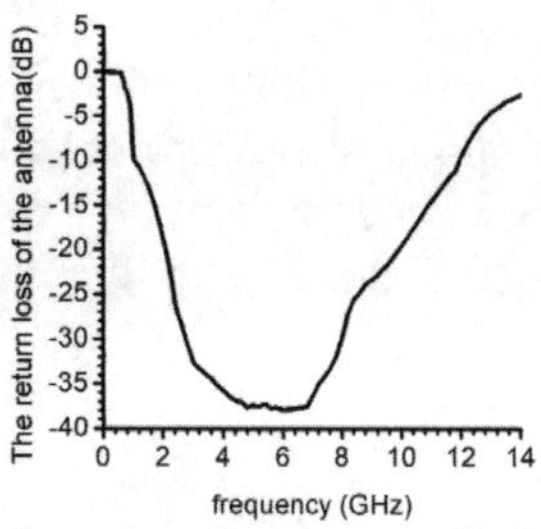

Fig.4 Measured results of antenna radiation performance

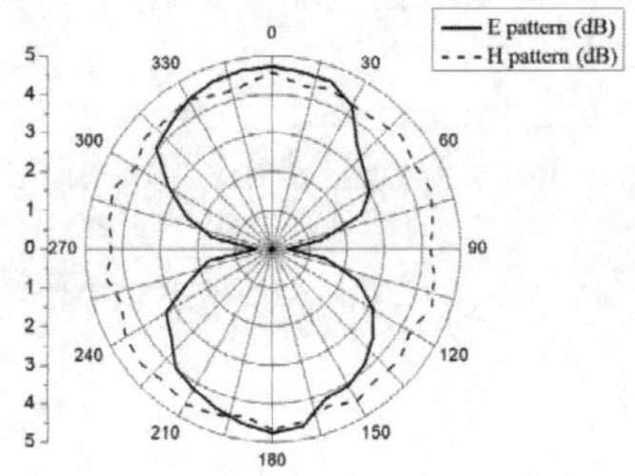

Fig.5 Measured results of antenna pattern

6. Conclusion

In this paper, according to the performance requirements of UAV for antenna system, the photosensitive resin of UAV shell material is selected as the antenna substrate material, and the antenna and UAV shell are fused together. The trigeminal monopole antenna is used as the radiation element antenna, and the antenna array is arranged according to the fractal law. The antenna array is designed with multi-band compatibility, which can cover the Bluetooth system frequency band, the second to fifth generation mobile communication frequency bands, the ultra-wideband system frequency band, the GPS system frequency band, and the Beidou system frequency band. The radiation performance is good, and the multi-band compatibility of trigeminal monopole array antenna with large performance redundancy is achieved. According to the actual test results, the antenna can conform to UAV shell, has excellent multi-band compatibility, strong radiation ability, good omnidirectional working ability, which has broad application prospects in the field of UAV.

Acknowledgments

The research work in this paper was funded by the university student innovation and entrepreneurship training planned project "UAV Multi-band Compatible Conformal Antenna Design".

References

[1] Chen S, Zhang M, Zhao J, et al. Reliable and Practical Bluetooth Backscatter With Commodity Devices [J]. IEEE/ACM Transactions on Networking, 2021, PP (99):1-13.

[2] Schwebel D C, Hasan R, Griffin R, et al. Reducing Distracted Pedestrian Behavior Using Bluetooth Beacon Technology: A Crossover Trial [J]. Accident Analysis & Prevention, 2021, 159(5):106253.
[3] Wisana I, Nugraha P C, Estiwidani D. The Effectiveness Obstructive Sleep Apnea Monitoring Using Telemedicine Smartphone System (TmSS) [J]. Journal of Biomimetics, Biomaterials and Biomedical Engineering, 2021, 50:113-121.
[4] Srilaxmi V, Babu S K. Bluetooth and LAN Enabled Printer Adapter using Raspberry Pi [J]. International Journal of Scientific and Engineering Research, 2021, 06(7):1370-1372.
[5] Kurri V, Raja V, Prakasam P. Cellular Traffic Prediction on Blockchain-Based Mobile Networks Using LSTM Model in 4G LTE Network [J]. Peer-to-Peer Networking and Applications, 2021, 14(3):1088-1105.
[6] Won S, Sang W C. Three Decades of 3GPP Target Cell Search through 3G, 4G, and 5G [J]. IEEE Access, 2020, 8:116914-116960.
[7] Gutierrez D, Gimenez F, Zerbini C, et al. Measurement of 4G LTE Cells with SDR Technology[J]. IEEE Latin America Transactions, 2020, 18(2):206-213.
[8] Dwairi M O, Soliman M S, Alahmadi A A, et al. Design and Performance Analysis of Fractal Regular Slotted-Patch Antennas for Ultra-Wideband Communication Systems [J]. Wireless Personal Communications, 2019, 105(3): 819-833.
[9] Zhu X M, Yi J J, Cheng J Y, et al. Adapted Error Map Based Mobile Robot UWB Indoor Positioning [J]. IEEE Transactions on Instrumentation and Measurement, 2021, 69(9): 6336-6350.
[10] Zhou N, Lau L, Bai R, et al. Novel Prior Position Determination Approaches in Particle Filter for Ultra Wideband (UWB)-Based Indoor Positioning [J]. Journal of the Institute of Navigation, 2021, 68(2): 277-292.

Spine Image Segmentation Based on U-Net and Atrous spatial pyramid pooling

Zhen Yang[1 a*], Long Chen[1 b*], Ting Fu[1 c*], Zhijian Yin[1 d*], Fan Yang[1 e*]

[1] Jiangxi Science and Technology Normal University, Nanchang 330013, China

[a*]yangzhenphd@aliyun.com,[b*]1020170701@jxstnu.com,[c*]criafting@aliyun.com, [d*]zhijianyin@aliyun.com, [e*] kooyang@aliyun.com

Abstract—In the past ten years, deep learning has achieved remarkable results in the area of natural image segmentation, and has gradually turned to the field of medical image segmentation. The precise segmentation of spine images can be used for early screening of spondylopathy, which is convenient for early detection and treatment of patients. Aiming at the segmentation of the spine image by the U-Net network, which structure will lead to large model calculation, network overfitting, image size, noise information and other issues. This paper introduces a new method based on spatial pyramid pooling module ASpp on U-Net network and applies the proposed network structure to the segmentation of the spine image. The model is first enhanced by the Spp module and Densenet data; secondly, the U-Net encoding and decoding structure is adopted, and the deep separable convolution is used. This method greatly reduces the complexity and computation of the model, and uses a rectangular convolution kernel to increase the computation of the model in a small amount. Based on the volume, the receptive field of the convolution operation is expanded; finally, in order to effectively enhance the segmentation area of the image, the SE-Net attention mechanism module is added to the lateral connection. The method proposed in this paper conducts ablation experiments on a public dataset of spine images and compares it with the U-Net network. The method proposed IOU in this paper can reach 72%, and the F1-score can reach 0.92. The comparison of the experimental results of spine medical image datasets shows that the method proposed in this paper can effectively improve the accuracy and accuracy of spine image segmentation.

1. Introduction

In the last decade, deep convolutional neural networks (CNNs) have been widely used in computer vision community, and have achieved great progress in a board range of tasks. e.g., image classification, object detection and instance segmentation[1]. Results of Deep Learning in Computer Vision was inspired by a lot of scholars, and applied it to medical image segmentation. However, the instance segmentation is still a challenging task in the computer vision, which requires an algorithm to predict a per-pixel mask with a category label for each instance of interest in an image[2]. U-Net network structure as the earliest deep learning medical image segmentation framework. However, the U-Net network loses the feature information of medical images when the encoder and decoder are connected laterally, resulting in a feature semantic gap [3].

The U-Net network is a fully convolutional neural network[4]. The input and output are images without fully connected layer. The shallower high-resolution layer is used to solve the problem of pixel positioning, and the deeper layer is used to solve the problem of pixel classification[5]. However, the U-Net still has some disadvantages and shortcomings, such as: effective convolution increases the difficulty, universality of model design, and the form of trimming is not symmetrical with the Feature Map[6].

Therefore, to overcome these problems in the process of image feature extraction, We propose a scheme for implementing image instance segmentation. The main strategies of this paper include the following three points[7]:

1) Inputting image through the Spp module. After the image passes through the Spp module, the final output image is the scale[8];

2) In the U-Net network, Atrous convolution is used to replace the original ordinary 3x3 convolution, which can expand the image receptive field without losing resolution. In this way, the U-Net network can simultaneously pay attention to the context information of the spine image area, and obtain more high-level semantic features of the image;

3) The attention mechanism SE-Net is added after each horizontally connected block of the U-Net network, which increases the accuracy and effectiveness of the network model for the segmentation of the target area while increasing the lightweight computation.

In section2, the paper will introduce recent work, and in section3 the paper will introduce their framework . In the last, we will show our experimental process and results, at the same time we will summarize our work and future prospects.

2. Related Works

In the computer vision, deep learning plays a key role in the application of face recognition, object detection, object tracking, defect detection and so on[9]. The precise segmentation of spine medical images is significant for the diagnosis of diseases such as spondylitis and cervical spondylitis. Therefore, the precise segmentation of spine images can be used for early screening of some spinal diseases, facilitating the early detection and treatment of patients.

Dilated convolution can expand the receptive field without loss of feature image resolution, which is very useful in target detection and segmentation tasks[10]. On the one hand, dilated convolution can detect and segment large targets in a large field, on the other hand, the high resolution can accurately locate target. Moreover, when different dilation rates are set for the dilated convolution, the receptive field will be different, that is, multi-scale information is obtained[11]. In the field of computer vision, multi-scale image information plays an important role in vision tasks.

One of the representative methods is squeeze-and-excitation network, which learns channel attention for each convolution block and provides a significant performance improvement for various deep CNN architectures[12].

Due to the variability of spine images, there are differences in size, brightness, and shape between patients, which makes it impossible to judge the boundaries. Therefore, the segmentation of many spinal images in hospitals mainly relies on the manual annotation of doctors and medical experts to segment the foreground and background information of spinal images. Yet, With the development of society, the number of patients is also increasing, manual labeling not only takes time and effort, but also has a high rate of misjudgment, which affects doctors' follow-up treatment methods. The method we propose in this paper improves the U-Net network, which can effectively segment medical spine images and greatly improve the accuracy and accuracy of segmentation.

3. Methods

The method proposed in this paper is mainly composed of four parts.: (1) Spatial Pyramid Pooling; (2) U-Net; (3) ASpp; (4) SE-Net module. In this section, we will describe our overall framework structure in detail, as shown in Figure 1, and describe the details of each branch in detail.

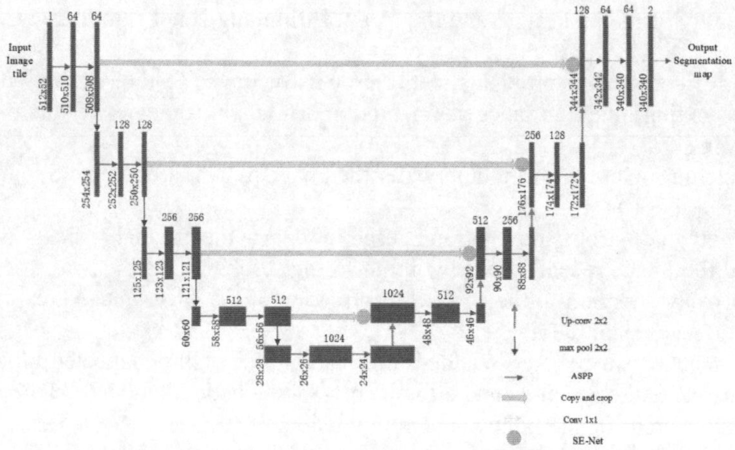

Figure 1. The network model structure of our method

1) Spatial Pyramid Pooling

As shown in Figure 2, the black picture represents the feature map after convolution, then we extract the features of the feature map with three blocks of different sizes, which are $4 \times 4, 2 \times 2, 1 \times 1$, and put these three grids below. Through feature extraction of three blocks of different sizes, on the input image feature map, we can get 16+4+1=21 different blocks (Spatial bins). From these 21 different blocks, we extract a feature from each block, which is exactly the 21 dimensions we want to extract. This process of pooling by combining different sizes of grids is called spatial pyramid pooling (Spp). For example, the maximum pooling of spatial pyramid is to calculate the maximum value of each block from 21 image blocks to obtain an output unit, and finally obtain a 21-dimensional feature output. In this way, the required number of features can always be obtained for different inputs, without fixing the input size of the convolution layer.

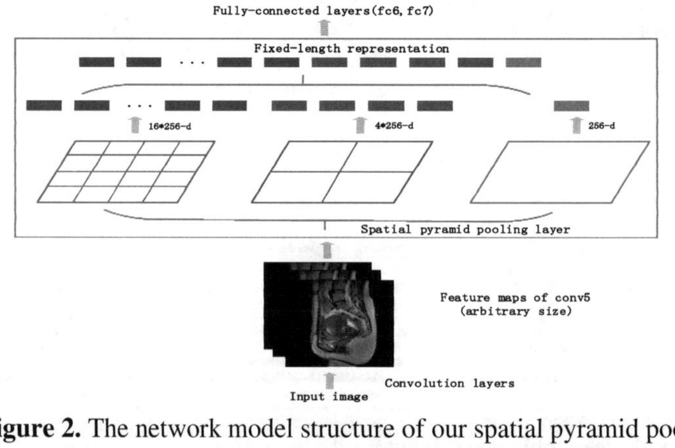

Figure 2. The network model structure of our spatial pyramid pooling

2)U-Net

As shown in Figure 3, in the structure of U-net, it includes a contraction path that captures context information and a symmetrical expansion path that allows precise positioning[13]. The U-Net network adopts an end-to-end training method, that is, the input is an image, and the output is also an image, and the best results can be obtained with very little data. The structure of U-Net is shown in the Figure 3. The left side of the U-Net network is the contraction path for encoding, and the right side is the expansion path for decoding. The encoder has four sub-modules, each containing two convolutional layers. After each sub-module, there is a down-sampling layer implemented by max pool[14]. The resolution of the input image is 572x572, and the resolutions of modules 1-5 are 572×572, 284×284, 140×140 , $68 \times 68$and 32×32 respectively. Since the convolution uses the valid mode, the resolution of the next sub-module here is equal to (resolution of the previous sub-module -4)/2. The decoder consists of four blocks, and through the up-sampling operation, the feature image resolution is restored in turn(since the convolution uses valid mode, the actual output is smaller than the input image). The U-Net network also uses lateral connections to fuse the features of the encoder and decoder, and fuse the results of the encoder network up-sampling and the decoder up-sampled results as the input of the next block.

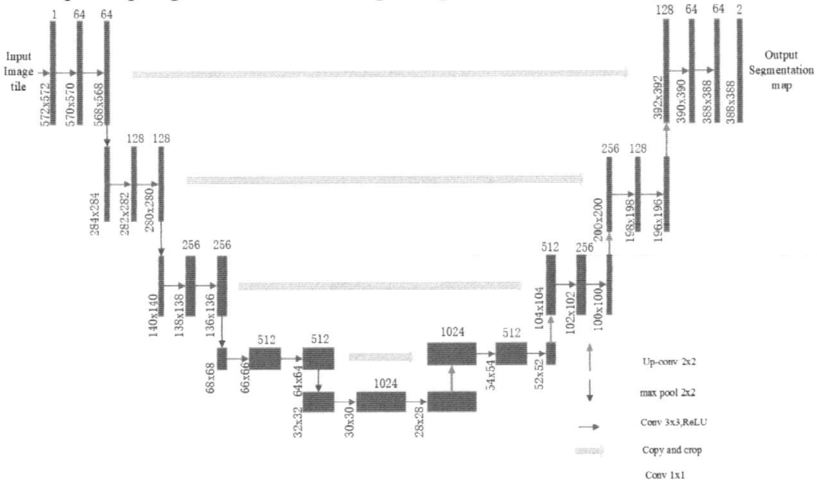

Figure 3. The network model structure of U-Net

3) ASpp

ASpp was proposed in deeplabv2. In the semantic segmentation task, we hope to have a larger receptive field for the features extracted from the images, and hope that the resolution of feature images will not decrease too much (too much resolution loss will lose a lot of detailed information about the image boundary), but these two are contradictory. If larger receptive fields are to be obtained, larger convolution kernels or larger steps are needed for pooling. For the former, the amount of calculation is too large, and the latter will lose resolution. Moreover, the dilated convolution is used to solve this contradiction, and it can get a larger receptive field without losing too much resolution. The dilated convolution is shown in Figure 4. Compared with ordinary convolution, atrous convolution enlarges the receptive field when the pooling loss information is small, so that each convolution output contains a wider range of information, ensuring that the output image features have a higher resolution. As shown in Figure 4, the ASpp module uses parallel convolution layers of holes with different dilation rates to capture multi-scale information on feature images, and fuse the output results to obtain image segmentation results. As shown in the Figure below, this is an improved ASpp in deeplab v3. A 1×1 convolution and three 3×3 dilated convolutions are used, and each convolution kernel has 256 channels and a BN layer. In fact, the 1×1 convolution is equivalent to a dilated convolution with a large rate, because the larger the rate, the smaller the effective parameters of the convolution kernel. Therefore, this 1×1 convolution kernel is equivalent to the large rate convolution kernel.

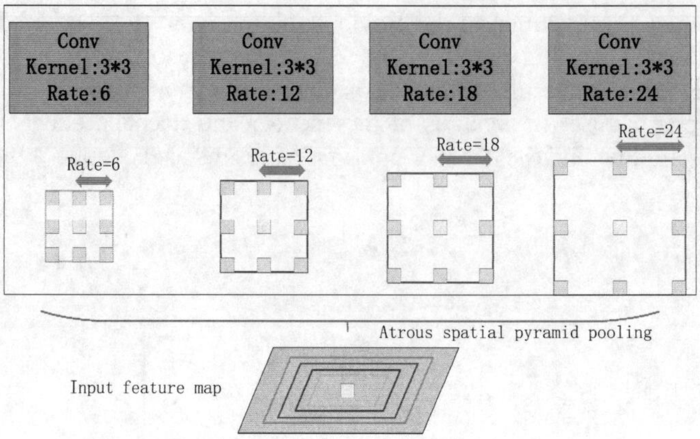

Figure 4. The network model structure of ASpp

4) SE-Net Attention

As shown in Figure 5, in the network structure of SE-Net, given an input x, the number of feature channels is C1. After convolution, compression, decompression, connection and other operations, the number of feature channels is finally obtained as C2. The SE-Net network first squeezes the operation, which uses a global pooling operation to synthesize the input features of size H*W*C into a feature description of C*1*1. Turn each two-dimensional feature channel into a real number with a global receptive field, and match the output dimension H*W*C with the input feature channel number. Second, the Excitation operation, which is a gate mechanism in a recurrent neural network, generates weights for each channel in the feature map by optimizing parameters, and sets different weights for different parts of the image, where the parameters are learned to explicitly construct features Correlation between channels. Finally, the Reweight operation weights the weight of the excitation output to the previous feature channel by channel by multiplication, Complete the re-calibration of the original features in the channel dimension and use them as the input data for the next level.

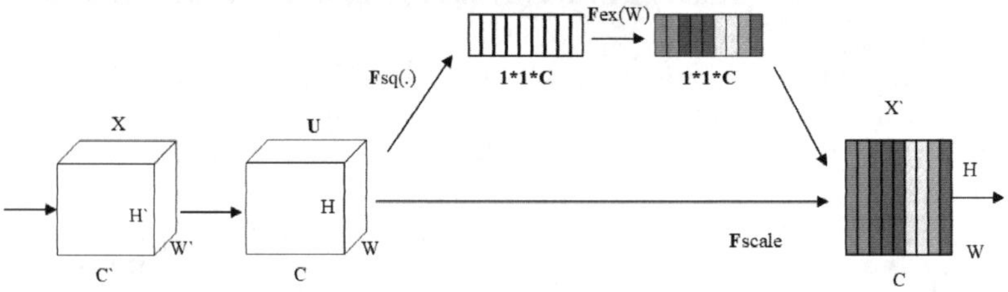

Figure5. Squeeze and Excitation block

4. Experiment

4.1. Implementation Details

The deep learning framework used in this experiment is Keras, which is based on the Tensorflow2.0 backend. The hardware configuration for network training is Intel Core i7-10700CPU@2.9GHz, GeForce GTX 1660Ti, Nvidia driver-450, Ubuntu18.04.5 LTS 64bit operating system, 16G RAM, 2.2TB Disk, and python3.6 and CUDA10 software configuration .0CUDNN7.0.5, image saving and other processing are based on the open-source software OpenCV.

In the model proposed in this paper, the size of the input image is uniformly 512*512*3, and in order to prevent distortion, the image is fine-tuned. The input image preprocessing input is normalized by dividing it by 255. When training the network, the batch_size is set to 4, and customized data enhancement methods were used, such as random direction translation, random contrast enhancement, random gamma transformation and random small-scale scaling, etc; each iteration steps_per_epoch is 50; the training epochs is 300 rounds; the early stopping method is used to control the training process. If the loss value of the verification does not decrease within 10 epochs, the training will be stopped to prevent overfitting. The optimizer is Adam, and the initial learning rate is both 0.0001; the learning rate training strategy is that if the loss does not decrease within 3 epochs during the training process, the learning rate is reduced to 0.1 times the original.

In this paper, the experimental data uses the public medical spine dataset, which is a spine image randomly selected from different ages and human bodies. The training set contains 1211 spinal images and corresponding masks; the remainder is in the test set. Real pixel labels are used for testing. The spine imageis in PNG format, the size of each image is 20kb, the resolution of each spine image is 120 pixels wide and 120 pixels high. The images and labels are shown in Figure 6:

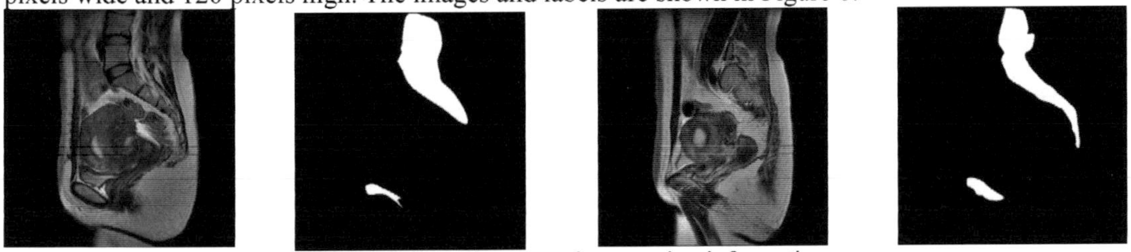

Figure6. Spine image and annotation information

The input image is uniformly fine-tuned to 512*512*3 size. The input image preprocessing input is divided by 255 for normalization. The normalization formula of Xmax-Xmin=255 is as follows:

$$X_{norm} = \frac{X - X_{min}}{X_{max} - X_{min}} \tag{1}$$

4.2. Evaluation Metrics

$$Iou = \frac{T_P}{T_P + F_N} \tag{2}$$

$$R_{call} = \frac{T_P}{T_P + F_N} \tag{3}$$

$$R_{Precision} = \frac{T_P}{T_P + F_P} \tag{4}$$

$$F_1 = \frac{2R_{Precision}R_{Precall}}{R_{Precision} + R_{recall}} \tag{5}$$

In the above formula, T_P represents the number of positive samples tested, and the test is correct, that is, true positive; T_N represents the number of negative samples tested, and the test is correct, that is true negative; F_P represents the number of positive samples tested, But the detection is wrong, that is false positive; F_N represents the number of negative samples detected, but the detection is wrong, that is false negative.

4.3. Loss Function

Our method uses a loss function with boundary weights:

$$Loss_{Dice} = 1 - \frac{2|X \cap Y|}{|X| + |Y|} \tag{6}$$

X is the Ground Truth segmented image, and Y is the predicted segmented image. Dice coefficient is a key evaluation index in medical image segmentation effect. The Dice loss function compares the intersection and union between the prediction result area and the ground truth area, and calculates the function loss by taking all the pixels in a category as a whole.

$$L = -y\log y - (1-y)\log(1-y') = \begin{cases} logy' & y = 1 \\ -\log(1-y') & y = 0 \end{cases} \tag{7}$$

Among them, γ is the key factor to increase the loss of cross entropy. When γ > 0, the loss of classified samples can be effectively reduced, and the problem that difficult samples are difficult to mine can be solved. α is a balance factor used to balance the problem of unbalanced proportion of positive and negative samples.

4.4. Experimental images and forecast results

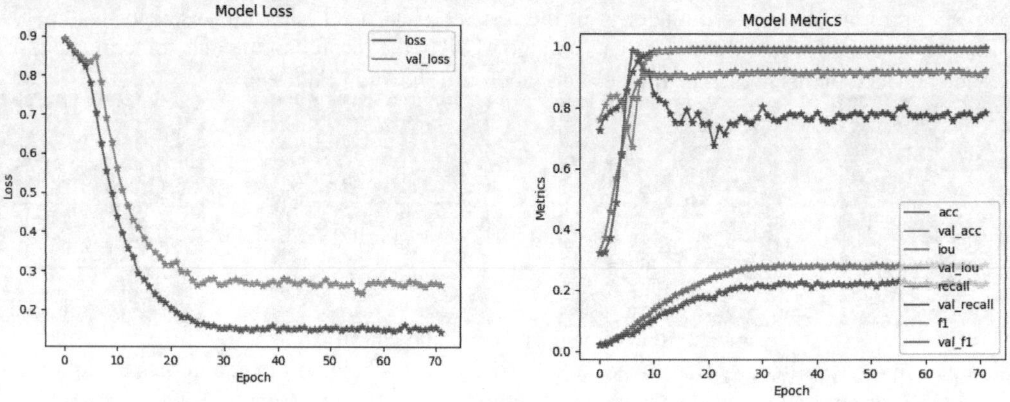

Training loss Training metrics

Algorithm	IOU /%	Recall /%	F1-score	Dice /%	Sensitivity /%	Specificity
U-Net	65.10	87.84	0.81	78.36	87.84	94.66
SE-UNet	66.82	88.32	0.84	79.74	88.49	95.56
Our Method	72.28	90.87	0.90	84.77	89.13	97.18

Figure7. Comparison results on three different methods
Table 1. A table with headings spanning two columns and containing notes

According to the comparison of ablation experimental results, the method proposed in this paper is the best for medical image segmentation, and our experimental results has obvious improvement in F1-score and IOU, which shows that ASpp and SE-Net network have some advantages in U-Net spine image segmentation.

4.5. Experimental images and forecast results

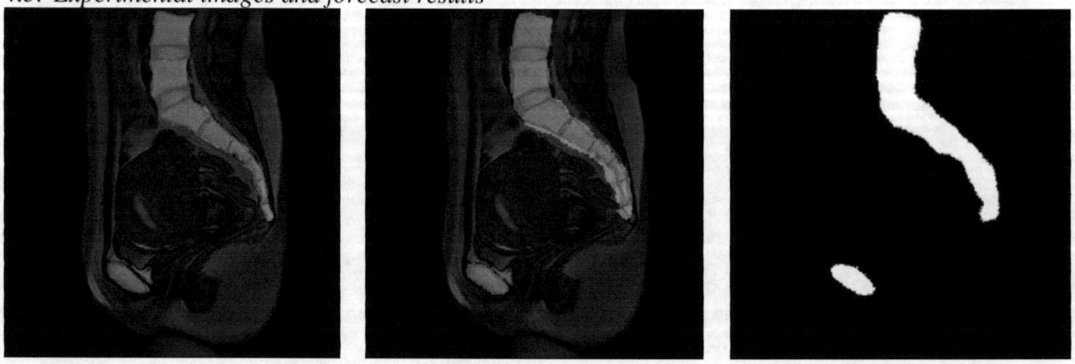

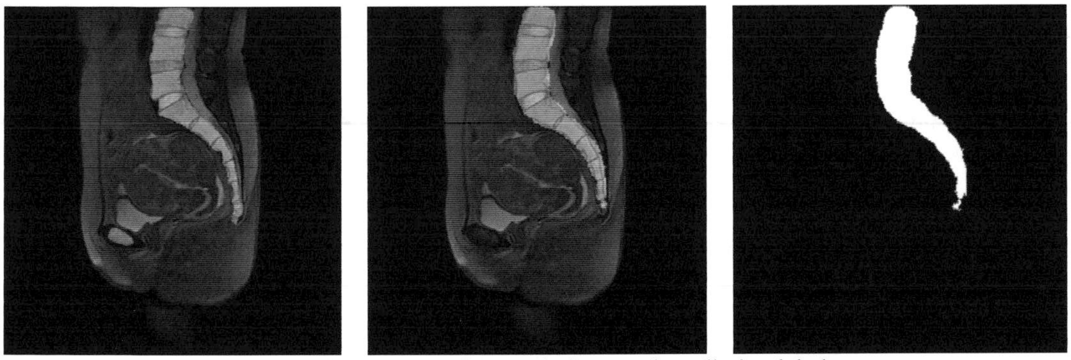

Figure8. Segmentation renderings and prediction labels

5. Conclusion

Aiming at the insufficient extraction of medical image features by the ordinary convolutional neural network in the U-Net network, the ASpp-UNet network proposed in this paper adds the Spp module to the U-Net network, and replaces the ordinary convolution with the hole convolution and the cavity convolution and the SE-Net attention mechanism to have stronger feature extraction for various complex spine images Ability and only a small number of calculations have been added. The public data set of spine images shows that the experiment has effectively increased the IOU and F1-score, and it can also effectively extract the key information of the spine image in the image. For the segmentation of spine images, this paper replaces the convolutional network with the ASpp network on the basis of the U-Net network model, and adds the attention mechanism SE-Net. Ablation experiment results show that our proposed new network structure can efficiently segment spine images and obtain higher segmentation results. Deep learning image segmentation is also a research direction that people will pay attention to.

Acknowledgement

This work was supported by the National Natural Science Foundation of China (61866016, 62061019), General Project of Jiangxi Natural Science Foundation (20202BABL202014), the Key Project of Jiangxi Education Department (GJJ201107), the Science Foundation of Artificial Intelligence and Bioinformatics Cognitive Research Base Fund of Jiangxi Science and Technology Normal University of China (2017ZDPYJD005).

References

[1] Wang Q , Wu B , Zhu P , et al. ECA-Net: Efficient Channel Attention for Deep Convolutional Neural Networks[J]. 2019.
[2] Tian Z , Shen C , Chen H . Conditional Convolutions for Instance Segmentation[M]. 2020.
[3] Zhou Z , Siddiquee M , Tajbakhsh N , et al. UNet++: A Nested U-Net Architecture for Medical Image Segmentation[C]// 4th Deep Learning in Medical Image Analysis (DLMIA) Workshop. 2018.
[4] Ronneberger O , Fischer P , Brox T . U-Net: Convolutional Networks for Biomedical Image Segmentation[J]. Springer International Publishing, 2015.
[5] X. Duan, K. Jia, B. Li, D. Guo, E. Zhang and C. Qin, "Reversible Image Steganography Scheme Based on a U-Net Structure," in IEEE Access, vol. 7, pp. 9314-9323, 2019, doi: 10.1109/ACCESS.2019.2891247.
[6] Zhang M , Yu F , Zhao J , et al. BEFD: Boundary Enhancement and Feature Denoising for Vessel Segmentation[J]. 2021.
[7] H. Wen, L. Chen, T. Fu, Z. Yang and Z. Yin, "Detecting the Surface Defects of the Magnetic- Tile Based on Improved YOLACT ++," 2021 IEEE 21st International Conference on Communication Technology (ICCT), 2021, pp. 1097-1102, doi: 10.1109/ICCT52962.2021.9658066.

[8] He K , Zhang X , Ren S , et al. Spatial Pyramid Pooling in Deep Convolutional Networks for Visual Recognition[J]. IEEE Transactions on Pattern Analysis & Machine Intelligence, 2014, 37(9):1904-16.
[9] Lei Junming,Wang Chenggang,Liu Xueyan.Retinal Vessel Image Segmentation Based on Improved U-Net[J].Information Technology and Information Technology,2019(10):168-169.
[10] Long J , Shelhamer E , Darrell T . Fully Convolutional Networks for Semantic Segmentation[J]. IEEE Transactions on Pattern Analysis and Machine Intelligence, 2015, 39(4):640-651.
[11] F Yu, Koltun V . Multi-Scale Context Aggregation by Dilated Convolutions[J]. 2016.
[12] Jie H , Li S , Gang S , et al. Squeeze-and-Excitation Networks[J]. IEEE Transactions on Pattern Analysis and Machine Intelligence, 2017, PP(99).
[13] Weng W , Zhu X . INet: Convolutional Networks for Biomedical Image Segmentation[J]. IEEE Access, 2021, PP(99):1-1.
[14] H. Wu, Y. Zhang, W. Fu, C. Zhang and S. Niu, "A Novel Pre-Processing Method for Neural Network-Based Magnetic Field Approximation," in IEEE Transactions on Magnetics, vol. 57, no. 10, pp. 1-9, Oct. 2021, Art no. 9200109, doi: 10.1109/TMAG.2021.3095590.

A method of multi-channel SAR moving target detection and imaging based on complex CNN

Zhao-Rong Deng[a], Xiang-Yang Zhang[b*], Wei-Min Gao[c], Hong-Ping Zhang[d], Yin-Qiang Xu[e]

(School of Information Engineering, Nanchang Hangkong University, Nanchang 330036, China)

[a]zhaorongdeng970621@163.com,[b*]xiangyangzhang@nchu.edu.cn,[c]gwm18855032560@163.com, [d]Evelyn20210131@163.com, [e]xuyinqiang1997@163.com.

Abstract: Aiming at the problem of the detecting difficulty and imaging the moving target submerged in clutter, a multi-channel Synthetic Aperture Radar (SAR) moving target detection and imaging method based on complex Convolutional Neural Network (CNN) is proposed. First, using Range Doppler algorithm (RDA) to image SAR data to obtain images with clutter, noise and defocused moving targets. Then, the SAR image is used as the input of complex CNN. A SAR image with the clutter and noise eliminated, and the moving target in focus is obtained after processing by neural network. The simulation experiment results show this method is effectiveness.

1. Introduction

Ground moving target indication (GMTI) and ground moving target indication imaging (GMTIm) are important applications of SAR. As for multi-channel SAR systems, classic algorithms for moving target detection include displaced phase center offset antenna technology (DPCA)[1], along-track interference technology (ATI[2]), space-time adaptive processing technology (STAP)[3] and other algorithms. However, the SAR imaging algorithm is originally designed for stationary targets. Moving target imaging will be offset and defocused due to its motion. The azimuth velocity causes the azimuth shift and azimuth defocus. In addition, the imaging of moving targets is also affected by clutter and noise, which pose challenges for SAR imaging.

In decades, deep learning has made outstanding achievements in some fields, including SAR recognition[4] and SAR target detection[5]. Except for classification and detection problems, CNN applied in the SAR field also solves many regression problems, for example, literature [6] proposed a SAR-GMTI method based on complex CNN. Literatures [7] and [8] used CNN for SAR imaging. Moreover, literature [9] also studied the estimation method of moving target range velocity and slant distance based on CNN. Currently, most of the CNNs used in the SAR field are classification and detection problems. The regression problems in SAR field based on complex CNN are still a novel and interesting direction.

This paper proposes a moving target detection and focuses imaging algorithm based on complex CNN. First, the complex image of the moving target after RDA imaging and the corresponding ideal scattering point is used as the training data pair. After the network training is completed, the verification complex image is input into the network to obtain moving target with clear clutter and well focus. The structure of this paper: Section 2 introduces the multi-channel moving target signal model. Section 3 describes the way to generate training data pairs, and contents of complex CNN via the signal model in detail. Section 4 conducts experimental simulation and analysis on the proposed algorithm. Section 5 draws conclusions and looks forward to future research directions.

2. Multi-channel moving target signal model

The geometric relationship between the airborne multi-channel radar system and the point target in the SAR imaging slant-range plane is shown in Figure 1. The airborne radar works in the front and side view mode, and the $y-axis$ is regarded as the distance direction. When the azimuth slow time of $\eta = \eta_c$, the moving target P is located at the center of the beam where the radar emits pulses. At this time, the coordinates (x_0, y_0) of the point target P are located in $(0, R_0)$, and the azimuth velocity and radial velocity are v_x and v_y, respectively. The radar flies with a constant speed along the $x-axis$ at a speed V_a, and the $x-axis$ is regarded as the azimuth direction. The working mode of the multi-channel SAR is single-transmit and dual-receive mode, channel 1 is used to transmit and receive signals, and channel 2 is used to receive signals.

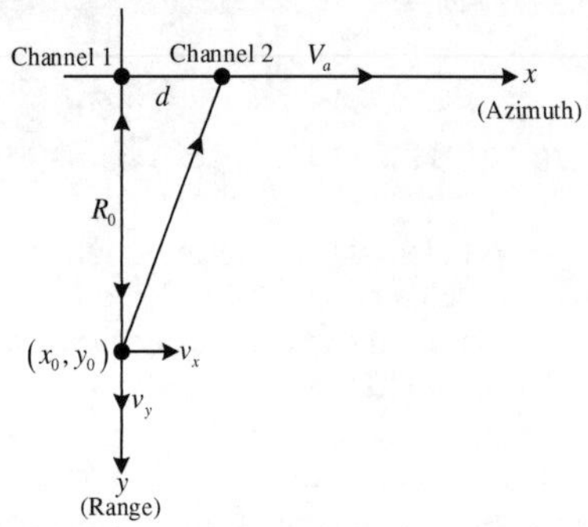

Figure.1 Multi-channel SAR slant-range plane geometric relationship diagram

The instantaneous slant distance between the moving point target and the mth receiving antenna is Equation (1), which is carried out by second-order Taylor expansion:

$$R_m(\eta) = \sqrt{\left[\left(R_0 + v_y\eta\right)^2 + (V_a\eta - v_x\eta + md)^2\right]} \approx R_0 + v_y\eta + \frac{(V_a\eta - v_x\eta + md)^2}{2R_0} \quad (1)$$

In Equation (1), η represents the azimuth slow time. For a single-transmit and dual-receiver SAR system, the signal expression after range compression received by receiving antenna of mth is:

$$s_{mrc}(\tau,\eta) = Ap_r\left(\tau - \frac{R_0(\eta)+R_m(\eta)}{c}\right)\omega_a(\eta - \eta_c)\exp\left(-j2\pi\frac{R_0(\eta)+R_m(\eta)}{\lambda}\right) \quad (2)$$

In Equation (2), A represents the constant related to the target backscattering coefficient, τ is the fast range time, $\omega_a(\cdot)$ represents the azimuth envelope, η_c represents the time when the beam center crosses, and $R_0(\eta)$ represents the instantaneous tilt between the transmitting antenna and moving target. $p_r(\cdot)$ represents the range of impulse response amplitude, as a sinc function.

After performing range migration correction and azimuth compression processing on the signal after distance compression in sequence, the Doppler frequency, Doppler center frequency of the moving target and the image domain signal expression of the moving target in the m-th channel is obtained, respectively, which expressed as follows[10]:

$$k_a' = \frac{2(V_a - v_x)^2}{\lambda R_0}$$
$$f_{mv} = -\frac{2v_y}{\lambda} \quad (3)$$

$$s_{mimg}(\tau,\eta) = Ap_r\left(\tau - \frac{2R_0}{c}\right)p_a(\eta - \eta_c + \eta_m)\exp\left(-j\frac{\pi(md)^2}{2\lambda R_0}\right)\exp(-j2\pi m f_{mv} t_d) \quad (4)$$

In equation (4), $p_a(\cdot)$ represents the amplitude of the azimuth impulse response, and $t_d = d/\big(2(V_a - v_x)\big)$ is the time domain sampling interval, $\eta_m = R_0 v_y/(V_a - v_x)^2 + md/2(V_a - v_x)$.

3. Complex convolutional neural network for moving target detection and imaging

This section discusses the method of using complex CNN to detect and image moving targets, where deep imaging learns the mapping from defocused SAR images to moving target images. The steps of the proposed method are as follows: Firstly, defocus moving target plus clutter and noise background image blocks are generated as input. With the help of training data, the network is trained by minimizing the loss function to obtain the optimal weights. When the training finished, the network is used as an imaging processor to achieve the task of eliminating clutter and SAR moving target imaging in focus. The data set generation and network architecture will be described in detail.

3.1. Training data set generation

Training imaging requires defocusing moving target image and corresponding static target image data pair. The moving target signal model is used for simulation owing to the lack of SAR moving target imaging data-set. For SAR-GMTI system, the parameters involved the speed and slant distance of each mission are different. According to Equation (3), the carrier speed and slant distance play an important role in moving target refocusing. Thus, in order to make the network suitable for all tasks, the speed and slant distance of the SAR-GMTI system are randomly selected with uniform distribution. In addition, multiple moving targets are randomly distributed in a given imaging area. The target motion parameters and reflection coefficients are randomly generated with uniform distribution. The main task of this network is to extract moving targets from clutter and focus imaging. If the radial velocity of the moving target is considered, the moving target will shift in the azimuth direction, which is not conducive to focusing and imaging, so the radial velocity is set to zero in this paper. The clutter used in the paper obeys the Rayleigh distribution, and the noise adds Gaussian white noise.

Since the input of the network is the imaging of complex image matrix after the moving target, whereas if the size of the complex image matrix is too large, it will take a lot of time to train the network. Moreover, many clutter regions excluding the moving target will not help the training of the network. Therefore, the data generated by the simulation should be as small as possible. Considering that the moving target is defocused in the azimuth direction, the azimuth sampling point of the image block should be set to be larger than the distance sampling point. And the phase is very important for the detection and imaging of moving target owing the moving target is a complex matrix after imaging, so the SAR complex image matrix is input into the complex CNN with the size $1 \times W \times H$. As for output, SAR images generally display amplitude information, so the output of the network is taken as an absolute value.

3.2. Complex convolutional neural network structure

Figure 2 shows the network structure used in this paper, where Complex Conv, Complex ReLU, and Complex BN represent complex convolutional layers, complex activation functions, and complex batch normalization functions, respectively. In this figure, 3×3×64, s=1 means that the parameters of the complex convolutional layer are that the size of the convolution kernel is 3, 64 feature maps, and the step size is 1. The detailed information of this network is given below.

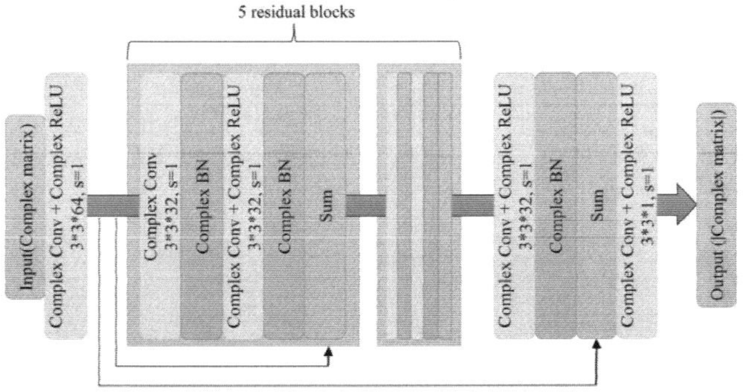

Figure.2 Complex network framework

1) Residual Blocks: The network includes 5 residual blocks, and each ResBlock contains two Complex Conv layers, with a convolution kernel and 64 feature maps, followed by Complex BN layer and Complex ReLU activation function. In addition, it also includes two jump connections. As shown in figure, ResBlocks can alleviate the degradation problem and facilitate the training of deeper networks, while the BN layer can solve the problem of gradient disappearance or gradient explosion. Therefore, these two structures are selected in this work.

2) Loss function: On account of importance of the amplitude and phase in the SAR image, similar to speech signal, the Euclidean distance loss function is commonly used in tasks of speech enhancement. And the common loss function based on CNN regression tasks also involves the Euclidean distance loss function. Thus, the function used is defined as follows:

$$L(I,O) = \frac{1}{W \times H} \sum_{w=1}^{W} \sum_{h=1}^{H} \left\| g_{\phi}(I_{w,h}) - O_{w,h} \right\|_2^2 \quad (5)$$

In Equation (5), I represents the input of the network, $g_{\phi}(I)$ represents the output through the network, and O represents the well-focused complex matrix of the stationary target corresponding to the defocused moving target. The network weights are updated through the optimization algorithm. After the optimal weights are obtained, the network determines the functional relationship between input and output, and then eliminates clutter and focuses imaging.

3) Network output: The image output size via network should be the same as the input image size, so the last layer uses a convolution kernel size of 3, 1 feature map and complex Conv layer with the step size of 1 to obtain a prediction image of $1 \times W \times H$.

4. Experimental verification

In this section, we use simulated point target data plus clutter and noise to verify the effectiveness of the method in this paper. The analog point target signal uses Equation (4). The dual-channel airborne radar works in the X-band, with carrier frequency $f_c = 5.3GHz$, pulse repetition frequency $PRF = 200$, antenna spacing $d = 0.5m$, and azimuth and range sampling points are 200 and 120, respectively. Carrier speed $V_a \sim U(150{,}250\, m/s)$ and slant distance $R_0 \sim U(15{,}20km)$, network input complex matrix size is $(W,H) = (200{,}120)$, the number of moving targets $N \sim U(1{,}10)$, moving target reflection coefficient $\sigma \sim U(0.8{,}1.2)$ and moving target azimuth speed $v_x \sim U(1{,}15\, m/s)$, range speed setting is 0. Adding the Rayleigh clutter with $SCR = 10dB$ and Gaussian white noise with $SNR = 18dB$, the network is trained on GTX1660Super GPU, the total of 20,000 training images and 20,000 test images are used, the optimizer choose Adam, the batch size is 4, the learning rate is 0.001, and a total of 10 Epochs are trained. In addition, the training time is about 10h.

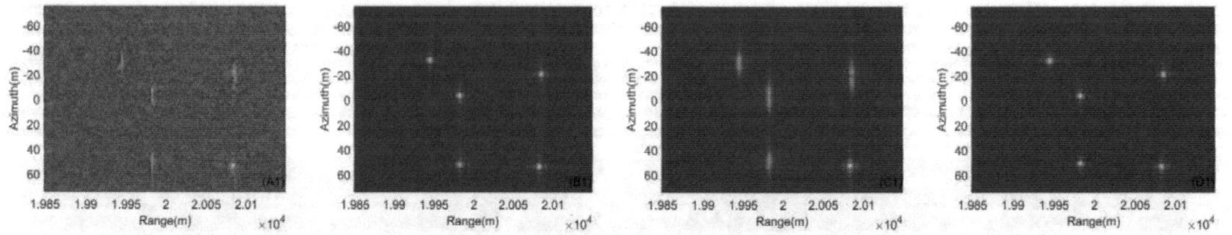

Figure.3 Experimental results of sports goals (A1) SAR Data (B1)Ideal scattering point (C1)DPCA detection results (D1)Complex CNN experimental results in this article

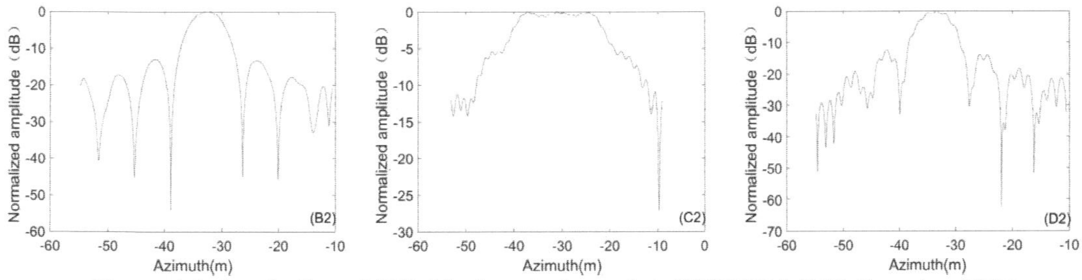

Figure.4 Azimuth slice of (B2) Ideal scattering points (C2)DPCA (D2) Complex CNN

The Figure 3 shows the verification result, and the Figure 4 shows the azimuth slice result of the Mark point in Figure 3. Among them, (A1) represents the RDA imaging result of the verification data, (B1) is the imaging result of the ideal scattering point model, and (B2) is the azimuth slice of the scattering point. (C1) shows the moving target signal obtained after DPCA detection, which displayed that the clutter has been eliminated, and only the moving target signal remains. (C2) shows the azimuth slice of the moving target, and the azimuth velocity causes the moving target to defocus in the azimuth direction. (D1) represents the imaging result obtained after the verification data by neural network. It shows that not only the clutter is eliminated, but the focusing effect of moving target is also perfect. As shown in (D2) and (B2) , the results show that the azimuth slice is relatively near, and the correlation coefficient between the experimental results and the ideal scattering point is 0.9631, which shows that our network is effective in eliminating clutter and focusing on moving targets.

5. Conclusion

This paper proposes a multi-channel SAR moving target detection and imaging method based on complex CNN. First, this paper uses RDA to perform imaging processing on the echo signal of moving target with clutter and noise, and then input the imaging result into the trained complex CNN for clutter elimination and moving target focusing imaging, and finally get a SAR image with excellent focusing effect. This paper uses simulated point targets and clutter to verify the algorithm, which proves that the algorithm can detect the moving targets and complete the refocusing imaging of moving targets.

Acknowledgments

This work was financially supported by the National Natural Science Foundation of China (61761031 & 61861033), National Aviation Science Foundation of China (20172056002 & 20142056005), Nanchang Hangkong University Teaching Reform Project (KCPY1179 & JY1625).

References

[1] Cerutti-Maori D, Sikaneta I. A generalization of DPCA processing for multichannel SAR/GMTI radars[J]. IEEE Transactions on Geoscience and Remote Sensing, 2012, 51(1): 560-572.
[2] Chiu S, Livingstone C. A comparison of displaced phase centre antenna and along-track interferometry techniques for RADARSAT-2 ground moving target indication[J]. Canadian Journal of Remote Sensing, 2005, 31(1): 37-51.
[3] Guerci J R. Space-time adaptive processing for radar[M]. Artech House, 2014.
[4] Min R, Lan H, Cao Z, et al. A gradually distilled CNN for SAR target recognition[J]. IEEE Access, 2019, 7: 42190-42200.
[5] Yang R, Wang R, Deng Y, et al. Rethinking the Random Cropping Data Augmentation Method Used in the Training of CNN-based SAR Image Ship Detector[J]. Remote Sensing, 2021, 13(1): 34.
[6] Mu H , Zhang Y , Jiang Y , et al. CV-GMTINet: GMTI Using a Deep Complex-Valued Convolutional Neural Network for Multichannel SAR-GMTI System[J]. IEEE Transactions on Geoscience and Remote Sensing, 2021, PP(99):1-15.
[7] Zjla C , Qi Q B , Hys B , et al. SAR moving target imaging based on convolutional neural network[J].

Digital Signal Processing, 106.
[8] Gao J , Deng B , Qin Y , et al. Enhanced Radar Imaging Using a Complex-valued Convolutional Neural Network[J]. 2017.
[9] Oveis A H , Giusti E , Ghio S , et al. CNN for Radial Velocity and Range Components Estimation of Ground Moving Targets in SAR[C]// 2021 IEEE Radar Conference (RadarConf21). IEEE, 2021.
[10] Zheng Hongchao. Research on moving target detection technology based on multi-channel synthetic aperture radar system [D].[Ph.D.dissertation],Shanghai Jiaotong University, 2018:40-54.

IBM Small-Scale Superconducting Quantum Computer Not Gate Error Test

Ming He[1], Xiaonan Liu[1*], Junchao Wang[1], Jiale An[1]

[1]State Key Laboratory of Mathematical Engineering and Advanced Computing, Information Engineering University, Zhengzhou, Henan, 450001, China

[*]Corresponding author's e-mail: prof.liu.xn@foxmail.com

Abstract—Quantum NOT gates play an important role in the process of quantum information conversion. However, when the X-gate operation is executed on a real quantum computer, there is a large deviation between the actual operation result and the theory, which will lead to inaccurate results when the quantum algorithm containing the X-gate operation is executed. In order to facilitate users to understand the error fluctuations of the X gate in time before executing the quantum algorithm containing the X gate operation on the IBM quantum cloud platform, this paper proposes a method to measure the X gate error. By measuring the X-gate error of four small-scale superconducting quantum computers on the back end of the IBM quantum cloud platform, analyze the degree of fluctuation of the actual measurement value of the quantum device; at the same time, under different execution times shots, the influence of the X-gate test is analyzed. The test results show that the measured value of each qubit of different quantum devices fluctuates to different degrees; the execution times of shots are different, and the test results of X-gate error will be affected to different degrees. This test method helps users select the optimal performance of the qubits before executing quantum algorithms with X gate on IBM's small-scale quantum computers.

1. Introduction

In recent years, quantum technology has developed rapidly, using the inherent parallelism of quantum to surpass supercomputers in terms of computing power [1]. However, compared with traditional computers, current quantum computers cannot be owned by every user like a classic computer due to the difficult manufacturing process, high cost, and extremely easy to be affected by the external environment. To this end, IBM opened the 5-qubit cloud platform in 2016, its name is IBM Quantum Experience [2]. Today, the IBM quantum computing cloud platform has provided users with a total of six 5-qubit superconducting quantum computers for free: ibmq_belem, ibmq_bogota, ibmq_lima, ibmq_manila, ibmq_quito, and ibmq_santiago [3]; at the same time, eight chip indicators [4] are provided to users: connectivity graph, energy relaxation time T1 [5], phase relaxation time T2 [6], device frequency, readout error, single qubit gate error, CNOT gate error and quantum volume [7]. It is designed to facilitate users to run quantum algorithms and test the performance indicators of quantum computers on real quantum computers.

Although the IBM quantum cloud platform provides great convenience for users to compile algorithms. However, as a precision instrument, a real quantum computer is easily affected by the inherent errors of the instrument and the noise of the external environment. When running the algorithm on the IBM quantum cloud platform, since the quantum gates that make up the quantum algorithm can be divided into single-qubit gates, double-qubit gates and multi-qubit gates according to the number of bits, there will be varying degrees of error in the implementation of different quantum gates. This leads to deviations between the actual operating results of the overall quantum algorithm and the theoretical results. Therefore, before executing the quantum algorithm, it is necessary to obtain the quantum gate operation error and the final state measurement error in the IBM quantum cloud platform in time.

For the test of quantum gates, Ref [8,9] proposed quantum process tomography, which can obtain the magnitude and type of the quantum system noise, but the number of measurements increases exponentially with the increase in the number of gate operations. The Ref [10-13] proposed a random test benchmark, which is mainly used to calculate the error probability of each gate operation. However, when the quantum computer system is large in scale, it is very difficult to characterize quantum noise. In order to solve the problem of the increase in the scale of the quantum system and the increase in the number of gate operations, the Ref [14] used the small-scale superconducting quantum computer supported by the back-end of the IBM quantum cloud platform to test only the CNOT gate error and the readout error of the quantum device. This test intuitively reflects the fluctuation of the CNOT gate error.

In order to facilitate users to obtain the error information in time before executing the single quantum gate algorithm, based on the Ref [14], this paper provides four small-scale superconducting quantum computers ibmq_belem, ibmq_lima, ibmq_manila and ibmq_quito (In alphabetical order) the single-qubit operation gate -- X gate's error floating degree is tested and analyzed. Simultaneously, verify whether there is an impact on the X-gate error test under different execution times shots.

2. Related Concepts

2.1. Superconducting quantum computer

At present, the more popular physical implementations of quantum computers mainly include ion traps, semiconductors, and superconductors. The ion trap quantum computer is a quantum computer that uses electric field trapping in a vacuum cavity and laser cooling of ions [15]. However, the computer integration of this physical realization method is difficult. The idea of semiconductor quantum computer [16] is derived from the semiconductor technology of classical computers, which can be integrated on a large scale, but there is charge noise between qubits.

Compared with the first two design schemes, the superconducting quantum computer is a computer scheme designed based on the principle of superconducting circuits. The core component is the superconducting Josephson junction [17], the middle is a thin layer of insulating material, and the two sides are superconductors, is a special LC resonator.

Superconducting quantum computers have the advantages of high integration, strong scalability, and mature processing technology. They are the focus of research by top teams at home and abroad. For example, in June 2021, the University of Science and Technology of China research team developed a 66-qubit quantum processor "*Zuchongzhi*" [18], which is a superconducting system; the four quantum computers tested in this article are also superconducting physics implementations.

2.2. Quantum not gate

Among single-qubit gates, quantum NOT gates (X gates) are widely used in algorithms such as Deutsch-Jozsa (D-J) [19], Quantum Phase Estimation (QPE) [20], and factorization *Shor* [21] , Is a common quantum operation gate, a kind of Pauli gate, sometimes it can also be called Puali-X gate [22]. On the Bloch sphere, the X gate rotates by π units around the x axis. Its function is similar to that of the classical NOT gate. It can convert the ground state $|0\rangle$ to the excited state $|1\rangle$ and the excited state $|1\rangle$ to the ground state $|0\rangle$.

2.3. Qubit Connected Graph

The connectivity of a quantum computer is similar to that of graph theory, which refers to the maximum value of the qubit dimension in a quantum topology graph. The four 5-qubit superconducting quantum computers in the IBM quantum cloud platform used in this experiment contain two connected graphs. Here, the *ibmq_manila* quantum computer connection diagram is shown in Figure 1; the *ibmq_belem*, *ibmq_lima*, and *ibmq_quito* quantum computer connection diagrams are the same, and the qubit Q1 is coupled with Q0, Q2, and Q3 respectively, as shown in Figure 2.

Figure 1. Connectivity of *ibmq_manila* device.

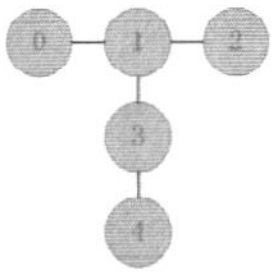

Figure 2. Connectivity of *ibmq_belem*, *ibmq_lima*, and *ibmq_quito* device.

Each circle in the quantum computer connection diagram represents a qubit, and the connecting line between the two circles indicates that the double qubit gate can directly act on the two qubits. If the double-qubit gate acts on two qubits that do not have a connecting line, the quantum compiler needs to add a series of switch gates (SWAP gate [23]) to realize the interaction between non-connected qubits. However, increasing the SWAP gate will introduce additional errors and reduce the depth of the quantum circuit. Therefore, the researchers proposed an all-to-all quantum connected graph, in which each qubit of the connected graph is connected to other qubits, that is, there is a coupling relationship. The higher the connectivity of the qubit, the stronger the hardware programmability and operability of the quantum computer.

2.4. *Fidelity*

In quantum computing, fidelity refers to the degree of fit between the real experimental results of a quantum computer and the ideal results. The mathematical expression is [24]

$$F = tr\left(\sqrt{\sqrt{\rho}\sigma\sqrt{\rho}}\right) \tag{1}$$

Here, F is the fidelity, and ρ is the experimental realization of ideal σ. In quantum circuits, fidelity can be divided into initial state preparation fidelity, gate operation fidelity and readout fidelity.

2.4.1. *Fidelity of initial state preparation*

Before executing the algorithm, the qubit needs to be initialized to reduce the energy level of the qubit to the ground state. However, due to the combined influence of the circuit noise inside the chip and the environmental noise outside the chip, the qubit cannot be completely initialized to the ground state. Therefore, before compiling the quantum algorithm, the initialization error needs to be measured. The fidelity of the initial state preparation is an index that measures the degree of fit between the actual ground state and the ideal ground state.

2.4.2. *Qubit gate operation fidelity*

The fault-tolerant law of quantum computers states that quantum computers can only work normally when the gate error rate is very low and the crosstalk is weak enough. Therefore, the engineering realization of quantum computers must reduce gate operation errors. Quantum gate fidelity can be used as an indicator of quantum gate operation to measure the degree of deviation between the gate operation performed in the experimental state and the ideal gate operation.

2.4.3. *Read fidelity*

In the quantum circuit, after the quantum gate operation is completed, the result of the operation needs to be measured. However, due to the inherent error of the readout instrument and the error caused by the external environment, the quantum state measurement value is not accurate. For this reason, it is necessary to read the index of fidelity. The readout fidelity, that is, the fidelity of the final state

Journal of Physics: Conference Series **2209** (2022) 012022 doi:10.1088/1742-6596/2209/1/012022

measurement, measures the degree of fit between the actual final state measurement value of the quantum computer and the ideal measurement value.

3. Test Method Design

The design steps of the quantum X-gate error test method in this paper include three aspects:

- Readout error test. Add measurement operations on the qubits q_0 - q_4 respectively, and measure the readout fidelity (measurement fidelity) of each qubit, according to the formula

$$E_r = 1 - F_r \tag{2}$$

Obtain the read error E_r. Among them, F_r is the read fidelity. As shown in Figure 3, it is the readout fidelity test circuit diagram of qubit q_0, and its measurement error pseudo code is shown in Table 1.

Table 1. Read error pseudo code

Algorithm: Readout error
Input: Qubit measurement circuit
Output: Qubit measurement error
1: import the toolkit
2: provider ← Visit Quantum Cloud Platform
3: Quantum Machine, qubit, shots ← Define the name of the IBM quantum computer, the number of qubits and the number of executions
4: backend ← Visit the Quantum Machine
5: qreg_q ← Define quantum register
6: creg_c ← Define classic registers
7: circuit ← Define quantum circuits
8: i ← 0
9: while i < qubit do
10: cq ← Measure qreg_q, creg_c
11: job ← Execute quantum circuit
12: R_error_results[i] ← Execution circuit result
13: print R_error_results

The readout error measured by this circuit includes the initial state preparation error.

- Add X gate error test. Using the method of adding X gates to qubits q_0 - q_4 and measuring, the probability of |00001>, |00010>, |00100>, |01000> and |10000> is the total of qubits q_0 - q_4. Fidelity, according to the formula

$$E_z = 1 - F_z \tag{3}$$

Find the total error E_z. Here, F_z is the overall fidelity. As shown in Figure 4, it is the circuit diagram of the fidelity test of adding X gate of qubit q_0. The pseudo code of the measurement error is shown in Table 2.

Table 2. Add X gate error pseudo code

Algorithm: Add X gate error
Input: Qubit plus X gate measurement circuit
Output: Qubit plus X gate measurement error
1: import the toolkit
2: provider ← Visit Quantum Cloud Platform
3: Quantum Machine, qubit, shots ← Define the name of the IBM quantum computer, the number of qubits and the number of executions
4: backend ← Visit the Quantum Machine
5: qreg_q ← Define quantum register
6: creg_c ← Define classic registers
7: circuit ← Define quantum circuits

```
8: i ← 0
9: while i < qubit do
10:   Circuit plus X gate operation
11:   cq ← Measure qreg_q, creg_c
12:   job ← Execute quantum circuit
13:   Add_X_results[i] ← Execution circuit result
14: print Add_X_results
```

The error measured by this circuit includes three aspects: initial state preparation error, X gate error and measurement error (reading error).

- Calculate the X gate error E_X. Using the difference between the X gate test error and the read error test error, the mathematical expression is

$$E_x = E_z - E_r \tag{4}$$

Calculate the X gate error E_X, the error pseudo code is shown in Table 3.

Table 3. X gate error pseudo code

Algorithm: X gate error
Input: Qubit measurement error, qubit plus X gate measurement error Output: X gate error result 1: i ← 0 2: While i < qubit do 3:   X_error_results[i] ← Add_X_results[i] - R_error_results[i] 4: print X_error_results

In order to prevent deviations in the indicator test results of real quantum computers in different time periods, and to ensure the accuracy of the X-gate error test results, this experiment parallelizes steps "Readout error test" and "Add X gate error test".

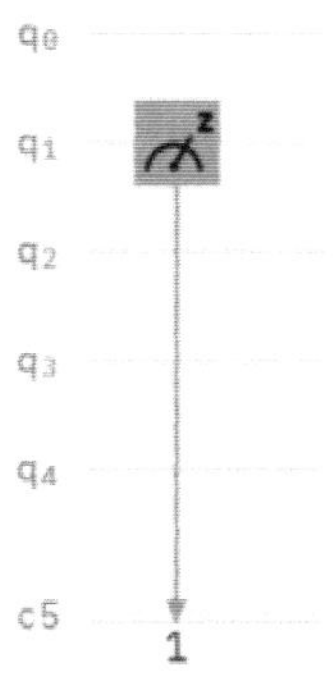

Figure 3. Readout fidelity test circuit.

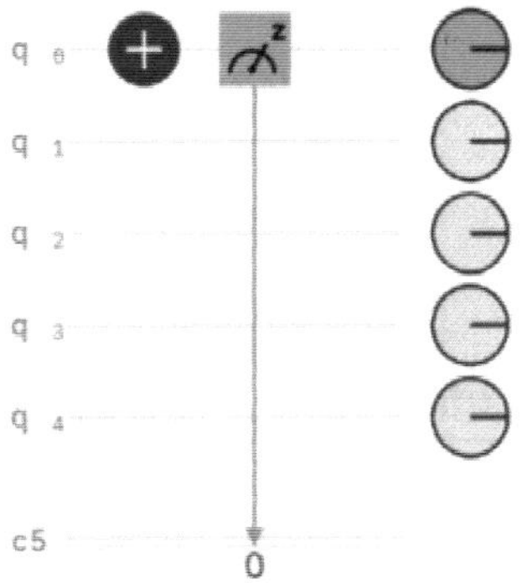

Figure 4. The fidelity test of adding X gate circuit.

4. Test Results

In this experiment, four X-gate error tests were performed on four 5-qubit superconducting quantum computers, *ibmq_belem, ibmq_lima*, *ibmq_manila* and *ibmq_quito*. The results show that in the four floating degree test experiments, the qubits contained in each quantum computer have different degrees of floating; In the shots comparison experiment, the *shots* were 8192, 4096, and 2048 respectively. By testing the X gate error of the bits in the quantum machine, it is proved that the test results of each quantum computer are affected to different degrees.

4.1. X gate error floating degree test

This experiment provides four X gate tests of quantum devices, analyzes the average, variance, maximum, minimum, and difference of the X gate error of each qubit, and observes the degree of fluctuation. Here, the average value, maximum value and minimum value of X-gate error analysis data of four quantum computers are shown in Figure 5-8, and the number of executions *shots* = 8192.

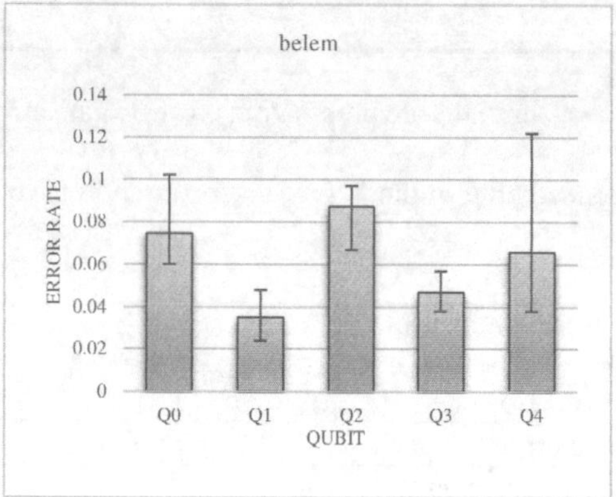

Figure 5. *ibmq_belem* device X-gate error data analysis.

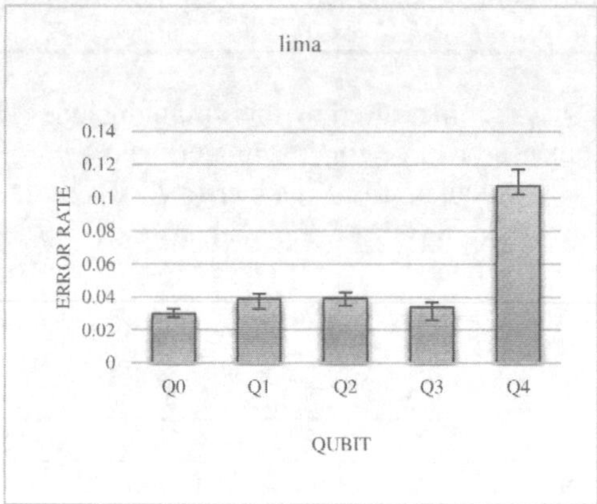

Figure 6. *ibmq_lima* device X-gate error data analysis.

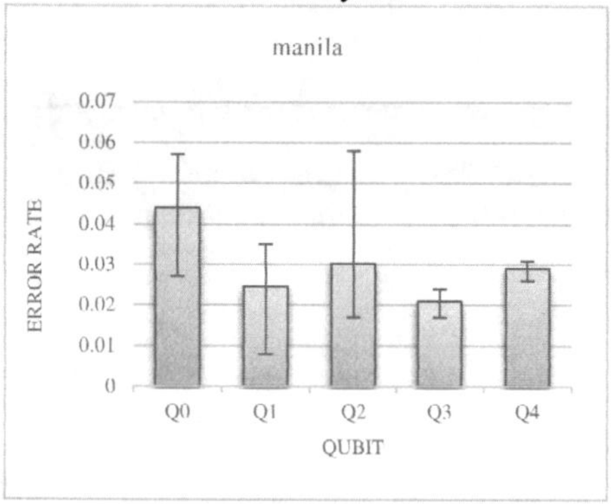

Figure 7. *ibmq_manila* device X-gate error data analysis.

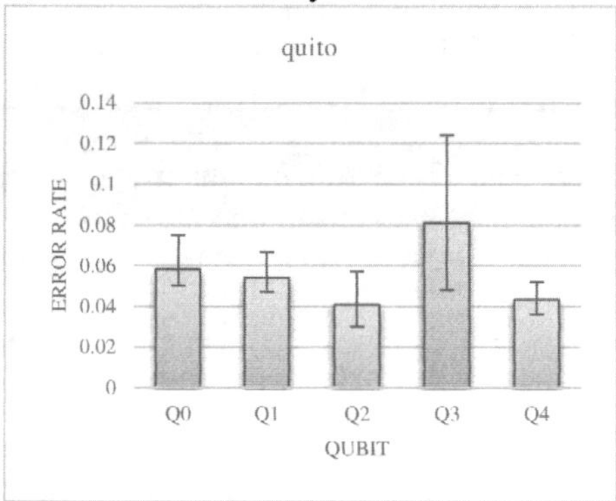

Figure 8. *ibmq_quito* device X-gate error data analysis.

Here, the upper end of the error line is the maximum value of the qubit X gate error, and the lower end of the error line is the minimum value. At the same time, the variance and difference of the test data are shown in Table 4-7.

- ibmq_belem device (Figure 5 and Table 4): The average value of qubits Q0-Q4 is 0.06195; the X gate error test value of Q1 is smaller than other qubits; the floating degree of Q3 is smaller than other qubits.

Table 4. *ibmq_belem* machine X-gate error floating

	Q0	Q1	Q2	Q3	Q4
Variance	3.50E-4	1.10E-4	1.88E-4	8.07E-5	1.49E-3
Difference	0.042	0.024	0.030	0.019	0.084

- ibmq_lima device (Figure 6 and Table 5): The average value of qubits Q0-Q4 is 0.04990; the X gate error test value of Q0 is smaller than other qubits; at the same time, the degree of fluctuation of Q0 is also smaller than other qubits.

Table 5. *ibmq_lima* machine X-gate error floating

	Q0	Q1	Q2	Q3	Q4
Variance	4.67E-6	1.67E-5	1.37E-5	2.87E-5	4.60E-5
Difference	0.005	0.009	0.008	0.011	0.015

- ibmq_manila device (Figure 7 and Table 6): The average value of qubits Q0-Q4 is 0.02975; the X gate error test value of Q3 is smaller than other qubits; the degree of floating of Q4 is smaller than other qubits.

Table 6. *ibmq_manila* machine X-gate error floating

	Q0	Q1	Q2	Q3	Q4
Variance	1.98E-4	1.35E-4	3.56E-4	1.00E-5	4.67E-6
Difference	0.030	0.027	0.041	0.007	0.005

- ibmq_quito device (Figure 8 and Table 7): The average value of qubits Q0-Q4 is 0.05550; the X gate error test value of Q4 is smaller than other qubits; at the same time, the floating condition of Q4 is also smaller than other qubits.

Table 7. *ibmq_quito* machine X-gate error floating

	Q0	Q1	Q2	Q3	Q4
Variance	1.39E-4	7.87E-5	1.32E-4	1.50E-3	5.16E-5
Difference	0.025	0.020	0.027	0.076	0.016

In summary, according to Figure 5-8 and Table 4-7, it can be seen that

- In the same quantum device: The bit with the smallest X gate error test value may not necessarily have the smallest floating degree.
- For quantum computers with the same connectivity (ibmq_belem, ibmq_lima, and ibmq_quito): the Q1 bit X gate error average of each quantum device is less than the Q0-Q4 bit X gate error average. Simultaneously, after comparison, the stability of the ibmq_lima device is higher than that of the other two devices.

4.2. The shots test

This test compares the impact of the X-gate error test on 4 superconducting quantum computers when the number of execution *shots* is 2048, 4096, and 8192. The experimental results are shown in Figure 9-12.

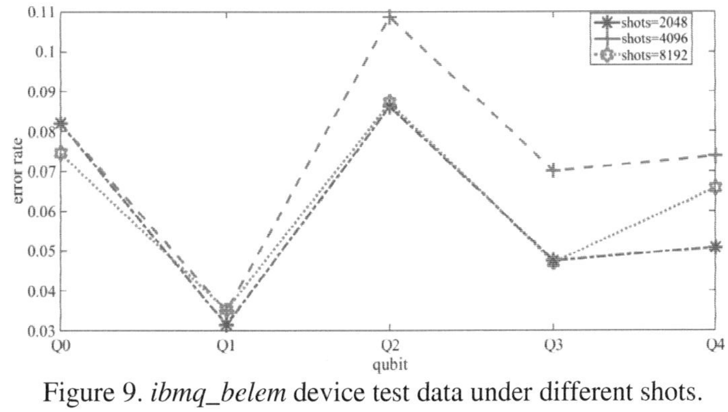

Figure 9. *ibmq_belem* device test data under different shots.

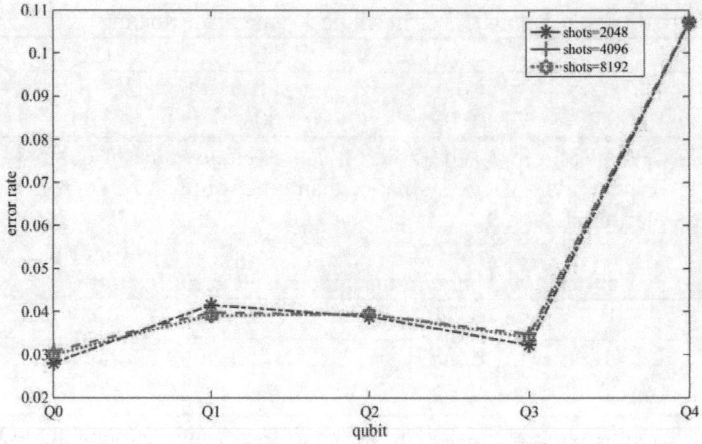

Figure 10. *ibmq_lima* device test data under different shots.

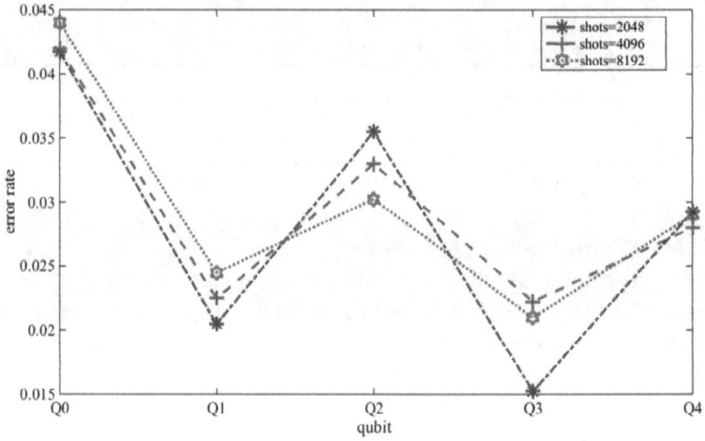

Figure 11. *ibmq_manila* device test data under different shots.

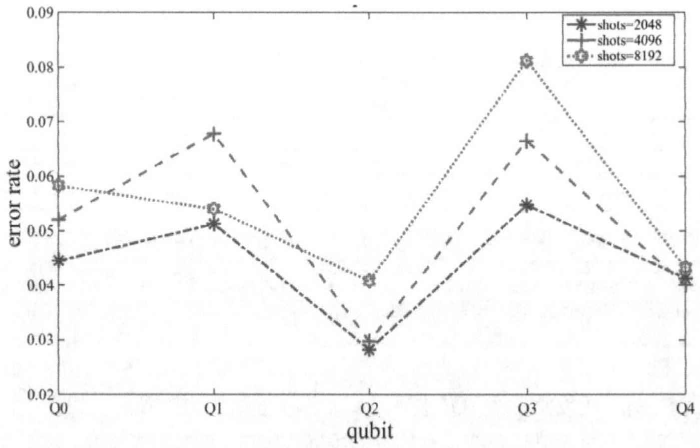

Figure 12. *ibmq_quito* device test data under different shots.

According to Figure 9-12, when the number of execution *shots* is 2048, 4096, 8192, it can be seen from the average value of the X gate error test of each qubit of the quantum computer,

- ibmq_belem device (Figure 9): It has the least impact on the X gate error of qubit Q1; when shots = 2048, the X gate error value of Q1 is the smallest, which is 0.0315.
- ibmq_lima device (Figure 10): It has the least impact on the X gate error of qubit Q0; when shots = 2048, the X gate error value of Q0 is the smallest, which is 0.0280.
- ibmq_manila device (Figure 11): It has the least impact on the X gate error of qubit Q4; when shots = 2048, the X gate error value of Q3 is the smallest, which is 0.0153.
- ibmq_quito device (Figure 12): It has the least impact on the X gate error of qubit Q4; when shots = 2048, the X gate error value of Q2 is the smallest, which is 0.0283.

In order to analyze the data in Figure 9-12 more accurately, this experiment calculates the variance value of the X gate test of the above four quantum devices under different *shots*, as shown in Figure 13-16.

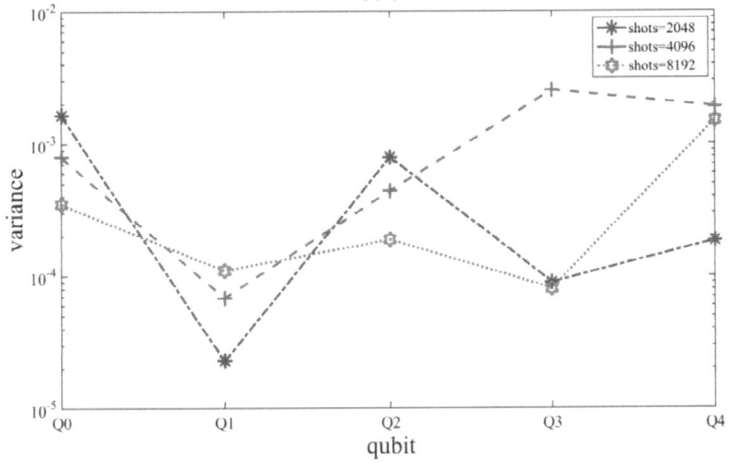

Figure 13. *ibmq_belem* device test variance under different shots.

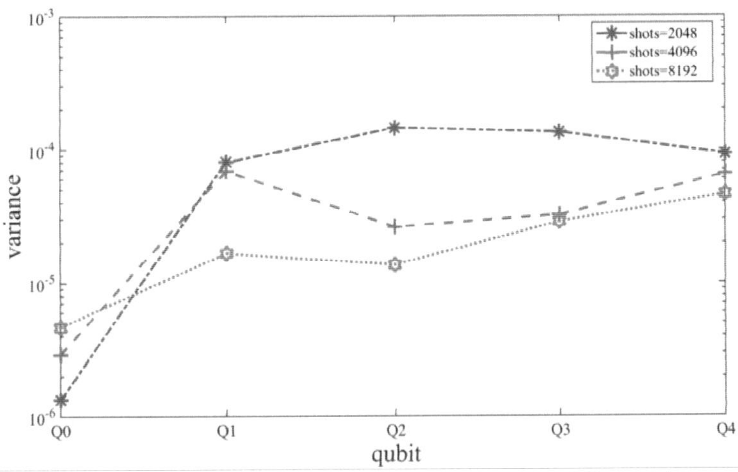

Figure 14. *ibmq_lima* device test variance under different shots.

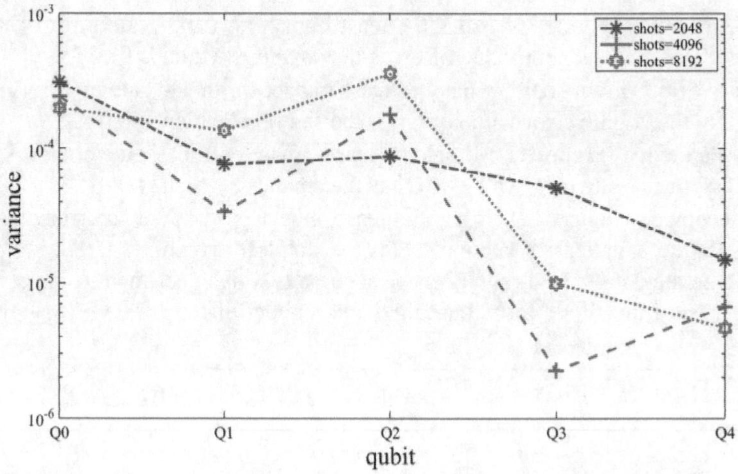

Figure 15. *ibmq_manila* device test variance under different shots.

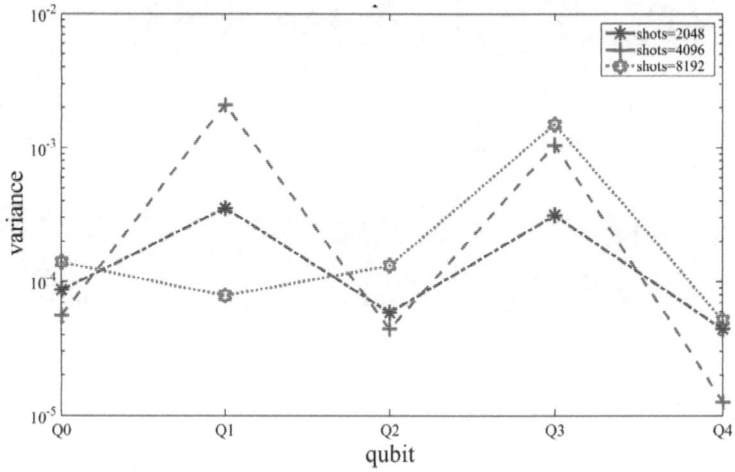

Figure 16. *ibmq_quito* device test variance under different shots.

Based on Figure 9-12 and Figure 13-16, we can see that for each qubit of the different devices in this experiment, when the *shots* are 2048, 4096, and 8192, there are different degrees of influence. On the other hand, for the quantum devices *ibmq_belem*, *ibmq_lima*, and *ibmq_quito*, under different execution times *shots*, the *ibmq_lima* device qubit X gate test variance average is smaller than the other two devices. Therefore, the X gate error test value of the *ibmq_lima* device has the least impact.

5. Conclusion

This paper proposes a method for testing quantum X-gate error, and uses four small-scale quantum computers provided by the IBM cloud platform as experimental objects to measure the X-gate error and analyze the degree of floating of each quantum computer. The experimental results show that through four error tests, each quantum device has different degrees of fluctuation. simultaneously, verify if the number of execution *shots* is 2048, 4096, 8192, whether there is any impact on the test data. The verification results show that, with different *shots*, the bits of each quantum device do have varying degrees of impact.

In the process of quantum information conversion, the quantum NOT gate plays an extremely important role. This article aims to test the quantum X-gate error in IBM's small-scale superconducting quantum computer, so that users can obtain the quantum computer's X-gate error fluctuation degree in time when executing quantum algorithms with X-gates, so that it is better to choose the quantum computer with the best performance. At the same time, the test method has scalability and portability, and can be used in quantum computers of larger scale and different physical realization systems. Therefore, the X-gate test method in this article has important practical significance.

Here, using the X-gate error test method to evaluate domestic quantum computers, while comparing the bit performance of domestic and foreign quantum computers will be a future work.

Reference

[1] David, D. (1985) Quantum theory, the church-turing principle and the universal quantum computer. J. Proceedings of the Royal Society of London A: Mathematical, Physical and Engineering Sciences., 400: 97-117.
[2] IBM. (2021) IBM Quantum. https://quantum-computing.ibm.com/.
[3] IBM Quantum Services. (2021) Services. https://quantum-computing.ibm.com/services?services=systems.
[4] IBM Quantum Services. (2021) ibmq_manila. https://quantum-computing.ibm.com/services?services=systems&system=ibmq_manila.
[5] Gonzalez, B., Palankovski, V., Kosina, H., Hernandez, A., Selberherr, S. (1999) An energy relaxation time model for device simulation. J. Solid-State Electronics., 43: 1791-1795.
[6] Stephen, J.N. (1979) T2 in coherent optics: Collision, dephasing time, or reciprocal linewidth. J. American Journal of Physics., 47: 987-991.
[7] Cross, A.W., Bishop, L.S., Sheldon, S., Nation, P.D., Gambetta, J.M. (2019) Validating quantum computers using randomized model circuits. J. Physical Review A., 100: 032328.
[8] Chuang, I.L., Nielsen, M.A. (1997) Prescription for experimental determination of the dynamics of a quantum black box. J. Journal of Modern Optics., 44: 2455-2467.
[9] Poyatos, J., Cirac, J.I., Zoller, P. (1997) Complete Characterization of a Quantum Process: The Two-Bit Quantum Gate. J. Physical Review Letters., 78: 390-393.
[10] Knill, E., Leibfried, D., Reichle, R., Britton, J., Blakestad, R.B., Jost, J.D., Langer, C., Ozeri, R., Seidelin, S., Wineland, D.J. (2008) Randomized benchmarking of quantum gates. J. Physical Review A., 77: 012307.
[11] Magesan, E., Magesan, Gambetta, J.M., Emerson, J. (2011) Scalable and robust randomized benchmarking of quantum processes. J. Physical review letters., 106: 180504.
[12] Magesan, E., Magesan, Gambetta, J.M., Emerson, J. (2012) Characterizing quantum gates via randomized benchmarking. J. Physical review A., 85: 042311.
[13] Moussa, O., Marcus, P., Silva, D., Ryan, C.A., Raymond, L. (2012) Practical experimental certification of computational quantum gates using a twirling procedure. J. Physical review letters., 109: 070504.
[14] He, M., Liu, X.N., Wang, J.C., Pang, J.M. (2021) Performance Test of IBM's Superconducting Quantum computer. J. Journal of Information Engineering University., 22: 175-184.
[15] Myatt, C., King, B., Kielpinski, D., Leibfried, D., Turchette, C., Wood, C., Itano, W., Monroe, C., Wineland, D. (1998) Trapped ions, entanglement, and quantum computing. J. Photonics West., 3270: 131-137.
[16] Johnston, I.D. (1992) Electrons in a crystal lattice: A simple computer model. J. American Journal of Physics., 60: 600-607.
[17] Yuriy, M., Gerd, S., Alexander, S. (2001) Quantum state engineering with Josephson - junction devices. J. Mesoscale and Nanoscale Physics., 73: 357-400.
[18] Wu, Y.L., Bao, W.S., Cao, S.R., Chen, F.S., Chen, M.C., Chen, X.W., Chung, T.H., Deng, H., Du, Y.J., Fan, D.J., Gong, M., Guo, C., Guo, C., Guo, S.J., Han, L.C., Hong, L.Y.,

Huang, H.L., Huo, Y.H., Li, L.P., Li, N., Li, S.W., Li, Y., Liang, F.T., Lin, C., Lin, J., Qian, H.R., Qiao, D., Rong, H., Su, H., Sun, L.H., Wang, L.Y., Wang, S.Y., Wu, D.C., Xu, Y., Yan, K., Yang, W.F.,Yang, Y., Ye, Y.S., Yin, J.H., Ying, C., Yu, J.L., Zha, C., Zhang, C., Zhang, H.B., Zhang, K.L., Zhang, Y.M., Zhao, H., Zhao, Y.W., Zhou, L., Zhu, Q.L., Lu, C.Y., Peng, C.Z., Zhu, X.B., Pan, J.W. (2021) Strong quantum computational advantage using a superconducting quantum processor. https://arxiv.org/abs/2106.14734.

[19] David, D., Richard, J. (1992) Rapid solutions of problems by quantum computation. J. Proceedings of the Royal Society of London, Series A., 439: 553.

[20] Richard, C., Artur, E., Chiara, M., Michele, M. (1998) Quantum algorithms revisited. J. Proceedings of the Royal Society A: Mathematical, Physical and Engineering Sciences., 454: 339-354.

[21] Harry, B. (1996) A short note on Shor's factoring algorithm. J. Association for Computing Machinery., 27: 89-90.

[22] Babazadeh, A., Erhard, M., Wang, F.R., Malik, M., Nouroozi, R., Krenn, M., Zeilinger, A. (2017) High-Dimensional Single-Photon Quantum Gates: Concepts and Experiments. J. Physical Review Letters., 119: 180510.

[23] Colin, W. (2011) A generalized quantum SWAP gate. https://arxiv.org/abs/0811.1684.

[24] Richard, J. (1994) Fidelity for mixed quantum states. J. Journal of modern optics., 41: 2315-2323.

Object detection based on Yolov4-Tiny and Improved Bi-directional feature pyramid network

Qi Liu[1 a*], Xiaoyu Fan[1 b*], Zhipeng Xi[1 c*], Zhijian Yin[1 d*], Zhen Yang[1 e*]

[1] Jiangxi Science and Technology Normal University, Nanchang, China

[a*]criafting@aliyun.com, [b*]1020170701@jxstnu.com, [c*]John_boy0618@163.com, [d*]zhijianyin@aliyun.com, [e*] yangzhenphd@aliyun.com

Abstract—In the field of small object detection, Yolov4-Tiny is inadequate in feature extraction and does not make best of multi-scale features. In this paper, an improved BiFPN framework is proposed based on Yolov4-Tiny to increase object detection precision. Moreover, the Yolov4-Tiny is taken as the backbone network and introduce spatial pyramid pooling (SPP) to connect and merge multi-scale regions. Finally, our method can achieve 79.53% map on Pascal VOC dataset, which is 2.12% higher than the original Yolov4-Tiny model.

1. Introduction

Object detection has become the foundation for solving more complex and advanced visual tasks, such as instance segmentation[1], scene understanding[2], target tracking[3] and so on. However, object detection methods are not only locating the different object in an image, also for classifying object category. Recently, these methods of object detection based on deep neuro network have been widely taken in pedestrian detection, worker's helmet detection, medical diagnosis et.al.

Early stage, object detection had low precision and less scope of application, so it is tough to achieve landing in practical application. The traditional object detection methods used many candidate regions by sliding window and extracting the features from the candidate regions, such as Haar-like feature[4], HOG[5], LBP[6] and DPM[7](Deformable Part Models). Various target detection methods based on handcrafted features had been proposed, and performed well on Pascal VOC dataset.[8] However, the above methods determine the candidate regions by window sliding that will lead to increase computational complexity.

More efficient methods begin to emerge in the field of image vision with the development of computer technology, especially the emergence of deep learning. Krizhevsky et al. proposed a convolutional neural networks (CNN) called AlexNet[9] in 2012 which got the best image classification accuracy in ILSRVC, It is proved that using CNN to extract features can classify images more accurately than traditional machine learning feature extraction methods. Meanwhile, in terms of target detection, AlexNet opens up a new method.

Since then, more and more researchers have begun to pay attention to deep learning, and people have gradually begun to try to use CNN to solve the problems in target detection[10]. In the field of target detection, traditional methods have been replaced, and convolutional neural networks have been rapidly developed and applied. Generally speaking, according to whether the network generates candidate frames, two mainstream methods have gradually formed in the field of target detection. One of them is a stage, and the other is one-stage. The main object detection methods based on two-stage are R-CNN [11], SPPnet [12], Fast R-CNN[13], and Faster R-CNN[14]. By using the selective search algorithm to generate 2000 candidate boxes on each image, and sending these 2000 candidate boxes into CNN to extract features and SVM for classification in R-CNN. Using these candidate boxes does improve the performance of detection, but it also slows down the running speed of the network, especially when the data set is relatively large. In response to the question of the slow detection speed of R-CNN, SPPnet and Fast R-CNN were proposed. SPPnet is a spatial pyramid pooling at the top of R-CNN last convolution layer for a fixed-length feature map. Although, the SPPnet detection speed has improved, the training speed is still slow. Fast R-CNN can not only improve training speed and detection speed but also improve accuracy via sharing convolutional computing regions. Faster R-CNN propose a region proposal network (RPN) to generate candidate regions to replace the traditional selective search algorithm, and the speed is improved. Object detection methods based on one-stage include Yolo [15] and SSD [16], etc. The core idea of Yolo is that target detection is regarded as regression problem. The size of the input image is cropped to the set size, and then separated into S×S grid units. Each grid cell predicts B bounding boxes, and the position information, confidence and category of bounding boxes are obtained by regression in the output layer. For the one-stage target detection method SSD, selecting VGG16 as the backbone network when extracting features, borrow the anchor idea in Faster R-CNN to detect in different receptive fields, and uses Yolo regression idea to output boundary boxes and categories.

Multiple versions have evolved due to the Yolo and its variants [17][18]being highly representative and the potential for improvement in a single-stage inspection framework. Among them, Yolov4[19] and Yolov4-Tiny have excellent performance in accuracy and speed respectively and have a large space for improvement.

The multi-feature map fusion method improves the detection accuracy by fusing the feature maps of different scales and the receptive field information to form a new feature map. Therefore, CNN models don't need to fix the size of the input images, but also have robustness to detect multi-scale objects by fusing multi-scale features. Therefore, the finer layer of the network can take advantage of the context information learned from the coarser layer of the network. Lin et al. Proposed FPN (feature pyramid network)[20] method based on the inherent multi-scale pyramid structure of deep convolution neural network which a top-down transverse connection structure was designed to improve the accuracy of multi-scale detection. In recent years, FPN has been widely cited in the image field by using the advantages of fusing multi-scale features to obtain global information. PANet[21], NAS-FPN[22] and BiFPN[23]More network structures have been developed for cross-scale feature fusion to obtain more discriminative information which aiming to fuse feature outputs from different network layers, to enhance feature information and improve target detection accuracy.

Yolov4-Tiny target detection framework cannot make best of multi-scale features and insufficient feature extraction ability inspired by these classic research methods. This paper uses Yolov4-Tiny as the backbone network, then, introduces SPP and improved BiFPN feature extraction network to enhance the feature extraction ability. This paper aims to maintain the fast detection speed and small network volume as the premise to improve the accuracy of target detection. The major work and contributions are as follows:

1) We introduce the SPP module. The SPP module is mainly composed of three maximum pooling layers of different sizes. The pooling layers of different sizes extract features from different angles to form feature maps of receptive fields of different sizes.

2) We embed an improved BiFPN into the structure of Yolov4-Tiny that can fuse the features between different scales.

3) The improvements above are introduced in Yolov4-Tiny, which effectively alleviate the problems of insufficient feature extraction ability and inadequate utilization of multi-scale features.

2. Yolov4-Tiny network architecture

As a simplified version of Yolov4-Tiny, which has a relatively simple network structure, a small amount of computing, and a lower requirement on hardware, so it is easy to run in mobile terminal or device terminal. The Yolov4-Tiny network architecture (Figure 1) consists of the backbone network, neck, and the head.

The main network of Yolov4-Tiny uses the network structure of CSPDarknet53-tiny, consisting mainly of two normal convolution processes and three CSP structure. The common convolution process adopts CBL structure, which is composed of the convolution layer, BatchNorm layer and LeakyRelu; BatchNorm layer can reduce the difference of range between different samples, speed up the training and convergence speed of the model on the computer, and prevent gradient explosion and gradient disappearance problems during the back propagation process. For the part of input less than zero, LeakyReLU can calculate the gradient, and avoid the phenomenon of neuron inactivation. This structure can reduce model parameters to reduce computational complexity while ensuring the accuracy of the network. CSPBlock compares the structural features of CSPNet and divides the features of base layer into two parts. The main function of the first part is to directly form the residual edge, and the final output can be obtained by convolution using the splicing of the second part. This network structure can reduce computational complexity while maintaining network performance.

In Neck part, Yolov4-Tiny adopts FPN structure, which can integrate features of different scales for implementing rich semantic information of deep network and geometric details of shallow network for strengthening the ability of extracting features. Yolo head used the feature map obtained by FPN to make the final prediction, and finally formed two prediction scales of 26×26 and 13×13.

In Yolov4-Tiny, the input image size is 416×416, and then multi-scale detection is carried out by fusing FPN idea, and two detection layers are output. The two detection layers divide the image into 26×26 and 13×13 respectively. Then each grid cell has predicted three bounding box, each boundary box contains (x, y, w, h, Ci) five elements. Among them, the x and y are referred to the width and height after normalization based on bounding box of the entire image corresponding to the cell w and h of the offset. Ci means the bounding box confidence score, reflects a box contains the possibility of detecting target, and the accuracy of the test box. In order to better predict bounding boxes, Yolov4-Tiny generates six anchor boxes through k-means clustering algorithm, which serves as suggestion boxes during detection. In addition, the loss function of Yolov4-Tiny consists of three parts: including bounding box regression loss, confidence loss and classification loss. The bounding box regression loss function adopts CIOU function, and the latter two use the same cross entropy function as Yolov3 [18].

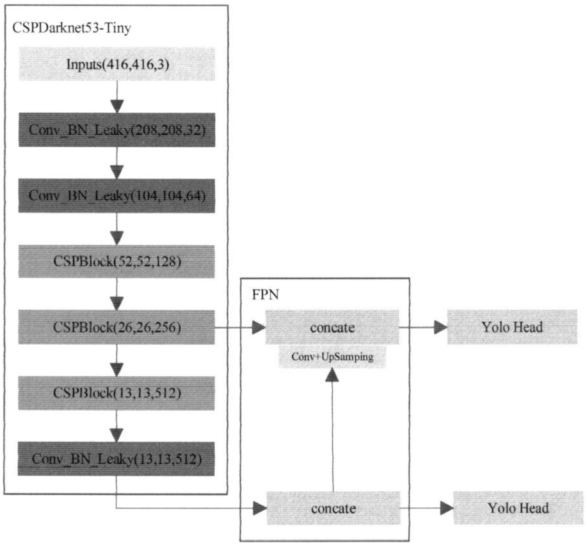

Figure 1 Network architecture of Yolov4-Tiny

3. methods

In this work, an improved target detection method based on Yolov4-Tiny is proposed. (hereinafter referred to as Yolov4-Tiny-Bi). The backbone network adopts CSPDarknet53-tiny, and the head uses a spatial pyramid pool and the improved BiFPN is introduced to make most use of the multi-scale information between different layers. Figure 2 shows the network structure.

3.1 Improved BiFPN

Yolov4-Tiny contains two prediction scales only, 13×13 and 26×26, two feature maps are output from the deeper network, while the deeper network is easy to lose the shallow edge information. The semantics of features can change low dimension to high dimension with the deepening of the network level. Each layer of the network will cause the loss of features to a certain extent. It is necessary to integrate features at different levels to enrich the semantic information of features. BiFPN make the most of the features between shallow network and deep network, and carries out bidirectional fusion of network features from bottom-up and from top-down. Therefore, the improved BiFPN is used to replace the original FPN.

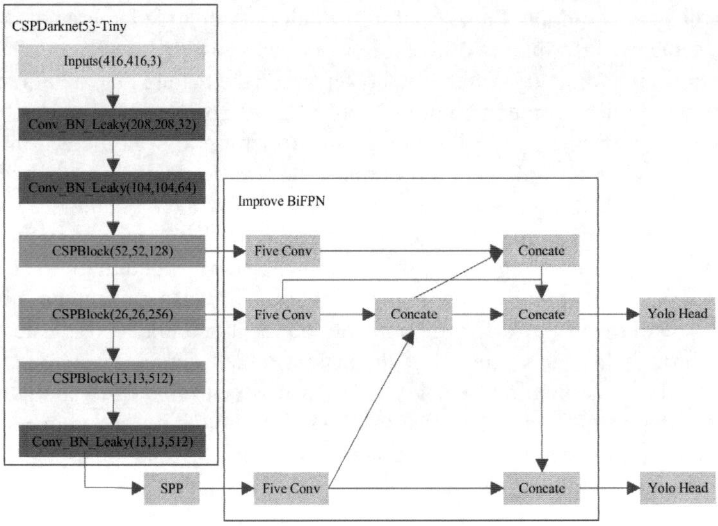

Figure 2 Network architecture of Yolov4-Tiny-Bi

BiFPN (Figure 3) use the idea of bidirectional fusion to construct a bottom-up and top-down bidirectional channel to carry out bidirectional fusion of information from diverse layers of the trunk network, so as to alleviate the loss of feature information caused by too many network layers.

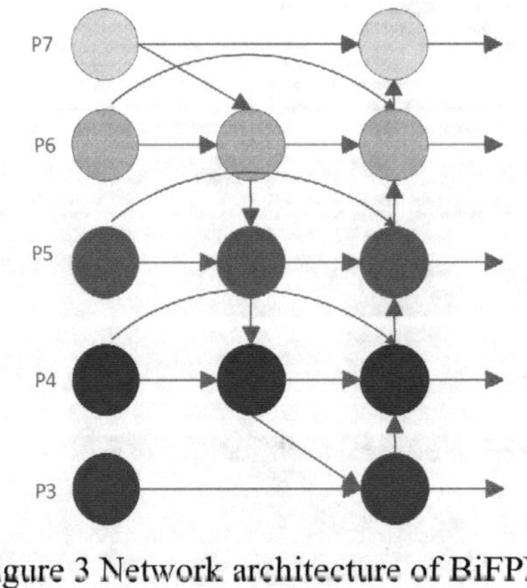

Figure 3 Network architecture of BiFPN

BiFPN is an improvement of PANet which removes the node with only one input edge, adding an edge from the primordial input to the last output node. Each bidirectional path is considered as a characteristic network layer and can be repeated several times. Bidirectional cross scale connections and the fast normalized fusion is integrated in our final BiFPN, the feature fusion method is shown in Equation (1):

$$P_5^{td} = conv\left(\frac{\omega_1 * P_5^{in} + \omega_2 * Resize(P_6^{td})}{\omega_1 + \omega_2 + \epsilon}\right) \quad (1)$$

$$P_5^{out} = conv\left(\frac{\omega_1' * P_5^{in} + \omega_2' * P_5^{td} + \omega_3' * Resize(P_4^{out})}{\omega_1' + \omega_2' + \omega_3' + \epsilon}\right) \quad (2)$$

In this formula P_5^{td} and P_5^{out} is the intellectual feature and the output feature at level 4, respectively.

The improved BiFPN is introduced between the backbone network and the head in Figure 4. We extract features of three different scales from the backbone network, we first carry out five convolution for these three feature layers of different scales, which consists of two 1×1 convolution layers and three depth separable convolution layers. We integrate the three different scale feature layers after five convolution into the improved BiFPN network, fusing the multi-scale features and setting the prediction branches of 13×13, 26×26 feature resolutions. In order to fully integrate the shallow features, we change the BiFPN bi-directional channel top down and bottom up. The improved feature fusion method is shown in Equation (2):

$$P_4^{td} = conv\left(P_4^{in} + Resize(P_5^{in})\right) \quad (3)$$

$$P_4^{out} = conv\left(P_4^{in} + P_4^{td} + Resize(P_3^{out})\right) \quad (4)$$

In this formula P_4^{td} and P_4^{out} is the intellectual feature and the output feature at level 4, respectively.

Different from the BiFPN proposed by Google et al. our new BiFPN in Yolov4-Tiny-Bi does not integrate the network features from bottom-up and top-down. Instead, we first convolved layers P3, P4 and P5 for five times, then carried out two-way fusion of deep and shallow layer features from bottom-up and top-down, and output only two prediction scales of 13×13, 26×26 at the end.

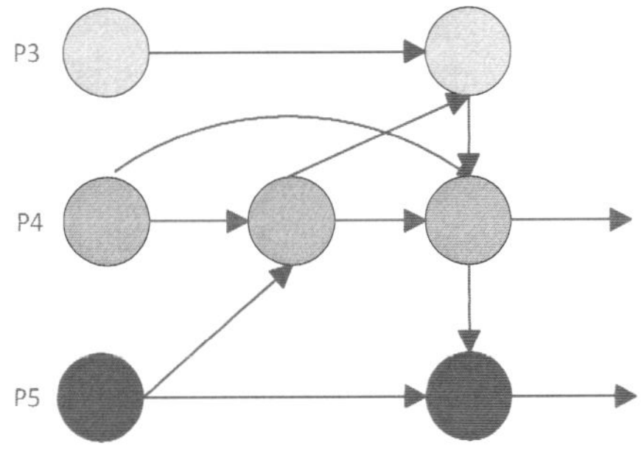

Figure 4 Network architecture of improved BiFPN

3.2 SPP

Spatial pyramid pooling (SPP) was proposed by He et al. SPP uses three sliding Windows of different sizes to maximize the pooling of the input feature graphs, the feature map is assembled by sliding windows whose size is the same as that of the plane elements, and one-dimensional feature vectors are obtained. Therefore, it is not feasible to apply it in full convolutional networks. Therefore, Redmon and Farhadi improve the SPP module to a series of maximum pool outputs with the kernel size of K × K, where k = {1, 5, 9, 13} and the span is equal to 1 in the design of Yolov3[18]. The receptive field of the backbone features is effectively increased by a relatively large K × K maximum pool. We add the SPP block to P5 because it effectively increases the receptive field, isolating above all contextual features that results in almost no reduction in network speed.

4. Experimental Analysis

4.1 Dataset

The PASCAL VOC [24] dataset is used in this article. This data includes 20 categories of objects. The 20 categories include planes, trains, people, potted plants, birds and so on. Yolov4-tiny is one of the most popular convolutional neural networks for embedded computing devices, its model size is small and their computational complexity is low, so they are used as a reference. The VOC2007 training set was used to train Yolov4-Tiny-Bi network, and average accuracy was calculated on the test set to evaluate the detection accuracy.

4.2 Model training

The detailed hardware configuration in this experiment is: the CPU of the computer is Intel(R) Core(TM) i7-10750H, frequency is 2.60GHz, the memory is 16GB, and the graphics card model is NVIDIA GeForce GTX 3060. The operating system for 64-bit Windows10 system, deep learning development platform uses Pytorch1.8. The parallel computing framework is CUDA10.0, and the deep neural network accelerator library is CUDNN V11.1.

When Pascal VOC2007 is trained, the initial learning rate is set to 0.001 and the batch size is set to 32.

4.3 Experimental results and analysis

The precision of Yolov4-Tiny-Bi is greatly improved due to its lightweight network structure by comparison with the Yolov4-Tiny. Compared with Yolov4-Tiny, MAP is 2.12 percentage points higher due to the introduction of improved BiFPN structure to enhance the feature fusion between different levels and make full use of shallow features. Meantime, the detection speed of the improved BiFPN structure has little change because the calculation amount is small. Finally, the map of Yolov4-Tiny-Bi algorithm using 416×416 resolution, SPP structure and improved BiFPN structure reached 79.53%, which was higher than that of Yolov4-Tiny, and achieved the expected effect.

Table 1. Comparison of the accuracy on the VOC2007 dataset

method	dataset	MAP/%
Yolov4-Tiny	VOC2007	77.41
Yolov4-Tiny-Bi	VOC2007	79.53

Figure 5 shows the comparison of detection results between Yolov4-Tiny and Yolov4-Tiny-Bi, indicating that Yolov4-Tiny has poor detection effect on occlusive objects and the problem of missing detection, while Yolov4-Tiny-Bi proposed in this paper detect most of the targets.

5. conclusion

Yolov4-Tiny algorithm has insufficient feature extraction ability in small target detection and can't take advantage of multi-scale features. In this paper, a bidirectional feature pyramid network target detection algorithm based on Yolov4-Tiny is proposed. An improved BiFPN is proposed and spatial pyramid pooling is introduced to improve the generalization ability of the model in actual detection scenes and make the model more geared for object detection tasks. Experimental results show that compared with the original algorithm Yolov4-Tiny, MAP improves by 2.12% on Pascal VOC dataset, and the detection rate reaches 72 frame/s.

Figure 5 Yolov4-Tiny and Yolov4-Tiny-Bi comparison of prediction results

Experimental results show the Yolov4-Tiny network model combining the improved BiFPN and SPP structure can meet the timeliness and accuracy of object detection, and has a good engineering application. The accuracy of the Yolov4-Tiny-Bi target detection algorithm in this article is less than that of some large-scale detection networks. In future research work, we will further study and optimize Yolov4-Tiny-Bi to make the model more suitable for more detection scenarios.

Acknowledgement
This work was supported by the National Natural Science Foundation of China (62061019, 61866016), General Project of Jiangxi Natural Science Foundation(20202BABL202014), the General Project of Jiangxi Education Department(GJJ190587), and the Youth Top Talent Foundation of Jiangxi Science and Technology Normal University (2018QNBJRC002).

References

[1] Peng S., Jiang W., Pi H., Li X., Bao H., & Zhou X. (2020). Deep snake for real-time instance segmentation. In Proceedings of the IEEE/CVF Conference on Computer Vision and Pattern Recognition. pp. 8533-8542.

[2] Pan X., Shi J., Luo, P., Wang, X., & Tang, X. (2018). Spatial as deep: Spatial cnn for traffic scene understanding. In Thirty-Second AAAI Conference on Artificial Intelligence.

[3] Bertinetto L., Valmadre J., Henriques J. F., Vedaldi A., & Torr P. H. (2016). Fully-convolutional siamese networks for object tracking. In European conference on computer vision. pp. 850-865.

[4] Papageorgiou C. P., Oren M., & Poggio T. (1998). A general framework for object detection. In Sixth International Conference on Computer Vision. pp. 555-562.

[5] Dalal N., & Triggs B. (2005). Histograms of oriented gradients for human detection. In 2005 IEEE computer society conference on computer vision and pattern recognition. pp. 886-893.

[6] Ojala T., Pietikainen M., & Maenpaa T. (2002). Multiresolution gray-scale and rotation invariant texture classification with local binary patterns. IEEE Transactions on pattern analysis and machine intelligence, 24(7): 971-987.

[7] Felzenszwalb P. F., Girshick R. B., McAllester D., & Ramanan D. (2009). Object detection with discriminatively trained part-based models. IEEE transactions on pattern analysis and machine intelligence, 32(9): 1627-1645.

[8] Zhang J., Huang K., Yu Y., & Tan T. (2011). Boosted local structured hog-lbp for object localization. In CVPR 2011.pp. 1393-1400.

[9] Krizhevsky A., Sutskever I., &Hinton G. E. (2012). Imagenet classification with deep convolutional neural networks. Advances in neural information processing systems, 25: 1097-1105.
[10] Girshick R. (2015). Fast r-cnn. In Proceedings of the IEEE international conference on computer vision. pp. 1440-1448.
[11] Girshick R., Donahue J., Darrell T., & Malik, J. (2014). Rich feature hierarchies for accurate object detection and semantic segmentation. In Proceedings of the IEEE conference on computer vision and pattern recognition. pp. 580-587.
[12] He K., Zhang X., Ren S., & Sun J. (2015). Spatial pyramid pooling in deep convolutional networks for visual recognition. IEEE transactions on pattern analysis and machine intelligence, 37(9):1904-1916.
[13] Girshick R. (2015). Fast r-cnn. In Proceedings of the IEEE international conference on computer vision. pp. 1440-1448.
[14] Ren S., He K., Girshick R., & Sun J. (2015). Faster r-cnn: Towards real-time object detection with region proposal networks. Advances in neural information processing systems, 28:91-99.
[15] Redmon J., Divvala S., Girshick R., & Farhadi A. (2016). You only look once: Unified, real-time object detection. In Proceedings of the IEEE conference on computer vision and pattern recognition. pp. 779-788.
[16] Liu W., Anguelov D., Erhan D., Szegedy C., Reed, S., Fu C. Y., & Berg A. C. (2016). SSD: Single shot multibox detector. In European conference on computer vision. pp. 21-37.
[17] Redmon J., & Farhadi A. (2017). YOLO9000: better, faster, stronger. In Proceedings of the IEEE conference on computer vision and pattern recognition. pp. 7263-7271.
[18] Redmon, J., & Farhadi, A. (2018). Yolov3: An incremental improvement. arXiv preprint arXiv:1804.02767.
[19] Bochkovskiy, A., Wang, C. Y., & Liao, H. Y. M. (2020). Yolov4: Optimal speed and accuracy of object detection. arXiv preprint arXiv:2004.10934.
[20] Lin T. Y., Dollár P., Girshick R., He K., Hariharan B., & Belongie S. (2017). Feature pyramid networks for object detection. In Proceedings of the IEEE conference on computer vision and pattern recognition. pp. 2117-2125.
[21] Liu S., Qi L., Qin H., Shi J., & Jia J. (2018). Path aggregation network for instance segmentation. In Proceedings of the IEEE conference on computer vision and pattern recognition. pp. 8759-8768.
[22] Ghiasi G., Lin T. Y., & Le Q. V. (2019). Nas-fpn: Learning scalable feature pyramid architecture for object detection. In Proceedings of the IEEE/CVF Conference on Computer Vision and Pattern Recognition. pp. 7036-7045.
[23] Tan M., Pang R., & Le Q. V. (2020). Efficientdet: Scalable and efficient object detection. In Proceedings of the IEEE/CVF conference on computer vision and pattern recognition. pp. 10781-10790.
[24] Everingham M., Van Gool L., Williams C. K., Winn J., & Zisserman A. (2010). The pascal visual object classes (VOC) challenge. International journal of computer vision, 88(2): 303-338.

Journal of Physics: Conference Series **2209** (2022) 012024 doi:10.1088/1742-6596/2209/1/012024

Finding Discriminative Regions with More Subtle Granularities for Fine-grained Image Classification

Qi Liu[1a*], Zhipeng Xi[1b*], Xiaoyu Fan[1c*], Zhijian Yin[1d*], Zhen Yang[1e*]

[1]Jiangxi Science and Technology Normal University, Nanchang, China

[a*]criafting@aliyun.com, [b*]John_boy0618@163.com, [c*]fanxiaoyu2015@163.com, [d*]zhijianyin@aliyun.com, [e*]yangzhenphd@aliyun.com

Abstract---Classifying subcategories of the same category (such as birds, cars, airplanes) is mainly to find discriminative features and accuracy regional positions in the fine-grained visual classification (FGVC). In this article, we propose to use global average pooling slice feature maps to find significant discriminative regions without complicated network designs or training operations and use Drop-Block mechanisms to solve the problem of network overfitting. Specially, we use the feature maps of the multi-branches as inputs, and average-pooling them with different sizes of convolution kernels to obtain feature maps that containing deeper and shallow information for getting more accuracy granularities. Our methods are called as More Accurate Multi-Granular Convolutional Neural Network (MAG-CNN). Compared with other networks with more complex designs, the network only requires common operations such as pooling and convolution to achieve higher accuracy. The MAG-CNN can be trained end-to-end, without any bounding-box, and our method has reached the most advanced performance on three common fine-grained image classification benchmark datasets (CUB-Birds, FGVC-Aircraft, and Stanford-Cars).

1. Introduction

Fine-grained image classification [14] refers to distinguishing identical sub-categories in the field of computer vision (for example: birds, cars, airplanes, etc.). Compared with other classification tasks, fine-grained classification is more difficult due to large differences between intra-class and small differences between inter-class. Finding more discriminative features from the global area have become the key points for solving the fine-grained classification problem and the well-known effective method. In addition, borrowing methods from other areas of computer vision have also become one of the directions such as object detection[1][2] and pedestrian re-identification[11][19]. Early classification methods mainly relied on self-supervised convolutional neural networks and manual part annotations[3]. However, this kind of human annotations would take lots of time and efforts but is easy to make mistakes including error-prone. Recently, with the rise and development of weakly supervised including half-supervised and unsupervised[7][10], some methods have been proved that the results perform better on some datasets without bounding-boxes because they can dig out distinguishable features and locate key areas. Now, more and more complex and deeper neural networks are used to obtain deeper-level feature maps. The main reason is that shallow-level features cannot play a decisive role. Up to now, solving the fine-grained classification problem is mainly divided into two methods:(1) using the idea of target detection to find the key areas;(2) using the attention mechanism[5][9] to ensure that the convolutional network could learn enough discriminative features. The latest work of the fine-grained classification has achieved the most advanced achievements by exploring these two parts at the same time and using their complementarity[4][21].

FIG 1 The main problems of FGVC, i.e., low inter-class variance and high intra-class variance.

In this article, we use the method of image slicing to get multi-granularity information including global features and local discriminative features as the basis of classification for the model which can pay more attention on discriminative regions. We use FPN to fuse shallow information and high-level semantic information for obtaining more informative feature maps. The main works can be summarized as follows:

1) Using FPN to fuse shallow features with deep features. The shallow features express the overall information of the object target, but the discrimination is not strong; the deep features represent the key features of the object better, but the problem is the amount of deep information is large which is easy to cause misjudgments for the nerves. We use the three-layer convolutional network of the fourth layer of the Resnet50 network as the deep features, and merge them with the second, third and fourth-level features so that the obtained feature map contains both deep information and shallow information.

2) Adopting global average pooling with different sizes[6]. We use three branches as inputs: the first branch is responsible for the extraction of global information features of the image, the second branch uses pooling[12] with different sizes to divide the feature map into two parts, and the third branch will be divided into three, this part can extract more granular information. The three branches have cooperation and division of labor, sharing low-level feature weights and high-level semantic information independence.

3) Regularized drop-block. In order to solve the problem of too many parameters, we decide to adopt drop-block regularization module. The traditional drop-out is widely used in the fully connected layer of the neural network, but it does not play a big role in the convolutional layer. We introduce drop-block which is a structured form of dropout and puts the units in adjacent areas of the feature map together and drops them. We find that applying drop-block can improve accuracy in jump connections in addition to the convolutional layer. Otherwise, during the training process, gradually increasing the number of dropped units will lead to better accuracy and robustness to hyper-parameter selection.

2. Related work

Before many neural networks with very deep convolutional layers were designed, the object detection frames were mainly used to train and test the datasets in the early stage. But the annotation frame was marked and the labor cost was greatly increased and errors were prone to occur. Recently, some articles such as NTS-Net [20] which is using target detection to automatically generate discriminative regions in a convolutional neural network without external manual annotation. In the NTS-Net, it can learn effective information self-supervised without bounding boxes, and adopting the method of multi-agent collaborative to identify the information area of images in the fine-grained image classification task accurately. The NTS network also introduces FPN[15] (Feature Pyramid Network) and anchors to focus on the model's most informative areas.

In the WSDAN[13] network, some attention-maps are generated to represent the discriminative regions through weakly supervised learning, then these attention maps are randomly selected for data enhancement (the main method is attention crop and attention[8] drop). The acquired feature maps and original image will be combined to crop and drop. In order to make the attention maps of the same channel focus on a specific area in the same category, the author proposes a method of attention regularization which like a sliding window method making the network pay attention to different discriminative areas in different categories; in the test phase, to realize the target positioning and image refinement of the test set, the attention map is generated through 1x1 convolution. By averaging all the attention maps, a new attention map is obtained, and then positioning to the overall position of the target.

In addition, the attention mechanism is also one of the commonly used methods for classification in recent years which including channel-attention mechanism and spatial-attention mechanism. Self-Attention is used to calculate the weight between different positions in the feature maps for achieving the effect of updating the feature in Transformer. In SE-Net, using Sigmoid to normalize for getting the channel attention map by applying the spatial dimension to adaptive average pool and learning the channel attention through two FCs, finally multiplying the Channel Attention Map with the original feature to get the weighted feature.

In MC-Loss, the cross-channel loss is introduced in the fine-grained image classification task.MC-Loss is composed of discriminative components and diverse components. Compared with other methods of SOTA, this paper has a small amount of code and it adds an auxiliary structure to the commonly used network structure. Moreover, the auxiliary structure does not introduce additional network parameters, only including regularization, pooling[16] and other operations which achieves the effect of SOTA.

Compared with the above classic and widely used methods, this paper draws on the idea of pedestrian re-identification slicing and improves it. We propose the use of neural network multi-layer features and only use average-pooling of different sizes to find more accurate multi-granularity discriminative areas.

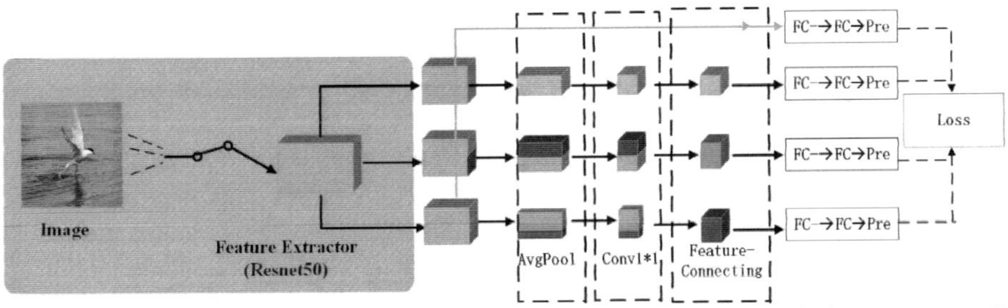

FIG 2 The framework of MAG-NET which consists of three parts. (A)Global Average-pooling of different sizes(B)Dimensionality reduction with 1×1 and Feature stitching. (C)CE-Loss

3. Methods

3.1 Shallow feature and deep feature fusion

In this paper, we use Resnet-50 as the fine-grained classification network's backbone in Fig 2. The feature maps are extracted by pre-trained extractor for an input image I that is expressed as $F \in R^{N\times W\times H}$, with the height H, the width W and the number of channels N. The feature map is expressed as $F \in R^{N\times H\times W}$, and we find that if only the fourth layer of the three-layer convolutional network is used as input, the final fine-grained classification effect is not ideal through lots of experiments. So, we fuse the second, third, and fourth-layer network of Resnet-50 with the fourth layer of the three-layers convolutional neural network by the FPN. In the FPN layer, we use the 1×1 and step size of 1 convolution layer on the feature maps so that all the numbers of channels are the same.

$$F_i \in R^{N\times 14\times 14}(i = 1,2, \ldots, n) \tag{1}$$

The feature maps not only contain rich shallow information but also has key discriminative features in this way. We have proved that part of the information will be missing which will reduce the accuracy of classification if only the deep feature map is used.

3.2 Global Average-pooling and Slicing

We choose a rectangular convolution kernel in order to achieve the purpose of slicing: the size of the feature map after the convolution layer operation is $F \in R^{N\times 14\times 14}$, and we choose three branches as inputs. The size of the first branch pooling convolution kernel is 14×14 which the size of the output feature map is $D_1 \in R^{N\times 1\times 1}$. But the second branch pooling size is 7×14, so we get Feature map size $D_2 \in R^{N\times 2\times 1}$ .The feature maps are divided into up and low parts through the slicing operation:

$$C_i = D_i[:, m{:}m+1, :, :](m \in 1,2,..n) \tag{2}$$

The feature map obtained by slicing contains discriminative features, which play a key role in the classification process. The feature map obtained by slicing is combined to form a new feature map. The classification is performed via two fully connected layers, the loss function can be expressed as:

$$L_{LOSS}(F) = L_{CE}\left(y, \frac{\left[e^{g(F0)}, e^{g(F1)}, \ldots, e^{g(Fn)}\right]^T}{\sum_{n=1}^{N} e^{g(Fn)}}\right) \tag{3}$$

Where g(·) is defined as:

$$G(Fi) = \frac{1}{WH}\sum_{n=1}^{N}[F_i] \tag{4}$$

3.3 Drop-Block

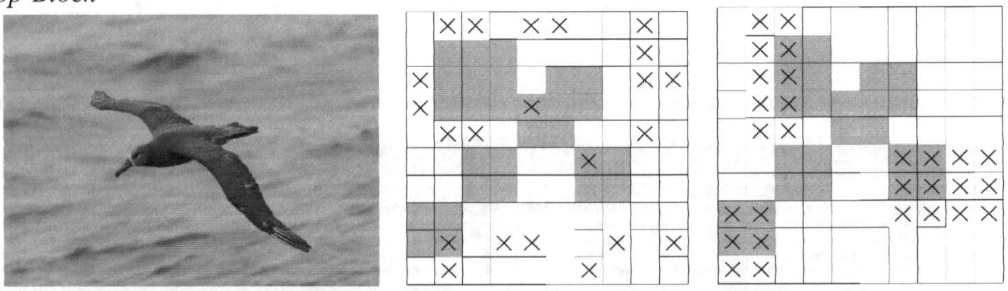

(a) (b) (c)

Fig 3: The green part represents the activated feature unit. (a) input image to the convolutional neural network. (b)Dropping out activations at random is less effective in removing information because nearby activations contain closely related information. (c)Dropping continuous regions can remove information and enforcing remaining units to learn other features for classifying.

In order to solve overfitting in the experiment, we adopt the regularized drop-block module in Fig 3. The traditional dropout played a small role in the convolutional layer: the figure b represents the random dropout activation unit, but the network will also learn the same information from nearby activations. The figure c represents the Drop-Block in this article which mainly has 2 parameters, block-size and γ. When block-size=1, Drop-Block degenerates to traditional dropout which can normally be 3, 5, or 7. γ: represents the probability:

$$\Gamma = \frac{1-\text{keepprob}}{\text{blocksize}^2}\frac{featuresize^2}{(featsize-blocksize+1)^2} \quad (5)$$

We find that drop-block can inactivate a part of the entire adjacent area (such as the bird's head and feet), then the network will focus on learning the characteristics of other parts of the object to achieve accuracy classification for showing better generalization.

4. Experiments

4.1 Dataset

In this section, we evaluate the performance of our designed framework on the fine-grained image classification benchmark datasets including CUB-200-2011, Stanford-Cars, and FGVC-Aircraft[18]. All datasets contain the same A set of subcategories of a super category. The following is a brief description of these data sets:

CUB-200-2011 has 11778 pictures from 200 categories, 5994 pictures are officially di-vided into training sets, and 5794 pictures are divided into test sets.

Stanford-Cars has 16,185 pictures from 196 categories, 8144 pictures are divided into training sets, and 8041 pictures are divided into test pictures.

FGVC-Aircraft has 10,000 images from 100 categories, officially divided into 6,667 training images and 3,333 test images.

Table1. Introduction to the benchmark data set

Dataset	#Class	#Train	#Test
CUB-200-2011	200	5,994	5794
Stanford Cars	196	8,144	8,041
FGVC Aircraft	100	6,667	3,333

4.2 Results

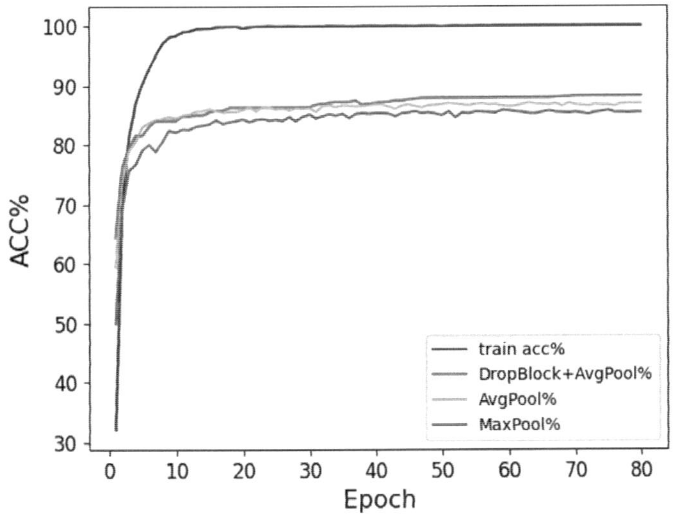

FIG 4 Training and test accuracy of AvgPool、DropBlock and MaxPool on CUB-200-2011 dataset.

We choose Resnet50 pre-processed in ImageNet for experiments and analysis. More specifically, we choose the second, the third, and the fourth layers of Resnet50 to perform feature fusion with the fourth layer of the three-layers convolution through the FPN feature pyramid network. The size of the input network image is 448×448 after flipping, cropping and other operations which are common basic image operations. In addition, we do not use additional bounding box annotations to facilitate comparison of our method with other weakly supervised methods. We use the open source Pytorch as our code base and train all models on a single GTX 2080Ti GPU. Use a stochastic gradient descent with a momentum of 0.9 and a mini-batch of 16 for optimization. The initial learning rate is set to 0.001 and is reduced to 0 using the cosine annealing schedule. All models have been trained for 200 iterations.

Table 2 shows the comparison between the neural network we designed and the current performance SOTA classification network. We also compare the global average pooling (GAP) and global maximum pooling (GMP)[17] on the Resnet50 classification network for slicing operations and show the effect of regularized Drop-Block on the performance of the convolutional neural network. As shown in Table 3, slicing with GAP improves the classification performance of the network compared with GMP. Compared with dropout, Drop-Block has a better performance in solving the network overfitting problem.

Table2. CUB-200-2011, Stanford Cars, FGVC-Aircraft data set comparison results. The best and the second are marked in bold and underlined.

Method	CUB-200-2011	Stanford Cars	FGVC Aircraft
RA-CNN	85.3	92.5	88.2
MA-CNN	86.5	92.8	89.9
MA-CNN	86.5	92.8	89.9
DFL-CNN	87.4	93.1	91.7
NTS-Net	87.5	93.9	91.4
TASN	87.9	93.8	-
B-CNN	84.1	91.3	84.1
MAG-Net	88.1	94.4	92.7

Table 3 With or without Drop-Block, GAP, and the comparison results of the two slicing methods of GAP and GMP in the CUB-200-2011 data set. Drop-Block: regularization module. GAP: global average pooling, GMP: global maximum pooling.

Method	Base Module	Accuracy(%)
Baseline	Resnet50	85.5
MaxPooling(Slicing)	Resnet50	86.1
GAP(Slicing)	Resnet50	87.4
Drop-Block+GAP(Slicing)	Resnet50	88.1

4.3 Visualization

In order to show the advantages of MAG-Net more intuitively, we use Grad-Cam to realize the channel visualization of the feature maps on CUB-200-2011, Stanford Cars and FGVC Aircrafts datasets. We can see that the network can learn the key areas of the target for classification, and focusing on different discriminative areas for the same class in the figure 5.

Fig 5 Some samples of feature maps visualization on CUB-200-2011, Stanford Cars and FGVC Aircrafts datasets

5. Conclusion

The problem of fine-grained image classification is still a challenging research topic in the field of computer vision for its goal is to identify sub-categories. Finding those subtle traits that fully characterize the object is not straightforward. To handle this circumstance, we propose to use the global average pooling method of slicing operation to replace the maxi-mum pooling for fine-grained classification without manual part annotations. The global information and local information are combined to form a multi-granular convolutional neural network, which provides a shallow Layer semantic information and deep discrimination information. Using drop-block regularization module, further improve the classification ability of the network. The effectiveness of our method is proved on the data sets of CUB-200-2011, Stanford Cars and FGVC Aircraft. Our network model is very effective for large goals but it is still somewhat insufficient for small goals, this is a future research direction for us.

Acknowledgement

This work was supported by the National Natural Science Foundation of China (62061019, 61866016), General Project of Jiangxi Natural Science Foundation(20202BABL202014), the General Project of Jiangxi Education Department(GJJ190587), and the Youth Top Talent Foundation of Jiangxi Science and Technology Normal University (2018QNBJRC002).

References

[1] Anelia Angelova and Shenghuo Zhu. Efficient object detection and segmentation for fine-grained recognition. In Proceedings of the IEEE Conference on Computer Vision and Pattern Recognition, pages 811–818, 2013.

[2] Steve Branson, Grant Van Horn, Serge Belongie, and Pietro Perona. Bird species catego-rization using pose normalized deep convolutional nets. arXiv preprint arXiv:1406.2952, 2014.

[3] Yuning Chai, Victor Lempitsky, and Andrew Zisserman. Symbiotic segmentation and part localization for fine-grained categorization. In Proceedings of the IEEE International Conference on Computer Vision, pages 321–328, 2013.

[4] Dongliang Chang, Yifeng Ding, Jiyang Xie, Ayan Kumar Bhunia, Xiaoxu Li, Zhanyu Ma, Ming Wu, Jun Guo, and Yi-Zhe Song. The devil is in the channels: Mutual-channel loss for fine-grained image classification. IEEE Transactions on Image Processing, 29:4683– 4695, 2020.

[5] Long Chen, Hanwang Zhang, Jun Xiao, Liqiang Nie, Jian Shao, Wei Liu, and Tat-Seng Chua. Sca-cnn: Spatial and channel-wise attention in convolutional networks for image captioning. In Proceedings of the IEEE conference on computer vision and pattern recog-nition, pages 5659–5667, 2017.

[6] Yin Cui, Feng Zhou, Jiang Wang, Xiao Liu, Yuanqing Lin, and Serge Belongie. Kernel pooling for convolutional neural networks. In Proceedings of the IEEE conference on computer vision and pattern recognition, pages 2921–2930, 2017.

[7] Cheng Deng, Erkun Yang, Tongliang Liu, Jie Li, Wei Liu, and Dacheng Tao. Unsuper-vised semantic-preserving adversarial hashing for image search. IEEE Transactions on Image Processing, 28(8):4032–4044, 2019.
[8] Yifeng Ding, Zhanyu Ma, Shaoguo Wen, Jiyang Xie, Dongliang Chang, Zhongwei Si, Ming Wu, and Haibin Ling. Ap-cnn: weakly supervised attention pyramid convolutional neural network for fine-grained visual classification. IEEE Transactions on Image Pro-cessing, 30:2826–2836, 2021
[9] Jianlong Fu, Heliang Zheng, and Tao Mei. Look closer to see better: Recurrent attention convolutional neural network for fine-grained image recognition. In Proceedings of the IEEE conference on computer vision and pattern recognition, pages 4438–4446, 2017.
[10] Weifeng Ge, Xiangru Lin, and Yizhou Yu. Weakly supervised complementary parts models for fine-grained image classification from the bottom up. In Proceedings of the IEEE/CVF Conference on Computer Vision and Pattern Recognition, pages 3034–3043, 2019.
[11] Weifeng Ge and Yizhou Yu. Borrowing treasures from the wealthy: Deep transfer learning through selective joint fine-tuning. In Proceedings of the IEEE conference on computer vision and pattern recognition, pages 1086–1095, 2017.
[12] Golnaz Ghiasi, Tsung-Yi Lin, and Quoc V Le. Dropblock: A regularization method for convolutional networks. arXiv preprint arXiv:1810.12890, 2018.
[13] Tao Hu, Honggang Qi, Qingming Huang, and Yan Lu. See better before looking closer: Weakly supervised data augmentation network for fine-grained visual classification. arXiv preprint arXiv:1901.09891, 2019.
[14] Alex Krizhevsky, Ilya Sutskever, and Geoffrey E Hinton. Imagenet classification with deep convolutional neural networks. Advances in neural information processing systems, 25:1097–1105, 2012.
[15] Tsung-Yi Lin, Piotr Dollar,´ Ross Girshick, Kaiming He, Bharath Hariharan, and Serge Belongie. Feature pyramid networks for object detection. In Proceedings of the IEEE conference on computer vision and pattern recognition, pages 2117–2125, 2017.
[16] Tsung-Yu Lin, Aruni RoyChowdhury, and Subhransu Maji. Bilinear cnn models for fine-grained visual recognition. In Proceedings of the IEEE international conference on com-puter vision, pages 1449–1457, 2015.
[17] Zhanyu Ma, Dongliang Chang, Jiyang Xie, Yifeng Ding, Shaoguo Wen, Xiaoxu Li, Zhongwei Si, and Jun Guo. Fine-grained vehicle classification with channel max pooling modified cnns. IEEE Transactions on Vehicular Technology, 68(4):3224–3233, 2019.
[18] Subhransu Maji, Esa Rahtu, Juho Kannala, Matthew Blaschko, and Andrea Vedaldi. Fine-grained visual classification of aircraft. arXiv preprint arXiv:1306.5151, 2013.
[19] Guanshuo Wang, Yufeng Yuan, Xiong Chen, Jiwei Li, and Xi Zhou. Learning discrimi-native features with multiple granularities for person re-identification. In Proceedings of the 26th ACM international conference on Multimedia, pages 274–282, 2018
[20] Ze Yang, Tiange Luo, Dong Wang, Zhiqiang Hu, Jun Gao, and Liwei Wang. Learning to navigate for fine-grained classification. In Proceedings of the European Conference on Computer Vision (ECCV), pages 420–435, 2018.
[21] Heliang Zheng, Jianlong Fu, Zheng-Jun Zha, and Jiebo Luo. Looking for the devil in the details: Learning trilinear attention sampling network for fine-grained image recognition. In Proceedings of the IEEE/CVF Conference on Computer Vision and Pattern Recogni-tion, pages 5012–5021, 2019.

Dynamic Analysis of the influence of the Rotor Centroid Offset of Gyrowheel

Xiaokun Liu[1,2,3*], Guanglie Yang[1,2], Chaoming Mo[1,2], Minglang Tan[1,2], Dongqing Gu[1,2], Bin Song[1,2], Shufan Wu[3]

[1]Aerospace System Engineering Shanghai, Minhang District, Shanghai, China

[2]Shanghai Engineering Research Center of Space Robotics, Minhang District, Shanghai, China

[3]School of Aeronautics and Astronautics, Shanghai Jiaotong University,No 800,DongChuan Road,Minhang District,Shanghai,China

*Corresponding author: *email: liuxiaokun@hit.edu.cn

Abstract. Gyrowheel is a new electromechanical servo device which can realize the integration of spacecraft attitude control and attitude measurement. Due to the existence of machining and assembly errors, there must be a deviation between the rotor centroid of the gyrowheel and the support center, that is, the centroid offset. Firstly, to solve this problem, the dynamic modeling of rotor center of mass offset is carried out from the perspective of vector mechanics by using D'Alembert principle. Secondly, the established dynamic model is simplified, and the influence of rotor center of mass offset on the micro vibration of gyrowheel is analyzed. Finally, the correctness of this method is verified by simulation.

1. Introduction

With the rapid development of spacecraft technology, Gyrowheel has great advantages in mass, power consumption, size and functional integration density, so its application in space is becoming more and more urgent[1]. At present, the domestic research on gyrowheel mostly focuses on the theoretical level. Although some research institutions have completed the development of engineering prototype, there is no on orbit application case[2].

The main reason is that there are a lot of non ideal factors in the processing and assembly process of gyrowheel, such as mismatch angle of torque converter, non coplanar and non vertical orthogonal support, non coincidence of rotor center of mass and orthogonal support center[3], etc. In view of these non ideal factors, some scholars at home and abroad have deduced the influence of the non ideal factors of the DTG[4], such as the internal and external torsion bars are not orthogonal, not vertical and the rotor mass is not balanced. Staley studies the characteristic polynomials of the equation of rotor system with internal and external torsion bar supporting scheme by the dynamic equation of DTG[5]. Since the gyrowheel works in a large number of non tilting states in the torque output process, its nonlinear characteristics are particularly critical.

Based on the above analysis, this paper first establishes the dynamic model of the gyro flywheel when the orthogonal support center and the center of mass are not coplanar. Second, it is properly simplified under engineering constraints, and the analytical formula of the disturbance torque caused by the rotor centroid offset characteristic is given, and the theoretical analysis is carried out by using the expressions. Finally, the effectiveness of this study is verified by simulation.

2. Dynamic modeling of the rotor centroid offset

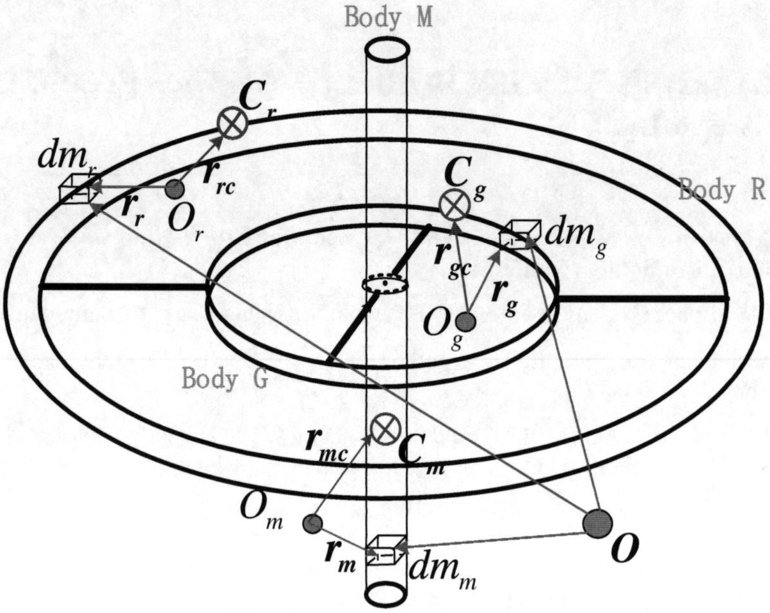

Fig. 1 The rotor centroid offset

Gyrowheel rotor is the main component to realize momentum exchange. In the process of machining and assembly, it is affected by process technology. Due to technical limitations, it is difficult for the rotor centroid to completely coincide with the support center, that is, the rotor centroid is offset as shown in Fig.1.

This error source will inevitably lead to changes in the dynamic characteristics of the gyrowheel. Therefore, in this section, the mathematical model of the gyrowheel in the presence of centroid offset is established from the perspective of vector mechanics by using the D'Alembert principle, which is used to provide a theoretical basis for the analysis of the dynamic characteristics of the gyrowheel under this nonideal factor. The support points O_r, O_g, O_m of each body coincide with the centroid. And the radius vector $R_{mc}=0$, $R_{gc}=0$, and the radius vector $R_{rc}\neq 0$, this is used to mathematically characterize the deviation of the rotor centroid C_r from the support point O_r .It is assumed that the vector $\boldsymbol{r}_{rc}$ can be expressed as $\boldsymbol{r}_{rc}^{(r)}=\left[R_{rx}^{(r)}\quad R_{ry}^{(r)}\quad R_{rz}^{(r)}\right]^T$ and $\boldsymbol{r}_{r}^{(c)}=\left[R_{rx}^{(c)}\quad R_{ry}^{(c)}\quad R_{rz}^{(c)}\right]^T$ in the rotor coordinate system and shell system, and according to the coordinate transformation relationship, the following equation can be obtained.

$$\boldsymbol{r}_r^{(c)}=\boldsymbol{C}_{mc}^{-1}\cdot\boldsymbol{C}_{gm}^{-1}\cdot\boldsymbol{C}_{rg}^{-1}\cdot\boldsymbol{r}_r^{(r)}=\begin{bmatrix} C_{\theta_z}\left(R_{rx}^{(r)}C_{\theta_y}+R_{rz}^{(r)}S_{\theta_y}\right)-S_{\theta_z}\left[R_{ry}^{(r)}C_{\theta_x}+\left(R_{rx}^{(r)}S_{\theta_y}-R_{rz}^{(r)}C_{\theta_y}\right)S_{\theta_x}\right] \\ S_{\theta_z}\left(R_{rx}^{(r)}C_{\theta_y}+R_{rz}^{(r)}S_{\theta_y}\right)+C_{\theta_z}\left[R_{ry}^{(r)}C_{\theta_x}+\left(R_{rx}^{(r)}S_{\theta_y}-R_{rz}^{(r)}C_{\theta_y}\right)S_{\theta_x}\right] \\ R_{ry}^{(r)}S_{\theta_x}-\left(R_{rx}^{(r)}S_{\theta_y}-R_{rz}^{(r)}C_{\theta_y}\right)C_{\theta_x} \end{bmatrix} \quad (1)$$

Using the above description, in order to derive the mathematical description of the gyrowheel model with centroid offset, the absolute velocity and acceleration of M body, G body and R body are required, firstly; Secondly, it is necessary to establish the moment balance equation by the D'Alembert principle[6].

Considering the rotation of R body to point O_r, the equilibrium equation of R body can be obtained, as shown in the formula: $\int_R \boldsymbol{r}_r \times \boldsymbol{a}_r dm_r = \boldsymbol{T}_{re}$,where $\boldsymbol{T}_{re}$ is the sum of the external torque vectors on the R body. Then the following expression can be obtained:

$$\int_R \boldsymbol{r}_r \times \boldsymbol{a}_r dm_r = \int_R \boldsymbol{r}_r \times \boldsymbol{a}_{ro} dm_r + \int_R \boldsymbol{r}_r \times \left({}^o\dot{\boldsymbol{\omega}}^r \times \boldsymbol{r}_r \right) dm_r + \int_R \boldsymbol{r}_r \times \left[{}^o\boldsymbol{\omega}^r \times \left({}^o\boldsymbol{\omega}^r \times \boldsymbol{r}_r \right) \right] dm_r = \boldsymbol{T}_{re} \quad (2)$$

According to the vector calculation relationship, the second item on the right in equation (2) can be obtained, and the third item can be sorted and written as shown in the following.

$$\int_R \boldsymbol{r}_r \times \left({}^o\dot{\boldsymbol{\omega}}^r \times \boldsymbol{r}_r \right) dm_r = \boldsymbol{e}_r^T \int_R \left(\boldsymbol{r}_r \cdot \boldsymbol{r}_r \mathrm{E}_r - \boldsymbol{r}_r \cdot \boldsymbol{r}_r \right) dm_r \boldsymbol{e}_r \cdot {}^o\dot{\boldsymbol{\omega}}^r = \boldsymbol{J}_r \cdot {}^o\dot{\boldsymbol{\omega}}^r \quad (3)$$

$$\int_R \boldsymbol{r}_r \times \left[{}^o\boldsymbol{\omega}^r \times \left({}^o\boldsymbol{\omega}^r \times \boldsymbol{r}_r \right) \right] dm_r = {}^o\boldsymbol{\omega}^r \times \int_R \boldsymbol{r}_r \times \left({}^o\boldsymbol{\omega}^r \times \boldsymbol{r}_r \right) dm_r = {}^o\boldsymbol{\omega}^r \times \boldsymbol{J}_r \cdot {}^o\dot{\boldsymbol{\omega}}^r \quad (4)$$

By introducing equation (3) and equation (4) into equation (2), we can get:

$$\int_R \boldsymbol{r}_r \times \boldsymbol{a}_r dm_r = \boldsymbol{r}_r \times \boldsymbol{a}_{ro} m_r + \boldsymbol{J}_r \cdot {}^o\dot{\boldsymbol{\omega}}^r + {}^o\boldsymbol{\omega}^r \times \boldsymbol{J}_r \cdot {}^o\dot{\boldsymbol{\omega}}^r = \boldsymbol{T}_{re} \quad (5)$$

Based on the derivation of equations (1) - (5), the inertia moments of G body to fixed point Og and M body to fixed point Om can be obtained, which can be sorted as follows:

$$\int_R \boldsymbol{r}_g \times \boldsymbol{a}_g dm_g = \boldsymbol{r}_{gc} \times \boldsymbol{a}_{go} m_g + \boldsymbol{J}_g \cdot {}^o\dot{\boldsymbol{\omega}}^g + {}^o\boldsymbol{\omega}^g \times \boldsymbol{J}_g \cdot {}^o\dot{\boldsymbol{\omega}}^g = \boldsymbol{T}_{re} \quad (6)$$

$$\int_R \boldsymbol{r}_m \times \boldsymbol{a}_m dm_m = \boldsymbol{r}_{mc} \times \boldsymbol{a}_{mo} m_r + \boldsymbol{J}_m \cdot {}^o\dot{\boldsymbol{\omega}}^m + {}^o\boldsymbol{\omega}^m \times \boldsymbol{J}_m \cdot {}^o\dot{\boldsymbol{\omega}}^m = \boldsymbol{T}_{re} \quad (7)$$

3. Establishment of moment balance equation of gyrowheel

(1) Taking the M + R + G body as a whole, the moment balance equation of the M + G + R body is established by using the D'Alembert principle, and the following can be obtained:

$$\int_R \boldsymbol{r}_r \times \boldsymbol{a}_r dm_r + \int_R \boldsymbol{r}_g \times \boldsymbol{a}_g dm_g + \int_R \boldsymbol{r}_m \times \boldsymbol{a}_m dm_m = \boldsymbol{T}_m + \boldsymbol{T}_r + \boldsymbol{T}_m^b + \sum_{i=m,g,r} \boldsymbol{r}_{ic} \times \boldsymbol{g} \cdot m_i \quad (8)$$

Taking the equation (5), equation (6) and equation (7) into equation (8), the following equation can be deduced:

$$\sum_{i=m,g,r} \boldsymbol{r}_{ic} \times \boldsymbol{a}_{io} \cdot m_i + \boldsymbol{J}_i \cdot {}^o\dot{\boldsymbol{\omega}}^i + {}^o\boldsymbol{\omega}^i \times \boldsymbol{J}_i \cdot {}^o\dot{\boldsymbol{\omega}}^i = \boldsymbol{T}_m + \boldsymbol{T}_r + \boldsymbol{T}_m^b + \sum_{i=m,g,r} \boldsymbol{r}_{ic} \times \boldsymbol{g} \cdot m_i \quad (9)$$

(2) Taking the G + R body as a whole, the moment balance equation of the G + R body is established by using the D'Alembert principle, as follows:

$$\int_R \boldsymbol{r}_r \times \boldsymbol{a}_r dm_r + \int_R \boldsymbol{r}_g \times \boldsymbol{a}_g dm_g = \boldsymbol{T}_r + \boldsymbol{T}_g^m + \sum_{i=g,r} \boldsymbol{r}_{ic} \times \boldsymbol{g} \cdot m_i \quad (10)$$

Taking equation (5) and equation (6) into equation (10) and sort it out to obtain:

$$\sum_{i=g,r} \boldsymbol{r}_{ic} \times \boldsymbol{a}_{io} \cdot m_i + \boldsymbol{J}_i \cdot {}^o\dot{\boldsymbol{\omega}}^i + {}^o\boldsymbol{\omega}^i \times \boldsymbol{J}_i \cdot {}^o\dot{\boldsymbol{\omega}}^i = \boldsymbol{T}_r + \boldsymbol{T}_g^m + \sum_{i=g,r} \boldsymbol{r}_{ic} \times \boldsymbol{g} \cdot m_i \quad (11)$$

(3) Taking R-body as a whole, the moment balance equation of R-body is established as follows:

$$\int_R \boldsymbol{r}_r \times \boldsymbol{a}_r dm_r = \boldsymbol{T}_r + \boldsymbol{T}_r^g + \boldsymbol{r}_{rc} \times \boldsymbol{g} \cdot m_i \quad (12)$$

Bring equation (5) into equation (12), and we can get:

$$\boldsymbol{r}_{rc} \times \boldsymbol{a}_{ro} \cdot m_r + \boldsymbol{J}_r \cdot {}^o\dot{\boldsymbol{\omega}}^r + {}^o\boldsymbol{\omega}^r \times \boldsymbol{J}_r \cdot {}^o\dot{\boldsymbol{\omega}}^r = \boldsymbol{T}_r + \boldsymbol{T}_r^g + \boldsymbol{r}_{rc} \times \boldsymbol{g} \cdot m_r \quad (13)$$

In order to obtain the radial dynamic equation of the gyrowheel, the vector equations (11) and (13) are transformed into the component form under the balance ring coordinate system by using the Newton Euler method from whole to part, as shown in equations (14) and (15):

$$\begin{aligned}&\boldsymbol{C}_{gc}\left(\boldsymbol{r}_{gc}^{(c)}\times\boldsymbol{a}_{go}^{(c)}\cdot m_g\right)+\left(\boldsymbol{J}_g^{(g)}\cdot\dot{\boldsymbol{\omega}}_g^{(g)}+\boldsymbol{\omega}_g^{(g)}\times\boldsymbol{J}_g^{(g)}\boldsymbol{\omega}_g^{(g)}\right)+\boldsymbol{C}_{gc}\left(\boldsymbol{r}_{rc}^{(c)}\times\boldsymbol{a}_{ro}^{(c)}\cdot m_r\right)\\&+\boldsymbol{C}_{gr}\left(\boldsymbol{J}_r^{(r)}\cdot\dot{\boldsymbol{\omega}}_r^{(r)}+\boldsymbol{\omega}_r^{(r)}\times\boldsymbol{J}_r^{(r)}\boldsymbol{\omega}_r^{(r)}\right)=\boldsymbol{C}_{gr}\boldsymbol{T}_r^{(r)}+\boldsymbol{T}_g^{m(g)}+\boldsymbol{C}_{gc}\left(\boldsymbol{r}_{rc}^{(c)}\times\boldsymbol{g}^{(c)}\cdot m_r\right)\end{aligned}\tag{14}$$

$$\boldsymbol{C}_{gc}\left(\boldsymbol{r}_{rc}^{(c)}\times\boldsymbol{a}_{ro}^{(c)}\cdot m_r\right)+\boldsymbol{C}_{gr}\left(\boldsymbol{J}_r^{(r)}\cdot\dot{\boldsymbol{\omega}}_r^{(r)}+\boldsymbol{\omega}_r^{(r)}\times\boldsymbol{J}_r^{(r)}\boldsymbol{\omega}_r^{(r)}\right)=\boldsymbol{C}_{gr}\boldsymbol{T}_r^{(r)}+\boldsymbol{T}_r^{g(g)}+\boldsymbol{C}_{gc}\left(\boldsymbol{r}_{rc}^{(c)}\times\boldsymbol{g}^{(c)}\cdot m_r\right)\tag{15}$$

Due to the rotor, balance ring, motor support point (O_r, O_g, O_m), motor and balance ring centroid(C_m, C_g) respectively coincide with the centroid, so the following can be shown:

$$\boldsymbol{a}_{ro}^{(c)}=\boldsymbol{a}_{go}^{(c)}=\begin{bmatrix}0&0&0\end{bmatrix}^T\tag{16}$$

After defining the centroid offset, the rotor moment of inertia tensor is expressed as:

$$\boldsymbol{J}_r^{(r)}=\begin{bmatrix}I_{rx}&-I_{rxy}&-I_{rxz}\\-I_{rxy}&I_{ry}&-I_{ryz}\\-I_{rxz}&-I_{ryz}&I_{rz}\end{bmatrix},\boldsymbol{J}_g^{(g)}=\begin{bmatrix}I_{gx}&0&0\\0&I_{gy}&0\\0&0&I_{gz}\end{bmatrix},\boldsymbol{g}^{(c)}=\begin{bmatrix}g_x^{(c)}\\g_y^{(c)}\\g_z^{(c)}\end{bmatrix}\tag{17}$$

Bring equations (16) and (17) into two-component equations (14) and (15) respectively, and take the X and Y-axis directions to obtain the radial dynamic equation of gyrowheel considering centroid offset, as shown in the following formulas:

$$\begin{aligned}&I_{gx}\dot{\omega}_{gx}-I_{gxy}\dot{\omega}_{gy}-I_{gxz}\dot{\omega}_{gz}+\left(I_{gz}-I_{gy}\right)\omega_{gy}\omega_{gz}+\left(I_{gxy}\omega_{gz}-I_{gxz}\omega_{gy}\right)\omega_{gx}+I_{gyz}\left(\omega_{gz}^2-\omega_{gy}^2\right)\\&+C_{\theta_y}\left[I_{rx}\dot{\omega}_{rx}-I_{rxy}\dot{\omega}_{ry}-I_{rxz}\dot{\omega}_{rz}+\left(I_{rz}-I_{ry}\right)\omega_{ry}\omega_{rz}+\left(I_{rxy}\omega_{rz}-I_{rxz}\omega_{ry}\right)\omega_{rx}+I_{ryz}\left(\omega_{rz}^2-\omega_{ry}^2\right)\right]\\&+S_{\theta_y}\left[I_{rz}\dot{\omega}_{rz}-I_{rxz}\dot{\omega}_{rx}-I_{ryz}\dot{\omega}_{ry}+\left(I_{ry}-I_{rx}\right)\omega_{rx}\omega_{ry}+\left(I_{rxz}\omega_{ry}-I_{ryz}\omega_{rx}\right)\omega_{rz}+I_{rxy}\left(\omega_{ry}^2-\omega_{rx}^2\right)\right]\\&=T_{gx}^{m(g_1)}+T_{gx}+\left(R_{ry}^{(c)}g_z^{(c)}-R_{rz}^{(c)}g_y^{(c)}\right)C_{\theta_z}+\left(R_{rz}^{(c)}g_x^{(c)}-R_{rx}^{(c)}g_z^{(c)}\right)S_{\theta_z}\\&I_{ry}\dot{\omega}_{ry}-I_{rxy}\dot{\omega}_{rx}-I_{ryz}\dot{\omega}_{rz}+m_r\cdot\left[\left(I_{rx}-I_{rz}\right)\omega_{rz}\omega_{rx}+\left(I_{ryz}\omega_{rx}-I_{rxy}\omega_{rz}\right)\omega_{ry}+I_{rxz}\left(\omega_{rx}^2-\omega_{rz}^2\right)\right]=T_{ry}^{g(r)}+T_{gy}\\&+m_r\cdot\left[-\left(R_{ry}^{(c)}g_z^{(c)}-R_{rz}^{(c)}g_y^{(c)}\right)C_{\theta_x}S_{\theta_z}+\left(R_{rz}^{(c)}g_x^{(c)}-R_{rx}^{(c)}g_z^{(c)}\right)C_{\theta_x}C_{\theta_z}+\left(R_{rz}^{(c)}g_x^{(c)}-R_{rx}^{(c)}g_z^{(c)}\right)S_{\theta_x}\right]\end{aligned}\tag{18}$$

4. The Influence analysis of rotor centroid offset

According to the mathematical description of rotor centroid offset given in the previous section, the disturbance torque generated by the rotor centroid offset of gyrowheel can be expressed as formula (19) in the shell system:

$$\boldsymbol{T}_{\boldsymbol{u}}^{(c)}=\boldsymbol{r}_{\boldsymbol{rc}}^{(c)}\times\boldsymbol{g}^{(c)}\cdot m_r=\begin{bmatrix}T_{ux}^{(c)}\\T_{uy}^{(c)}\\T_{uz}^{(c)}\end{bmatrix}=\begin{bmatrix}R_{ry}^{(c)}g_z^{(c)}-R_{rz}^{(c)}g_y^{(c)}\\R_{rz}^{(c)}g_x^{(c)}-R_{rx}^{(c)}g_z^{(c)}\\R_{rx}^{(c)}g_y^{(c)}-R_{ry}^{(c)}g_x^{(c)}\end{bmatrix}\cdot m_r\tag{19}$$

Where

$$\begin{bmatrix}R_{rx}^{(c)}\\R_{ry}^{(c)}\\R_{rz}^{(c)}\end{bmatrix}=\begin{bmatrix}C_{\theta_z}\left(R_{rx}^{(r)}C_{\theta_y}+R_{rz}^{(r)}S_{\theta_y}\right)-S_{\theta_z}\left[R_{ry}^{(r)}C_{\theta_x}+\left(R_{rx}^{(r)}S_{\theta_y}-R_{rz}^{(r)}C_{\theta_y}\right)S_{\theta_x}\right]\\S_{\theta_z}\left(R_{rx}^{(r)}C_{\theta_y}+R_{rz}^{(r)}S_{\theta_y}\right)+C_{\theta_z}\left[R_{ry}^{(r)}C_{\theta_x}+\left(R_{rx}^{(r)}S_{\theta_y}-R_{rz}^{(r)}C_{\theta_y}\right)S_{\theta_x}\right]\\R_{ry}^{(r)}S_{\theta_x}-\left(R_{rx}^{(r)}S_{\theta_y}-R_{rz}^{(r)}C_{\theta_y}\right)C_{\theta_x}\end{bmatrix}\tag{20}$$

$R_{rx}^{(r)}, R_{ry}^{(r)}, R_{rz}^{(r)}$ and $R_{rx}^{(c)}, R_{ry}^{(c)}, R_{rz}^{(c)}$ are the centroid offset of rotor in body coordinate system and shell coordinate system, respectively. $g_x^{(c)}, g_y^{(c)}, g_z^{(c)}$ are respectively the projection of gravity in the triaxial direction of the shell. θ_x, θ_y are the inner and outer torsion bar angle. θ_z is the motor rotation angle.

By the approximate assumption $\sin\theta_i \approx \theta_i, \cos\theta_i \approx 1, i = x, y$, the second-order small quantity is kept, and equation (20) is converted to shell coordinate system ,we can get:

$$\begin{bmatrix} R_{rx}^{(c)} \\ R_{ry}^{(c)} \\ R_{rz}^{(c)} \end{bmatrix} \approx \begin{bmatrix} \phi_y R_{rz}^{(r)} + R_{rx}^{(r)} C_{\theta_z} - R_{ry}^{(r)} S_{\theta_z} + \frac{1}{4} R_{rx}^{(r)} \left[\left(\phi_x^2 - \phi_y^2\right)\left(C_{\theta_z} - C_{3\theta_z}\right) + 2\phi_x\phi_y\left(S_{\theta_z} - S_{3\theta_z}\right) \right] \\ -\phi_x R_{rz}^{(r)} + R_{rx}^{(r)} S_{\theta_z} + R_{ry}^{(r)} C_{\theta_z} + \frac{1}{4} R_{rx}^{(r)} \left[\left(\phi_y^2 - \phi_x^2\right)\left(S_{\theta_z} + S_{3\theta_z}\right) + 2\phi_x\phi_y\left(C_{\theta_z} + C_{3\theta_z}\right) \right] \\ R_{rz}^{(r)} + R_{ry}^{(r)}\left(\phi_x C_{\theta_z} + \phi_y S_{\theta_z}\right) - R_{rx}^{(r)}\left(-\phi_x S_{\theta_z} + \phi_y C_{\theta_z}\right) \end{bmatrix} \tag{21}$$

where ϕ_x, ϕ_y is the rotation angle of the gyrowheel in the radial direction in the shell coordinate system.

According to formula (19), there are three frequency components in the formula: DC, frequency doubling and frequency doubling.

Since m_r and $g_i^{(c)}, i = x, y, z$ in equation (19) are constants, when equation (21) is brought into equation (19), the radial disturbing moments $T_{ux}^{(c)}$, $T_{uy}^{(c)}$ caused by centroid offset must have three torque components: DC, frequency doubling and frequency doubling

When these torques act on the rotor along the radial direction of the shell, the DC disturbance torque will cause gyro drift of the gyrowheel. Because the first and third frequency belong to high frequency, it will not cause gyro drift of the gyrowheel, but will stimulate and produce high-frequency vibration of one and three times the speed frequency as the rotor tilting motion.

5. Simulation

The gyrowheel rotor mass is set to $m_r = 1kg$, , the center of mass offset $R_r^{(c)} = \begin{bmatrix} 1 & 1 & 1 \end{bmatrix}^T mm$, and the initial inclination angles of the x-axis of the gyrowheel are $\phi_{x0} = 3^\circ, \phi_{y0} = 0^\circ$, the rotating speed is $\omega = \dot{\theta}_z = 398 rad/s$, and other simulation conditions remain unchanged relative to the ideal model of gyrowheel. The analysis of the tilting motion frequency of the rotor along the x-axis of the shell can be obtained as shown in the figure below.

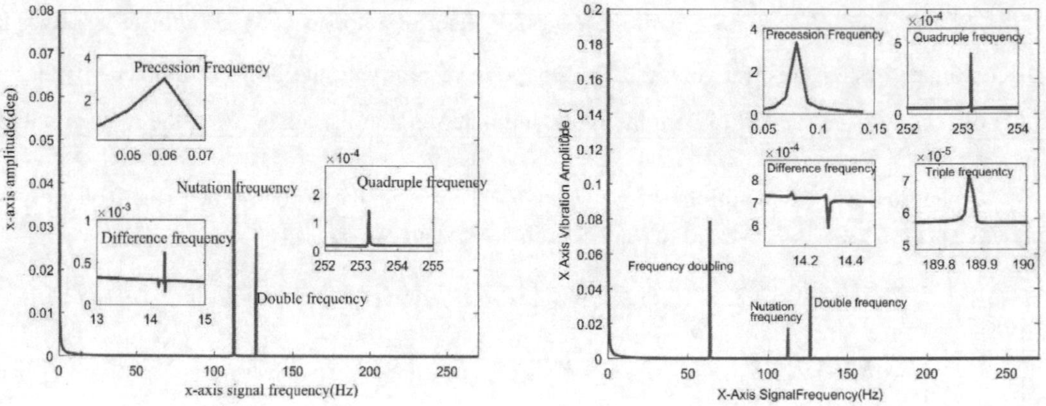

a. Ideal support b. Rotor centroid offset

Fig. 2 FFT Analysis of Gyrowheel

According to the analysis of the above figure, compared with the ideal state, when the rotor has centroid offset, the frequency component of the tilt motion response of the gyrowheel rotor increases the frequency of one and three times of the speed. Before and after the existence of centroid offset, the precession frequency of the system changes from 0.061hz to 0.064hz, which is caused by the DC torque component caused by the rotor centroid offset. It can be seen that this part of DC torque will drift the rotor, which will affect the accuracy of spacecraft attitude angular velocity measurement based on gyrowheel. The double frequency, nutation frequency, difference frequency and quadruple frequency all change slightly in the same order of magnitude before and after the centroid shift.

6. Conclusion

In this paper, two major contributions of this paper can be summarized as follows: (1) the model of gyrowheel considering the rotor centroid offset is established by Newton-Eular method; (2) The non-coplanarity of the rotor centroid offset has evident impact on vibration frequency, which means that research on the compensation of this non-ideal factor for gyrowheel is needed in the future work.

Acknowledgments

This work is supported by Shanghai Natural Science Foundation under Grant 19ZR1455600.

References

[1] Wang Qirui, Liu Qiang, Li Zhuang. (2021)Novel Implicit Lorentz Force Type Magnetic Bearing for High-Precision Agile Maneuver of Magnetically Suspended Gyrowheel. Transactions of Nanjing University of Aeronautics and Astronautics, 38:57-68.

[2] Lv Shuxin, Zhang Guojiang, Liu Xiaokun, Zhao Qing, Zhao Hui, Yao Yu. (2017)Modeling and Simulation Analysis of Gyrowheel System. In: Proceedings of the 36th Chinese Control Conference. DaLian.10353-10358.

[3] Cain, J S. (1999)Investigation of the crossed pivot and the dynamics of the momentum management system spacecraft control component and the dynamically tuned gyroscope. In:Carleton University. Ottawa.23-69.

[4] Xin H , Liu H , Guo Z. (2017)Modeling and analysis of unbalance for GyroWheel system. In: IEEE Control & Decision Conference. Chongqing.1556-1561.

[5] Staley D, Cain J S. (1997)Dynamics of a Tuned Rotor Gyroscope. Canadian Aeronautics and Space Journal, 43: 106-110.

[6] Izadi M,Sanyal A K. (2015)Rigid Body Pose Estimation based on the Lagrange-d'Alembert Principle. In: 54th IEEE Conference on Decision and Control. Osaka. 3688-3704.

Journal of Physics: Conference Series **2209** (2022) 012026 doi:10.1088/1742-6596/2209/1/012026

Improvement and Research of Collaborative Filtering Algorithm Based on Penalty Factor

Panying Li[1], Yuxuan Han[1], Xiumei Wen[2*] and Fanxing Meng[1*]

[1] Computer Science and Technology, Hebei Institute of Civil Engineering and Architecture, Zhangjiakou City, Hebei Province, Zip Code: 075000,China

[2]Zhangjiakou Big Data Technology Innovation Center, Zhangjiakou City, Hebei Province, Zip Code: 075000,China

[*]Corresponding author's e-mail:xiumeiwen@163.com.cn, sir363@163.com

Abstract-The rapid development of the Internet has pushed society into the era of information explosion, and people are faced with more and more information screening and choices. A recommendation system is an effective way to process massive amounts of information, and it is also a tool that can make recommendations based on user behavior. Traditional collaborative filtering algorithms generally use the cosine similarity formula to calculate the similarity between users or items to make recommendations. Due to the popularity of the Internet, more and more popular items have appeared. The appearance of popular items not only affects the recommendation results, but also fails to reflect the real needs of users. This paper proposes an improved formula for cosine similarity with a penalty factor, which can restrain the influence of popular items on the recommendation result. Finally, the Movie Lens data set is used to verify that the recommended performance indicators have been improved to a certain extent.

1. Introduction

In 1994, researchers from the Group Lens project research group at the University of Minnesota developed the first Group Lens system based on collaborative filtering. The basic idea of this system design is to analyze and mine users who have the same interests and preferences as the target user through the user's evaluation of the project, and finally give a recommendation list. The essence of the system is to use a user-based collaborative filtering algorithm. Collaborative filtering algorithm has many advantages in application. This technology is easy to implement, easy to operate, and has a good recommendation effect. Therefore, this technology has made great progress in both academic research and practical application. Domestic research on recommender systems is not very early, but some domestic researchers have also achieved certain results in recent years. At present, it mainly focuses on collaborative filtering algorithm and hybrid recommendation algorithm. In terms of commercial applications, many domestic companies have participated in research related to recommender systems. Among domestic companies, Percentage is the first to research and recommend technology. In order to quickly promote recommendation technology in China, the company has carried out a lot of technology sharing, provided personalized solutions, and developed China's recommendation technology. At the same time, domestic Internet company giant Alibaba also uses many recommendation techniques on Taobao and other websites. Therefore, recommendation systems have important applications in all walks of life. Many algorithms have emerged in the field of recommendation systems. These algorithms have their own application scenarios, but at the same time, these algorithms also have certain advantages and disadvantages. Commonly used recommendation algorithms mainly include content-based recommendation, collaborative filtering recommendation and model-based recommendation algorithm, etc. Collaborative filtering recommendation algorithms are further divided into user-based collaborative filtering and item-based collaborative filtering. Among them, the most commonly used recommendation algorithm is the collaborative filtering recommendation algorithm. The operation steps of the algorithm are simple and clear, easy to understand, and have strong applicability. However, the collaborative filtering algorithm still has some problems, such as cold start, sparse data, poor scalability, etc. These shortcomings will affect the performance of the recommendation system in practical applications.

This article starts with the shortcomings of the collaborative filtering recommendation algorithm. Aiming at the traditional collaborative filtering recommendation algorithm that uses the cosine similarity calculation formula for recommendation, the impact of popular items cannot be suppressed, so the idea of adding a penalty factor to the cosine similarity formula is proposed. Finally, the Movie Lens data set is used for experimental verification and comparative analysis of the results. The idea and implementation steps of user-based and item-based collaborative filtering recommendation algorithms are relatively similar. Therefore, the following uses user-based collaborative filtering recommendation algorithm as an example for experimental analysis and demonstration.

2. Overview of collaborative filtering recommendation algorithm

2.1 Traditional collaborative filtering algorithm

The user-based collaborative filtering recommendation algorithm generally needs to obtain the user's behavior information, and then find the neighbor set of the target user, and finally select the items in the neighbor set that are of interest to the target user but have not generated behaviors for recommendation. The specific implementation steps are as follows:

2.1.1 Collect user behavior data and generate user-item rating matrix

The simplest form of user behavior data on the website is the log[1], of course, there will be other forms as well. For these behavior data, preprocessing is required first. Preprocessing is to convert the user's historical behavior described in natural language into digital information [2]. After processing, the user's behavior information on the item can be transformed into a matrix R(m, n), which is called a user-scoring matrix.

2.1.2 Establish a set of neighbors with similar interests of target users

The key to this step is to calculate the interest similarity between the target user and other users, and generate the target user's neighbor set according to the user similarity. The similarity between users is calculated by the similarity formula. Common similarity calculation formulas include Pearson correlation coefficient, Euclidean similarity, and cosine similarity calculation formulas. The more commonly used is the cosine similarity calculation formula, the formula is as follows:

$$Wuv = \frac{|N(\mathrm{u}) \cap N(v)|}{\sqrt{|N(u)||N(v)|}} \tag{1}$$

Among them, N(u) represents a collection of items with positive feedback that user u has ever had, and N(v) represents a collection of items with positive feedback that user v has ever had. Positive feedback can also be called explicit feedback behavior, which includes the behavior of users expressly expressing their preferences for items [3].

2.1.3 Generate recommendations

After obtaining the interest similarity between users, the user-based collaborative filtering recommendation algorithm will recommend the most similar k favorite items to the user. The following formula measures the user u's interest in items in the user-based collaborative filtering recommendation algorithm:

$$P(u,i) = \sum_{v \in S(u,k) \cap N(i)} Wuv * Rvi \tag{2}$$

Among them, S(u,k) contains k users whose interests are closest to user u, N(i) is the set of users who have acted on item i, Wuv is the similarity of interest between user u and user v, and Rvi represents user v is interested in item i, because it uses implicit feedback data of a single behavior, and the value is 1 here. Implicit feedback data sets refer to behaviors that cannot clearly reflect user preferences, and the typical representative is user page browsing behavior [3].

2.2 Maintaining the Integrity of the Specifications

The algorithm flow of the recommendation system is generally divided into the following parts, namely the collection of user behavior data, the establishment of a user-item score matrix, the calculation of user similarity, and the generation of a user's neighbor set, and finally a recommendation list. The flowchart of collaborative filtering recommendation algorithm is shown in Figure 1.

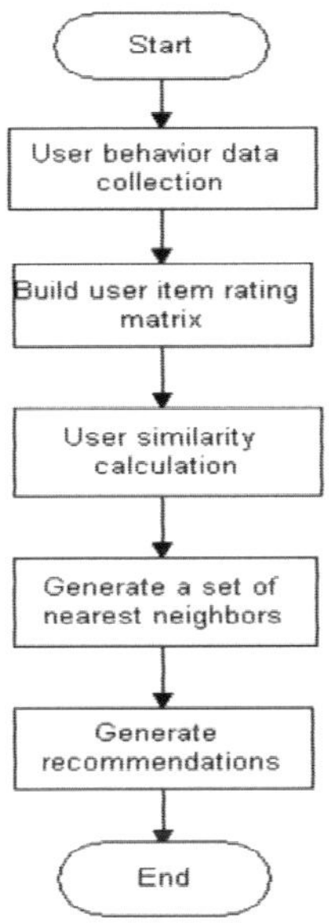

Figure1. Collaborative filtering recommendation algorithm flow

3. Research on collaborative filtering recommendation algorithm based on penalty factor

3.1 Penalty factor

Take books as an example. Two users have bought "Xinhua Dictionary". This does not mean that their interests are similar. If both users have bought "Introduction to Data Mining", then they can be considered to have similar interests, because Only people who study data mining will buy this book. Two users who have taken the same behavior on unpopular items can better illustrate the similarity of their interests.

In a recommendation system, if there are too many popular items, the recommendation results will be relatively single, without novelty, and the final recommendations are all items that appear on the popular rankings, which is very different from the actual needs of users. In order to reduce the impact of popular items, this paper considers adding a penalty factor as a weighting coefficient when calculating user similarity, which plays a role in restraining popular items from affecting the recommendation results. The more an item appears, the less the item contributes to the similarity of user interest. The revised formula can attenuate the impact of popular items. The revised formula is as follows:

$$Wuv = \frac{\sum_{i \in N(u) \cap N(v)} \frac{1}{\log(1+\sqrt{|N(i)|})}}{\sqrt{|N(u)||N(v)|}} \tag{3}$$

N(i) in the formula represents the number of occurrences of product i, and the impact of popular items on similarity is penalized by the number of occurrences of the item.

3.2 Improved algorithm flow

In the above process step of calculating user similarity, a penalty factor is added, and the penalty factor can be used to reduce the impact of popular items. The more items appear, the greater the effect of the penalty factor, and the smaller the impact of popular items on the calculation of user similarity, which has a good recommendation effect. The flow chart of the improved recommendation algorithm is shown in Figure 2.

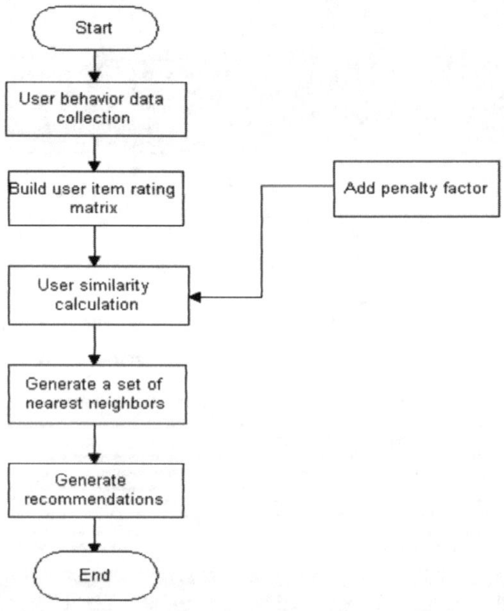

Figure2. Improved algorithm flow

4. Experimental results

4.1 Data set analysis

This article uses the classic Movie Lens (ml-100k) data set in the recommendation system to verify some performance indicators of the traditional user-based collaborative filtering recommendation algorithm and the improved user-based collaborative filtering recommendation algorithm. There are a total of three tables in the data set. Users, Movies and Ratings. The data set stores more than 100,000 pieces of user rating data for movies, of which the data rating value is 1 to 5. The data set is shown in Table 1.

Table 1 Introduction to the data set

Category	Quantity
Subscribers	A total of 610, uid from 1 to 610
Project volume	A total of 1682, mid from 1 to 1682
Score value	100k, an integer of 1-5
Sparsity	100k/(610*1682)=0.097

When the system provides recommendation services, it generally provides users with a personalized recommendation list. This type of recommendation is called Top-N recommendation [4]. The prediction accuracy recommended by Top-N is generally measured by precision and recall.

R(u) is a recommendation list made to the user based on the user's behavior on the training set, and T(u) makes the user's behavior list on the test set. The accuracy rate describes how much of the final recommendation list is the user-item rating record that has occurred, The recall rate describes the percentage of user-item rating records included in the final recommendation list.

4.2 Experimental design

The offline experiment of the collaborative filtering algorithm is generally designed as follows: First, the user behavior data set is randomly divided into M parts according to a uniform distribution (M=8 in this experiment), one part is selected as the test set, and the remaining M-1 parts are used as the training set. Then build a user interest model on the training set, predict user behavior on the test set, and calculate the corresponding evaluation indicators. In order to ensure that the evaluation index is not the result of overfitting, M experiments are required, and a different test set is used each time. Finally, the average value of the evaluation indicators measured in M experiments is used as the final evaluation indicator.

4.3 Experimental results and analysis

In the verification experiment, this paper uses the classic cosine similarity algorithm and the improved cosine similarity calculation method for comparative analysis. First, the recommended number in the algorithm is set to 25, and the number of neighbors of the target user is 2, 3, 5, 10, 20, 40, 50. The calculation accuracy rate and recall rate (the calculation result retains four decimal places) are shown in Table 2.

Table 2 Traditional cosine similarity algorithm

Number of similar users	Accuracy	Recall rate
2	0.1622	0.0822
3	0.1755	0.0888
5	0.2063	0.1043
10	0.2298	0.1162
20	0.2500	0.1273
40	0.2604	0.1302
50	0.2615	0.1322

The improved cosine similarity algorithm is used to calculate the accuracy and recall rates. As shown in Table 3.

Table 3 Improved cosine similarity algorithm

Number of similar users	Accuracy	Recall rate
2	0.1670	0.0836
3	0.1816	0.0915
5	0.2083	0.1046
10	0.2325	0.1171
20	0.2566	0.1295
40	0.2636	0.1326
50	0.2662	0.1342

Set the input parameters (the number of user neighbors) to 2, 3, 5, 10, 20, 40, 50, and the recommended number to 25. The evaluation results are shown in Figure 3 and Figure 4.

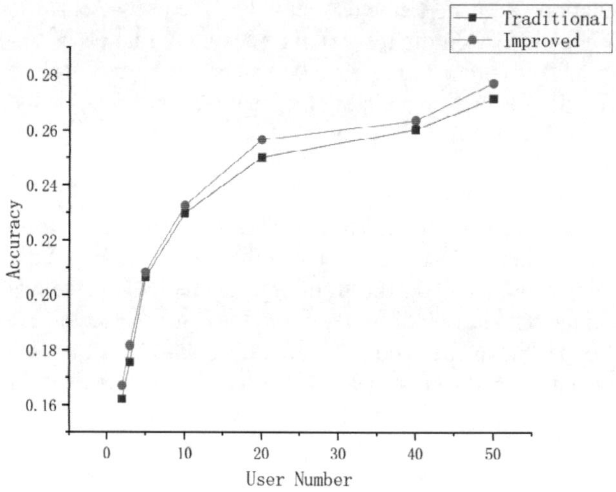

Figure3. Comparison of accuracy of traditional and improved recommendation algorithms

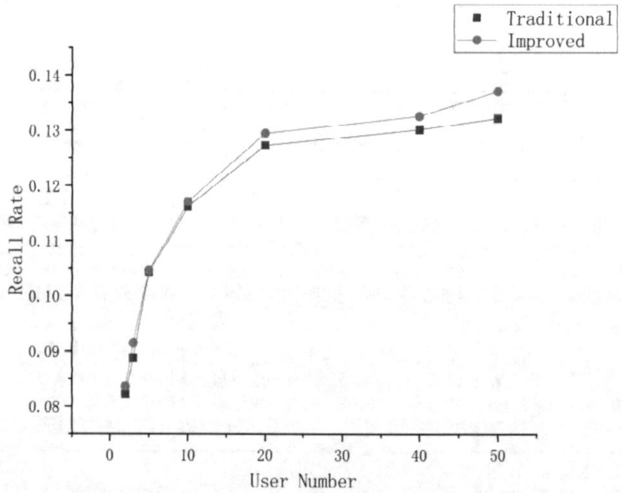

Figure4. Comparison of the recall rate of traditional and improved recommendation algorithms

The figure above shows the comparison between the traditional collaborative filtering recommendation algorithm and the collaborative filtering recommendation algorithm with penalty factor added. It can be seen from the figure that the improved collaborative filtering recommendation algorithm has a higher accuracy and recall rate than the traditional collaborative filtering recommendation algorithm.

5. Conclusion

This article first analyzes that when the traditional collaborative filtering recommendation algorithm calculates user similarity, popular items will affect the final recommendation result. Using traditional cosine similarity algorithm for recommendation, the final recommended items are almost all popular items. Therefore, This paper proposes to add a penalty factor to the cosine similarity calculation formula to reduce the impact of popular items. Finally, it is verified by experiments that the improved algorithm can improve the accuracy and recall rate. It is concluded that the collaborative filtering algorithm based on the penalty factor proposed in this paper can improve the recommendation efficiency and make the recommendation result more accurate.

Acknowledgments

Thank you to the initiator of this research-"Hebei Provincial Higher Education Fundamental Scientific Research Business Expenses Research Project 2021XSTS04". In addition, I would like to thank Zhangjiakou Big Data Technology Innovation Center for providing experimental platforms and equipment.

References

[1] Xiang,L.(2012) Recommended system practice.People Post Press,Beijing.

[2] Meng,XW. Hu,X. Wang,LC.(2013) Mobile recommendation system and its application.ournal of Software .24(1):91-108.

[3] Meng,XW. Ji,WY. Zhang,YJ.(2015) Recommendation system in big data environment. Journal of Beijing University of Posts and Telecommunications. 38(2):1-15.

[4] Leng,YJ. Lu,Q. Liang,CY.(2014) Overview of Collaborative Filtering Recommendation Technology. Pattern recognition and artificial intelligence. 27(8):720-734.

Data enhancement method for object detection

Zhijian Yin[1], Lei Xu[1], Hanqing Yu[1], Zhen Yang[1*]

[1]School of Communication and Electronics, Jiangxi Normal University of Science and Technology, Nanchang, Jiangxi, China

*Corresponding author's e-mail: yangzhen@jxstnu.edu.cn

Abstract—A data enhancement method for object detection is proposed to address the problem that there are few data enhancement means for target detection. This method has obvious advantages for small targets and scenarios where the target background is lacking. It is also a method to solve the problem of small and unbalanced samples. In this paper, the object is subtracted from the bounding box, and after processing, it is fused into a new image using various image classification improvement means such as random brightness, random contrast, random rotation, random cropping, etc. Using mixup-like fusion methods, and XIOU is proposed to guarantee the t intersection ratio of the target synthesis during synthesis. An improvement of 2% is obtained on yolov3 over the VOC dataset.

1. Introduction

The fusing map strategy to separate target and background is proposed to improve the performance of target detection. It has been shown to be effective for guidance, especially when there are only a few hundred or thousand records on small datasets with small data samples, such as COCO128 so that the network can be better promoted and provide better object positioning. Train the image by dividing the target away from the entire large graph, by inserting it into other images after transformation, or by increasing the random noise. This approach is not desirable and leads to a decrease in the accuracy of the early training compared to before the enhancement. Therefore, we propose a strategy to improve the map[1]: the entire cut target in the training image, where the real ground label is also proportionally mixed with the patch area. By creating rotation changes, transparency changes, stretching, subtly unifying the use of training pixels, and maintaining regional dropouts, CIFAR and image network classification tasks and image network localization tasks with weak supervision are always superior to the most advanced enhancement strategies. In addition, unlike previous enhancement methods, our way of training image network classifiers can achieve consistent performance improvements and MS-COCO image caption benchmarks in Pascal detection when used as a pre-training model[2]. We also show that this approach improves model robustness for input corruption and its out-of-distributed detection performance.

2. Related Works

Regional dropout: A method for removing random areas in an image is proposed to improve the overall performance of CN. Object positioning methods also use regional drop-out techniques to improve the localization of CNs. Our method is similar to these methods, and the key difference is that the removed area is filled with patches from another training image. DropBlock summarizes regional dropouts as functional spaces and shows enhanced replicability[3]. Our method can also be performed in the functional space.

Synthesizing training data: Some work explores integrated training data for further promotion. Create new training samples through the modeling image network to make the model focus more on shape than texture for better classification and object detection performance. Our approach also improves performance on many computer vision tasks by cutting and pasting patches in small batches to generate new samples; Unlike our approach, our approach incurs only negligible additional training costs. For object detection[4], the object insertion method is recommended as a background object composition method.

Mixup: Our approach is similar to Mixup's in that both samples combine two samples, where the ground authenticity label of the new sample is given by a linear interpolation of a thermal label. As we will see in our experiments, the Mixup samples are affected by local ambiguity and unnaturalness, which confuses the model, especially localization. Recently, hybrid variants have been proposed: they perform feature-level interpolation and other types of transformations. However, the above works often lack an in-depth analysis of localization capabilities and transfer performance[5].

Tricks for training deep networks: The efficient training of Deep networks is one of the most important issues in the computer vision community because they are very computationally intensive and require a lot of data. Methods such as weight attenuation[6], dropout, and batch normalization are widely used for efficient training depth networks. Recently, it has been proposed ways to add noise to the internal features of the CNN or to incorporate additional paths in the architecture to improve image classification performance.

3. Method

In each random batch, all targets are deducted. After random enhancement transformation, it is randomly combined with the original image to form a new group of images. Each original image and target is randomly combined with 2 to 5 new targets to form an image. When combining, you need to check the frame to avoid the target being completely covered by the other targets. In addition, you need to ensure a certain coverage ratio, which can enhance the robustness of the model[7].

3.1 XIOU Method

In processing the training image, the crucial step is to insert the target frame into the original image. Without considering the relationship between the original image and the size of the target image, direct mapping leads to the problem that the target object in the original image will be occluded in a large area if the size of the target object is larger than the size of the target object in the original image. Due to the excessive occlusion, the object in the original image loses the crucial part of the information. The trained model cannot learn the critical features of the object in the original image. Based on this, this experiment proposes a new method to calculate the object overlap degree IOU, called XIOU. This approach considers three relationships between the mapping target and the original target. The specific mapping steps are as follows: First, a central coordinate is randomly generated within the original map[8], and the target is moved to the generated central coordinate. Second, a simple coordinate analysis is performed to determine whether there is an overlap between the mapped target and the coordinates of each target on the original map. If there is no overlapping area, it means that every two targets in the map are in a relationship away from each other, then there is no need to calculate XIOU.

When the mapping target and the target in the original map have intersections or overlaps, it is necessary to calculate XIOU according to formula 1. Equation 1 consists of two parts: 1)the first half determines the intersection between the target objects;2)the second half determines the separation of the two objects. Here is an extreme example, if the two targets overlap and the overlap area is half of the larger target area. In this case, the first half of the equation yields 0.25, and the second half is also 0.25, so the final XIOU is 0.5. We use a value between 0.5 and 1 as a threshold to control the degree of overlap between two bounding boxes. to control the effect of compositing. If the overlapping area of two objects is bigger than half of the larger target area, then the calculated XIOU will be bigger than the threshold value. If the two targets are separated or intersect (not including the case of overlap), then XIOU will be less than this threshold. Control XIOU can control the target's position in the composite image, the target in the original image will not be covered entirely.

$$XIOU = (\frac{S_{\cap}}{S_1} + \frac{S_{\cap}}{S_2}) * 0.25 + (\frac{S_1}{S_{\Delta}} + \frac{S_2}{S_{\Delta}}) * 0.25 \quad (1)$$

$$S_{\cap} = S_1 \cap S_2 \quad (2)$$

$$0 < XIOU < 1 \quad (3)$$

Suppose there are two targets in the figure, target 1 and target 2, and their areas are S_1 and S_2。The intersection of S_1 and S_2 is $S_{\cap}$. The coordinates of the lower left corner of target 1 are (x1, y1), the coordinates of the lower left corner of target 2 are (x2, y2), and $x1 < x2$, $y1 < y2$。the area of the region enclosed by the point (x1, y1) and the point (x2, y2) is S_{Δ}, as shown in the diagonal part of the figure 1.

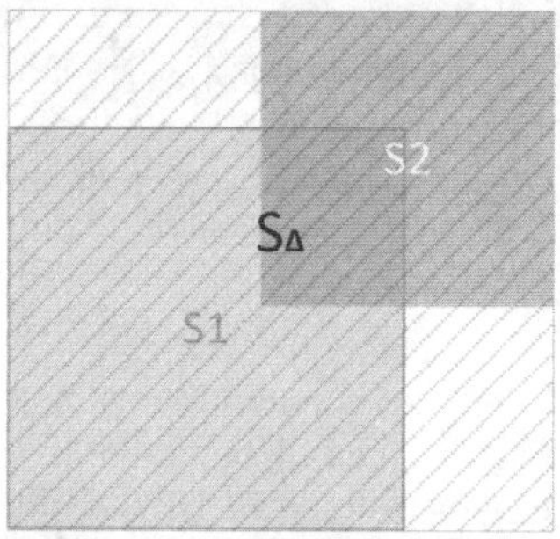

Figure 1. XIOU schematic diagram

4. Experimental analysis

4.1 Environment

The hardware configuration for the experiments in this work is Intel (R) Core (TM) i7-8700, the GPU model is NVIDIA GeForce GTX 1050Ti with 8 GB video memory, 8 GB of RAM, and the operating system is ubuntu 16.04. The number of global iterations is 300, the momentum constant $\beta$is set to 0.9, the weight attenuation is 0.0005, and the initial learning rate is 0.001. The initial experiment was conducted on yolov5s using COCO128 dataset and VOC dataset, so the experiments in this paper are all conducted on YOLOv5s[9].

4.2 Datasets

The VOC2007 and VOC2012 datasets are divided into four broad categories: humans, common animals, traffic vehicles, and interior furniture supplies, which are mainly used for image classification, target detection recognition, and image segmentation[10], and the VOC2007 and VOC2012 datasets are mutually exclusive. In this paper, the training set and verification set of VOC2007 and VOC2012 are used as the training set of experiments, and the test set of VOC2007 is used as the test set, of which there are 16551 pictures in the training set and 4952 in the test set.

4.3 Experiments baseline

Training the YOLOv5s model on the VOC dataset without adding any data enhancement strategy[11], epoch number is 50, and the obtained map@.5 reached 59%.

4.4 Black pixel filling method

The target of detection is first keyed out of the image. The blank position of the image after keying out the target is filled with black pixels, and then this background image is saved in the background folder[12].

The keyed-out detection targets are stored in different folders by category. Randomly select n targets from the detection target folder, where the value of n is ranging from 3 to 5. Then, a background image is randomly selected from the background folder, and the selected targets are pasted onto the background image as new training data, see figure 2. The new data is expanded to the original dataset for model training[13]. The results obtained from the training are shown in Table 1, from which it can be seen that the map@.5 obtained from training on the expanded dataset decreases by 2%. It is presumed that the direct violent keying to fill 0 pixels destroys the correlation between the original target and the background leading to a reduced recognition of a part of the targets that need to be recognized by the background.

Table 1. Black pixel filling method result

Method	Map@.5
Unused data enhancement	59.45%
Black pixel fill	57.13%

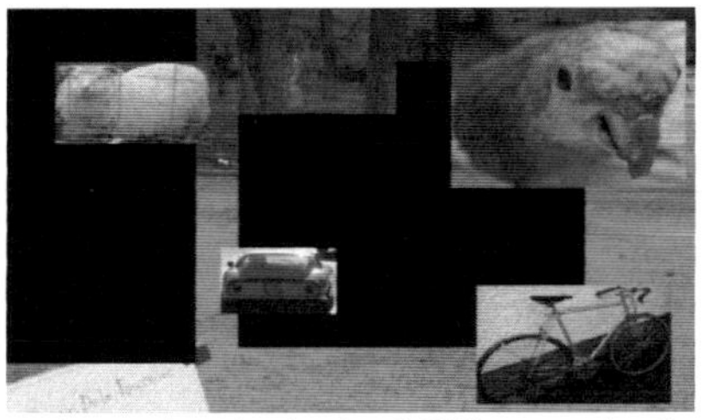

Figure 2. Black pixel filling method

4.5 Random noise filling method

The black pixel padding method in the previous experiment led to a decrease in model accuracy. Therefore, this experiment uses random noise to explore the effect of the difference infill on the model performance. First, the targets in the dataset are keyed and placed in different directories according to the target classification. The blank parts of the keyed-off target objects are filled with random noise and saved as images separately[14]. Then, n targets are randomly selected, and the targets are pasted on the ultimately deducted target images to generate new training images. The newly obtained training maps are blended into the original dataset for training, see Figure 3. The trained map@.5 is reduced by 25% from the baseline, see table 2. Based on the experimental phenomenon, we infer that the random noise reaches the bottom layer of the network through the convolutional network layer by layer and affects the network performance. In addition, since the mapping keeps the original size, the excessive difference between the background map and the mapping scale when pasted onto the background map also degrades the training results.

Table 2. Random noise filling method result

Method	Map@.5
Unused data enhancement	59.45%
Random noise filling	26.24%

Figure 3. Black pixel filling method

4.6 Adding targets to the background

The earlier experiments were all about keying the target out of the background and recombining it to compose a new image. Neither the black background fill nor the random noise fill can improve the model accuracy[15]. So this experiment improves the method of synthesizing images. The original image without keying operation and the target are saved separately. When synthesizing, targets are randomly selected and pasted onto the original image with targets, and 2~5 additional targets are added to each image to generate a new training set. Each target undergoes a simple stretching and scales transformation operation before being pasted onto the original image. When pasting, we follow the principle that the newly added targets do not intersect with the targets in the original image, as shown in figure 4. This strategy also leads to some problems, for example, when the target in the original image occupies a large proportion of the original image. Since the newly added targets cannot intersect with the targets in the original map, then the newly added targets can only be reduced to a relatively small size[16], leading to more small targets in the generated dataset, which affects the network's performance.The experiment results are in table 3.

Table 3. Random noise filling method result

Method	Map@.5
Unused data enhancement	59.45%
Adding targets to the background	58.52%

Figure 4. Black pixel filling method

4.7 Coordinate control with XIOU

We introduce the XIOU introduced in the previous method based on Experiment 6.4. This method calculates the IOU for the newly added target and the target in the original image, as shown in equation(1). We mapped with a threshold value less than 0.5. In this way, the targets in the original map can be in three states: away from each other, intersecting, and overlapping. However, it is essential to note that the area of intersection or overlap here will not exceed half the size of the larger scale target[17]. We can control the size of the threshold to control the mapping form, including controlling the maximum occlusion area of the overlapping part[18]. The following figure 5 shows the detail.

Figure 5. Coordinate control with XIOU

4.8 Mapping target scale transformation

Targets are keyed from the target dataset and stored in category directories, saving the original image and the keyed targets separately. When compositing, the targets are randomly selected and added to the original image[19], and 2~5 additional targets are added to each image. When adding targets, the previously mentioned XIOU method is introduced to calculate XIOU for the newly added targets and the original targets, with a specified threshold value less than 0.5, and the relative positions of the two targets are controlled by the threshold size. The experiment results are in table 4. The spliced new graph is expanded to the original dataset to generate a new training set[20], as shown in Figure 6.

Table 4. Coordinate control with XIOU result

Method	Map@.5
Unused data enhancement	59.45
Adding targets to the background	61.97

Figure 6. scale transformation

5. Conclusion

Most of the current enhancement methods for target detection still use the whole image as the enhancement object. In this paper, based on the idea that most target detection does not require background information, the target and background are separated. The separated target and the background of the keyed target are enhanced separately by transformations such as random brightness, saturation and random cropping. The transformed targets are then stitched back into the original image so that a new target detection dataset is synthesized. This dataset can enhance the generalization ability of the model. The map@.5 of the model after data enhancement improves by xx percentage points over the original model. However, the hyperparameters in this enhancement method still have a great influence on the experimental effect, and the next step will be to further explore the influence of hyperparameters on the model.

ACKNOWLEDGMENTS

This work was supported by the National Natural Science Foundation of China (62061019, 61866016), the Key Laboratory of System Control and Information Processing, Ministry of Education (Scip202106); the Key Project of the Ministry of Education of Jiangxi (GJJ201107) and the General Project of the Ministry of Education of Jiangxi Province (GJJ190587), the Youth Top Project of Jiangxi Science and Technology Normal University (2018QNBJRC002);General Project of the Natural Science Foundation of Jiangxi Province (20202BABL202014).

References

[1] Liang-Chieh Chen, George Papandreou, Iasonas Kokkinos. Deeplab: Semantic image segmentation with deep convolutional nets, atrous convolution, and fully connected crfs.IEEE transactions on pattern analysis and machine intelligence, 40(4):834–848, 2018.

[2] Junsuk Choe and Hyunjung Shim. Attention-based dropout layer for weakly supervised object localization. InProceedings of the IEEE Conference on Computer Vision and Pattern Recognition, pages 2219–2228, 2019.

[3] Terrance DeVries and Graham W Taylor. Improved regularization of convolutional neural networks with cutout.arXiv preprint arXiv:1708.04552, 2017.

[4] Nikita Dvornik, Julien Mairal, and Cordelia Schmid. Modeling visual context is key to augmenting object detection datasets. InProceedings of the European Conference on Computer Vision (ECCV), pages 364–380, 2018.

[5] Debidatta Dwibedi, Ishan Misra, and Martial Hebert. Cut, paste and learn: Surprisingly easy synthesis for instance detection. InProceedings of the IEEE International Conference on Computer Vision, pages 1301–1310, 2017.

[6] M. Everingham, L.Van Gool, C.K.I. Williams, J. Winn, and A.Zisserman. The pascal visual object classes(VOC)challenge. International Journal of Computer Vision, 88(2):303-338,June 2010.

[7] Robert Geirhos, Patricia Rubisch, Claudio Michaelis, Matthias Bethge, Felix A Wichmann, and Wieland Brendel. Imagenet-trained cnns are biased towards texture; increasing shape bias improves accuracy and robustness.arXiv preprint arXiv:1811.12231, 2018.

[8] Golnaz Ghiasi, Tsung-Yi Lin, and Quoc V Le. Dropblock: A regularization method for convolutional networks. In Advances in Neural Information Processing Systems, pages 10750–10760, 2018.
[9] Ian J Goodfellow, Jonathon Shlens, and Christian Szegedy. Explaining and harnessing adversarial examples. InInternational Conference on Learning Representations, 2015.
[10] Hongyu Guo, Y ongyi Mao, and Richong Zhang. Mixup as locally linear out-of-manifold regularization.arXiv preprint. arXiv:1809.02499, 2018.
[11] Dongyoon Han, Jiwhan Kim, and Junmo Kim.Deep pyramidal residual networks. InCVPR, 2017.
[12] Kaiming He, Xiangyu Zhang, Shaoqing Ren. Deep residual learning for image recognition. InCVPR, 2016.
[13] Dan Hendrycks and Kevin Gimpel. A baseline for detecting misclassified and out-of-distribution examples in neural networks. InInternational Conference on Learning Representations, 2017.
[14] Jie Hu, Li Shen, Samuel Albanie, Gang Sun, and Andrea V edaldi. Gather-excite: Exploiting feature context in convolutional neural networks. InAdvances in Neural Information Processing Systems, pages 9423–9433, 2018.
[15] Jie Hu, Li Shen, and Gang Sun. Squeeze-and-excitation networks. InarXiv:1709.01507, 2017.
[16] Gao Huang, Zhuang Liu, and Kilian Q Weinberger. Densely connected convolutional networks. InCVPR, 2017.
[17] Gao Huang, Y u Sun, Zhuang Liu, Daniel Sedra, and Kilian Weinberger. Deep networks with stochastic depth. InECCV, 2016.
[18] Sergey Ioffe and Christian Szegedy. Batch normalization:Accelerating deep network training by reducing internal covariate shift. InICML, 2015.
[19] Hanjoo Kim, Minkyu Kim, Dongjoo Seo. NSML: meet the mlaas platform with a real-world case study.CoRR, abs/1810.09957, 2018.
[20] Alex Krizhevsky, Ilya Sutskever, and Geoffrey E Hinton. Imagenet classification with deep convolutional neural networks. InNIPS, 2012.

Research on face tracking Algorithm Based on Detection and Supervision Tracking

Gang Yuan[1], Fengkun Lv[1], Zhihao Xu[1*], Lv Luo[1], Yi Rong[1], Yanan Zhang[1], Zheng Tian[1]

[1] School of Mechanical and Electrical Engineering, Nanchang Institute of Technology Nanchang, China

[*]Corresponding author's e-mail: zhxuhi@whu.edu.cn

Abstract. Abstract: Accurate face detection and tracking is widely used in many social life scenes. However, the uncontrollable background noise and random illumination change in the application scene will reduce the detection accuracy of the tracked target, and the rotation, occlusion and overlap of the tracked target will also affect the accuracy and success rate of the tracking algorithm. In order to solve the above problems, this paper proposes a method that uses deep learning detection method to supervise the correlation filter tracking algorithm to improve the success rate of de-tection and tracking. First of all, in the first frame of the picture, we use the deep learning SSD (Single Shot MultiBox Detector) algorithm to detect the face, and take the detected face as the tracking target, and use the correlation filtering algorithm DSST (Discriminatiive Scale Space Tracker) Tracking in the process of tracking, face detection is con-tinuously carried out on the tracking target, and the detection results are used to monitor the tracking results, so as to reduce the target drift caused by the boundary effect, thus improving the accuracy of the tracking algorithm. The algorithm is tested and verified on OTB100 data set. The final experimental results show that the accuracy of this algorithm is obviously better than the mainstream classical algorithm, and the frame rate meets the real-time re-quirements.

1. Introduction

Face tracking is a classic problem in computer vision, which is applied in many fields, such as human-computer interaction, intelligent video surveillance and autonomous driving, etc. In recent years, face tracking has been very popular in the field of computer vision.Face tracking is to predict the position and scale of a given face target in the following frames.Generally speaking, the tracking framework consists of four modules, which are feature extraction, appearance modeling, motion estimation and target location.Feature extraction is the process of mapping the k-dimensional feature subset of the target to a new feature set. It can also be expressed as the method and process of extracting characteristic information from images by using computers.Appearance modeling is a process of detecting target features and constructing a mathematical model with the ability to discriminate.Motion estimation is the process of inferring and tracking the position of the target in subsequent frames. At this time, the predicted position may be multiple.Target location is the process of determining target location by non-maximum suppression algorithm or greedy algorithm.Face tracking is faced with several difficulties as well as target tracking, such as the deformation of target appearance, environmental lighting, motion blur and jitter, high speed movement and background interference.The commonly used face tracking methods include generation model method and discriminant model method.The generation model method is to use the current frame to model the target region, and the next frame will look for the region with the largest model response according to the model. Common methods include Kalman filter algorithm[1], particle filter algorithm[2] and means-shift algorithm[3].The discriminant model method is mainly to extract image features and then perform machine learning on the features. Common methods include TLD algorithm[4] and Struct algorithm[5] based on online learning, KCF[6] and DSST algorithm based on correlation filtering, and SiamFC algorithm[7] based on the deep learning.

Correlation filtering algorithms have incomparable advantages in speed, especially MOSSE (Visual Object Tracking using Adaptive Cor-relation Filters) algorithm[8].In the OTB50 data set, the FPS of MOSSE algorithm reaches 615, but the average accuracy only reaches 43.1%[9]. This is the first time that correlation filtering method has been applied in the field of target tracking, showing the amazing potential of correlation filtering algorithm in response speed.MOSSE is a correlation filtering based on single channel gray features.On the basis of MOSSE, Henriques Joao F et al proposed CSK (Circulant Structure of Tracking-by-detection with Kernel) algorithm[10] and KCF ((high-speedTracking with Kernelized Correlation Filters (CSK) algorithm, based on MOSSE, adds dense sampling, cyclic displacement and closed solution of computational kernel function.KFC algorithm adds HOG feature of multi-channel gradient on the basis of CSK algorithm. Although this reduces THE FPS of KCF to 292, the average accuracy reaches 72.8%, which is higher than most tracking algorithms at that time.Martin Danelljan et al extended CSK with multi-channel color features to obtain the adaptive Color Attributes for real-time Visual Tracking (CN) algorithm[11].Yang Li from Zhejiang University proposed a SCALe-Adaptive and Multi Feature Integration Tracker (SAMF) algorithm based on KCF, which used HOG+CN Feature to search targets with position filters on multi-scale images.The position at the maximum response output and its scale are taken as the target of the current frame[12].SAMF can be regarded as the first Scale adaptive target Tracking algorithm, but Martin Danelljan proposed DSST (Discriminative Scale Space Tracking) algorithm to carry forward the target Tracking algorithm, DSST algorithm only uses HOG feature.DCF is used for translational filtering, and scale filter is introduced to build 33 layers of pyramid feature vector to respond to the target scale.Although correlation filtering can solve the problem of scale change of targets and improve the accuracy compared with traditional methods, it is not very ideal when tracking fast targets and fast deformation targets.

Firstly, the Deep Learning Tracker (DLT) algorithm proposed by Naiyan Wang et al.[13] combines Deep Learning with target tracking. This algorithm first uses the stacked denoting autoloader,SDAE) to obtain the object representation ability, and then the sigmoID part was trained online to get the classifier, and then the particle filter was used to extract a batch of candidate regions, and the candidate regions were input into the classification network to get the prediction target with the highest confidence.With the development of deep learning, MA et al proposed the HCFT (Hierarchical Convolutional Features for Visual Tracking) algorithm[14], which uses VGGNET-19 as the feature extraction device.After the feature extractor, three feature layers are obtained. After training the three feature graphs, responses are obtained through filters. Finally, the three responses are added with different weights to obtain the final response graph.Bertinetto et al proposed the target tracking method SiamFC (Fully Convolutional Siamese Networks)[15], which uses the network model of offline training to obtain a faster speed, but the positioning accuracy is lower.Although most deep learn-based algorithms improve the tracking accuracy compared with the correlation filtering method, most of them fail to meet the requirement of real time in terms of speed[16].

In the method of correlation filtering, DSST algorithm can obtain fast and relatively accurate effect, but when the fast movement of the target produces boundary effect, it will lead to the failure of tracking.On the basis of literature[17], this paper firstly extracts the multidimensional HOG feature of the target region and retains more target information. By introducing the weight coefficient matrix, it makes a centralized response to the central position of the target region and obtains the coarse positioning of the target.At the same time, the image is input into the SSD network to obtain the supervised position of the target, and the target template of the filter is updated by comparing the Euclidean distance between the coarse positioning position and the supervised position and the threshold value.The algorithm in this paper refers to the long time tracking algorithm and the tracking algorithm based on deep learning, and adds a neural network detector, which reduces the tracking failure and drift caused by the boundary effect, and realizes accurate positioning.Experimental results show that the proposed algorithm can significantly improve the tracking accuracy on the basis of real-time performance.

2. Detection and monitoring tracking algorithm

The algorithm idea in this paper is Detection and supervised Tracking, namely Detection SU-Pervise Tracking, which is abbreviated as DST algorithm in this paper.First of all, the input image sequence is judged, if it is the first frame, directly into the detection module to detect the face, the detected face as the target to create a filter.If it is not the first frame, it is filtered and detected.In the filtering part, the HOG feature of the frame is extracted, and the feature input filter is filtered to get the response of position and scale, and then the non-maximum suppression algorithm is applied to the response to get the optimal solution of position and scale.Detection part to an initialization of the input image, adjust its 300 x300x3 three-dimensional matrix, according to the model, in turn, to extract the image feature of the network structure, the characteristics of the layer to test for the input, get more target detection box, the use of the maximum suppression algorithm to filter results, get the optimal output.

After the above process to get the detection and tracking box, calculate the center of the center and tracking test box, both subtraction, get the relative error of the detection and tracking, the error compared with the adaptive threshold, if it is greater than the threshold, will detect the target image as the input updating filter, namely to detect the target as a new target tracking.If the value is lower than the threshold, the tracking result is accurate and is displayed.The detector uses the relatively advanced SSD algorithm, and the tracker uses the classical correlation filter tracking algorithm DSST algorithm.

3. Face detection based on SSD

3.1. Formatting the title

The target detection algorithm of SSD (Single Shot MultiBox Detector)[18] was proposed by Wei Liu et al. VGG was used as the basic network. The full connection layer of VGG was changed into the convolution layer and several additional convolution layers were added.

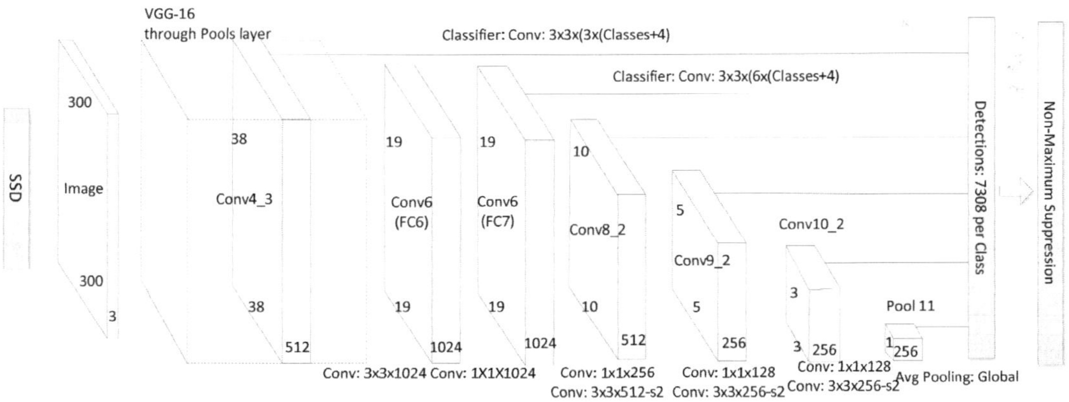

Fig.1 SSD network model

SSD network structure: (1) The network structure of SSD300 is shown in Figure 1.The original input image is scaled to a fixed size as the input of the convolutional network model, whose base network uses VGG-16.(2) SSD to VGG16 FC6 layer, FC7 layer full connection mode to convolution layer, and then continue to add a number of convolution layer to Conv10_2, the last full connection layer.(3) Save the feature map obtained after convolution. SSD adopts the feature pyramid structure for regression detection of objects, that is, feature maps generated by multiple different convolution layers are used in detection.(4) Extract the feature layer as the input of the detection layer.Generate multiple Prior boxes on feature maps of different sizes, and perform classification and order regression on the features generated by these feature maps.(5) Non-maximum suppression algorithm is used to filter the final results.

3.2. SSD network loss function

The loss function of SSD is defined as the weighted sum of position error and confidence error:

$$L(x,c,l,g)=\frac{1}{N}(L_{conf}(x,c)+\alpha L_{loc}(x,l,g)) \tag{1}$$

Where N is the number of positive samples of the prior box, L_{conf} is the confidence error, and L_{loc} is the position error.For position error, Smooth $L1$ loss is adopted, which is defined as follows:

$$L_{loc}(x,l,g)=\sum_{i\in Pos}^{N}\sum_{m\in\{cx,cy,w,h\}} x_{ij}^{k} smooth_{L1}(l_i^m-\hat{g}_j^m) \tag{2}$$

$$\text{smooth}_{L1}(x)=\begin{cases}0.5x^2 & if\ |x|<1\\ |x|-0.5 & \text{otherwise}\end{cases} \tag{3}$$

For confidence error, surtax Loss is adopted:

$$L_{conf}(x,c)=-\sum_{i\in Pos}^{N} x_{ij}^{p}\log(\hat{c}_i^p)-\sum_{i\in Neg}\log(\hat{c}_i^0),\hat{c}_i^p=\frac{\exp(c_i^p)}{\sum_p \exp(c_i^p)} \tag{4}$$

Here, $X^p{}_{ij}\in\{1,0\}$ (this symbol should correspond to Formula 1) is an indicator parameter. When $X^p{}_{ij}$ =1, it means that the in prior box matches the JTH real target, and the category of the real target is p, c is the predicted value of class confidence, L is the predicted value of the corresponding boundary box of the prior box, and g is the positional parameter of the real target[1].Due to the existence of $X^p{}_{ij}$, the position error is calculated only for positive samples. The weight factor α was set to 1 by crossing validation.

4. Face tracking based on DSST

DSST algorithm is an efficient correlation filter tracking algorithm, including position filter and scale filter. The position filter can locate the target of the current frame, and the scale filter can estimate the target scale of the current frame. The two filters are relatively independent. So different feature types and eigenvalue calculation methods can be selected for the design.

4.1. Position filter

Assuming the size of the image block P where the target is located is M*N, fog feature of the target P is extracted to obtain the feature f with the size of $M*N*d$, where the dimension of the feature is d dimension. According to the Gaussian function, the response g is constructed with the size of $M*N$, and the intermediate response value of the response is the largest, which decreases successively in all directions.

Take the two-dimensional Fourier transform of the features of each dimension of f to get F, and take the two-dimensional Fourier transform of g to get G. Set the regular term as γ, and take the minimum value of ε, then the optimal filter H should satisfy the following formula:

$$\varepsilon=\left\|\sum_{l=1}^{d}h^l\star f^1-g\right\|^2+\lambda\sum_{l=1}^{d}\left\|h^l\right\|^2 \tag{5}$$

Where the asterisk represents cyclic correlation, h^l and pl are the l-th dimension of f, and l ranges from 1 to d. Minimize the above equation to obtain the formula (6), where the overline represents the complex conjugate.

$$H^l=\frac{\overline{G}F^l}{\sum_{k=1}^{d}\overline{F^k}F^k+\lambda} \tag{6}$$

Divide H into numerator A and denominator B, and carry out iterative update respectively. The update method is as follows[2]:

$$A_t^l = (1-\eta)A_{t-1}^l + \eta \bar{G}_t F_t^l \quad , \quad B_t = (1-\eta)B_{t-1} + \eta \sum_{k=1}^{d} \overline{F_t^k} F_t^k \tag{7}$$

Where, η is the learning rate, and the new image, its feature is z. Take the two-dimensional Fourier transform of each dimension to get Z. The corresponding response y can be obtained according to the following formula:

$$y = \mathcal{F}^{-1}\left\{\frac{\sum_{l=1}^{d} \overline{A^l} Z^l}{B+\lambda}\right\} \tag{8}$$

The position of the maximum value of y is the position of the new target center.

4.2. Scaling filter

After the position of the target is obtained through the position filter, the scale information is determined by the one-dimensional scale filter. The target sample scale selection principle for scale evaluation is as follows:

$$\mathrm{a}^n P \times a^n R \qquad n \in \left\{\left\lfloor -\frac{S-1}{2} \right\rfloor, \ldots\ldots, \left\lfloor \frac{S-1}{2} \right\rfloor\right\} \tag{9}$$

Where, P and R are the width and height of the target in the previous frame, a =1.02 is the scale factor, and S =33 is the total scale series[2]

Since the scale factor is not a linear function, the growth of each scale is not the same.For the scale larger than the current scale, a larger stride length is adopted, and for the scale smaller than the current scale, a smaller step length is adopted, so as to achieve coarse detection of large targets and fine detection of small targets, so as to achieve higher accuracy.

Each sample was normalized to a unified fixed size, and 31-dimensional flow features were extracted respectively. All frog features of each sample were connected in series into a feature vector to form a 33-layer pyramid feature, which was multiplied by an one-dimensional hand window as input Z.

The corresponding output response y can be obtained by substituting the input Z into the equation below:

$$y = \mathcal{F}^{-1}\left\{\frac{\sum_{l=1}^{d} \overline{A^l} Z^l}{B+\lambda}\right\} \tag{10}$$

The scale of the maximum value of y is the final scale estimation result of the current frame.

5. The experimental results

The experimental environment in this paper is Ubuntu18.04, the CPU is Intel Core I5 7300H, the GPU is NVIDIA GTX 1050, the regularization parameter is set as 0.01, the scale factor A =1.02, and the total weight scale coefficient =0.75.In this paper, DST algorithm is compared with TLD algorithm, CT algorithm, CXT algorithm, MS-V algorithm, KCF algorithm and DSST algorithm. The video used in the experiment is selected from the FaceOcc2 video of OTB100 data set, including the background environment of lighting change, occlusion, rotation inside and outside the plane, etc.

This paper compares the above seven algorithms from three aspects: success rate, accuracy and frame rate .Where, the Euclidean distance between the center position of the target box output by the tracking algorithm and the center position of the actual target is compared with the threshold value, and the ratio of the number of frames less than the threshold value to the total number of frames of the video is the accuracy of the video at this distance. Overlap degree between the output box of the tracking algorithm and the actual box is calculated, and the overlap degree is compared with the threshold value. The ratio between the number of frames greater than the threshold value and the total number of frames is called the success rate of the video under the overlap rate.

Figure 2 for face recognition and tracking test under different condition video capture, figure 3 to test the success rate of video figure, figure 4 to test the precision of the video, the red represents the average test results of the algorithm in this paper the success rate and the average accuracy of algorithm is superior to the rest to table 1 for seven kinds of algorithm is the average success rate, average precision and frame rate of the comparison table.

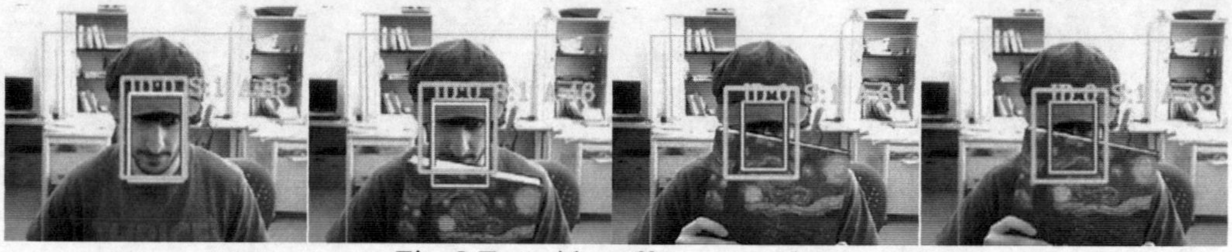

Fig. 2 Test video effect screenshot

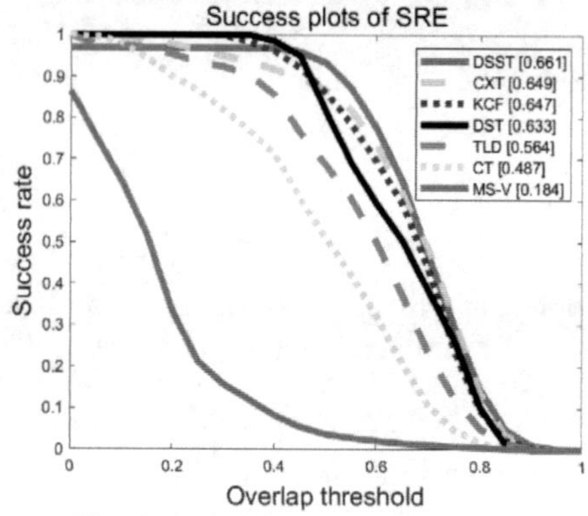

Fig. 3 Test video success rate charts

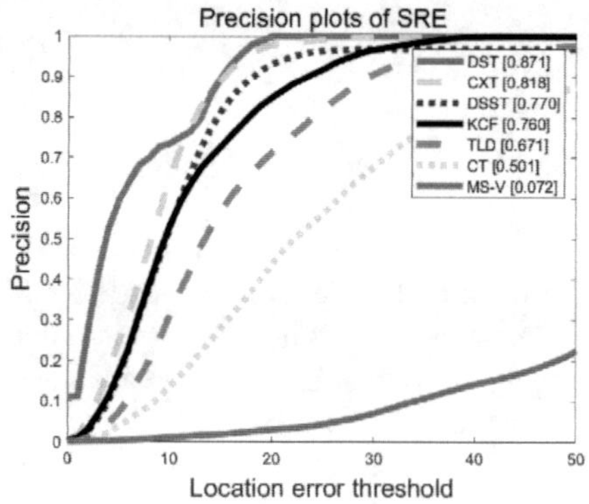

Fig. 4 Test video accuracy chart

Table 1 Comparison table of average success rate, average precision and frame rate

The test algorithm	Average success rate	The average accuracy
DST	0.633	0.871
CXT	0.649	0.818
KCF	0.647	0.76
TLD	0.564	0.67
CT	0.487	0.5
MS-V	0.184	0.072
DSST	0.661	0.77

6. conclusion

DSST algorithm is presented in this paper with the SSD algorithm by SSD detection algorithm is proposed on the basis of supervision DSST update and tracking algorithm based on monitoring results of face tracking algorithm, namely the inspection supervision and tracking algorithm, experiments show that the presented algorithm in serious background interference and object shelter and deformation under the condition of real-time tracking can still be good, The accuracy of the algorithm is improved compared with other algorithms.

As the detection and monitoring tracking algorithm adopt deep learning SSD algorithm as the detection part, the calculation amount of the algorithm is improved. The frame rate is lower than DSST, but still higher than CXT algorithm, and meets the real-time requirements. The follow-up research work can focus on solving the problem of speed decrease caused by large amount of computation, such as introducing faster target detection algorithms such as YOLO, etc., or focusing on using depth features to replace manual features, etc., so that the improved algorithm can take into account the requirements of accuracy, success rate and speed.

References

[1] Kalman R E. A New Approach to Linear Filtering and Prediction Problems[J]. Journal of Fluids Engineering, 1959, 82D:35-45.
[2] Numbers K,Koller-Meier E,Gool L V.An adaptive color-based particle filter[J].Image and Vision Computing,2003,21(1):99-110.
[3] D. Comanche and P. Meer, "Mean shift: A robust approach toward feature space analysis," IEEE T. PAMI, vol. 24, no. 5, pp. 603-619, May 2002
[4] All Z, Mikolajczyk K, Matas J. Tracking-Learning-De- section[J]. IEEE Transactions on Software Engineering, 2011, 34(7):1409-1422.
[5] S. Hare, A. Safari,and P. H. S. Torr, "Struck: Structured output tracking with kernels," in Proc. IEEE Int. Conf. Comput. Vis., 2011, pp. 263–270
[6] Henriques João F,Caseiro Rui,Martins Pedro,Batista Jorge. High-Speed Tracking with Kernelized Correlation Filters.[J]. IEEE transactions on pattern analysis and machine intelligence,2015,37(3).
[7] Bertinetto L, Valmadre J, Henriques J F, et al. Fully- convolutional siamese networks for object tracking[C]// European Conference on Computer Vision (ECCV), October, 8-16, 2016, Amsterdam, the Netherlands, 2016. 9914:850-865.
[8] D. S. Bolme, J. R. Beveridge, B. A. Draper, and Y. M. Lui, "Visual object tracking using adaptive correlation filters. Computer Vision & Pattern Recognition, vol. 119, no. 5, pp. 2544-2550, 2010.
[9] Wu Y, Lim J, Yang M H. Online objects tracking: A benchmark [C]/ CVPR, 2013.
[10] Henriques J F, Caseiro R, Martins P, et al. Exploiting the Circulant Structure of Tracking-by-Detection with Kernels[C]/ Proceedings of the 12th European conference on Computer Vision - Volume Part IV. Springer, Berlin, Heidelberg, 2012.
[11] Danelljan M,Khan F S,Felsberg M,et al.Adaptive color attributes for real-time visual tracking[C] // Proc of 2014IEEE Conference on Computer Vision and Pattern Recognition,2014:1090-1097.
[12] Li Y, Zhu J. A Scale Adaptive Kernel Correlation Filter Tracker of Feature Integration[C]/ European Conference on Computer Vision. Springer, Cham, 2014.
[13] Wang N, Yeung D Y. Learning a deep compact image representation for visual tracking[J]. Advances in Neural Information Processing Systems, 2013.
[14] MA C,HUANG J B,YANG X K,et al. Hierarchical convolutional features for visual tracking[C]/Proceedings of the IEEE International Conference on Computer Vision. 2015:3074-3082.
[15] Bertinetto L, Valmadre J, Henriques J F, et al. Fully- convolutional siamese networks for object tracking[C]// European Conference on Computer Vision (ECCV), October, 8-16, 2016, Amsterdam, the Netherlands, 2016. 9914:850-865.
[16] Li Xi,Zha Yufei,Zhang Tianzhu,et al. Survey of visual objects tracking algorithms based on deep learning[J]. Journal of Image and Graphics2019,24(12):2057-2080.
[17] Danelljan M,Hager G,Khan F S,et al.Discriminative scale space tracking[J].IEEE Transactions on Pattern Analysis and Machine Intelligence,2017,39(8):1561-1575.
[18] LIU W,ANGUELOV D,ERHAN D,et al.SSD:Single Shot Multibox Detector[C]/ European Conference on Computer Vision.Springer,Cham:Springer,2016:21-37.

Overview of Energy Consumption Optimization in Mobile Edge Computing

Bingyi Hu[1], Jixun Gao[2], Yanxin Hu[3a], Huaichen Wang[3b], Jialei Liu[3c*]

[1] School of Computer Science and Technology, Henan Polytechnic University, Jiaozuo 454000, China

[2] School of Computer, Henan University of Engineering, Zhengzhou 451191, China

[3] School of Software Engineering, Anyang Normal University, Anyang 455000, China

[1] hby1260464202@126.com , [2] gaojixun@haue.edu.cn , [3c*] jialeiliu@aynu.edu.cn

Abstract—The energy consumption of traditional mobile network is mainly reflected by task calculation and data transmission. The edge server in mobile edge computing is placed at the edge of the network closer to the mobile terminal device. When the computing resources of the mobile terminal device are insufficient, the task can be offloaded to the edge server for processing. This paper discusses the energy consumption optimization in the Mobile Edge Computing (MEC) system, which is composed of the edge server and mobile devices at the edge of the network. Firstly, the system model of MEC is elaborated. Secondly, the energy consumption optimization for mobile devices and base stations is studied. Which oriented to the mobile device energy consumption optimization is divided into multiple users - single MEC server scenarios of energy consumption optimization and multiple users - the energy optimization of MEC server scenario, and enumerate the oriented to the base station and a new method for the optimization of energy consumption for mobile equipment energy consumption optimization method, in addition to the insufficiencies of these methods and the prospect of the future are also explained.

1. Introduction

With the continuous and rapid development of Internet of Things technology and 5G mobile communication technology [1], time-sensitive and computation-intensive computing tasks, including audio recognition, augmented reality and collaborative driving and other applications [2], have been gradually applied in our daily life, providing more convenient services for mobile users. However, such applications have problems such as large amount of data, time-delay sensitivity and excessive energy consumption. Cloud computing with centralized processing mode has been difficult to meet the demands of massive user data and long-distance transmission [3, 4]. Therefore, Mobile Edge Computing [5] (MEC) was proposed to solve these problems. MEC places servers close to mobile users, which can significantly improve the quality of the mobile device experience and help reduce power consumption [6, 7]. MEC has been regarded as one of the key technologies in the 5G era by the European 5GPPP organization [8].

Nowadays, with the development of MEC technology, more and more attention has been paid to the energy consumption of MEC system. How to reduce the energy consumption of MEC system has become the goal pursued by many scholars. In this paper, the energy consumption optimization for mobile device terminal and base station is studied. In MEC system, users have high requirements for delay, so we combine delay to optimize the energy consumption of MEC system, so that the energy of MEC system can be fully utilized.

2. System Model

With the rapid development of computer technology, the types and functions of mobile terminal devices are becoming more and more diversified. If all the computing tasks on the mobile terminal devices are offloaded to the traditional cloud computing center for processing, it is easy to cause network congestions [9]. If the computing tasks are all processed on the mobile terminal device, the battery life of the mobile terminal device will decrease, which cannot meet the user's demand for the battery of the mobile device [10]. The introduction of MEC has solved these problems well.

The network model of the MEC system is shown in Figure 1. The MEC system includes a base station and an edge server and mobile devices deployed in the base station. From the figure, we can see that compared with the cloud data center, the edge server deployed on the base station and the base station is the distance of users is closer, which effectively reduces the delay of task transmission [11]. The MEC system offloads the computing tasks on the mobile terminal equipment to the edge server for computing, which greatly saves the computing and storage resources of the mobile equipment, reduces the energy consumption of the mobile equipment, and greatly improves user satisfaction. The reasonable deployment of edge servers on base stations and the reasonable utilization of edge server computing resources can effectively reduce the energy consumption of the servers [12]. Nowadays, with the development of MEC technology, people are paying more and more attention to energy consumption optimization in the MEC system [13].

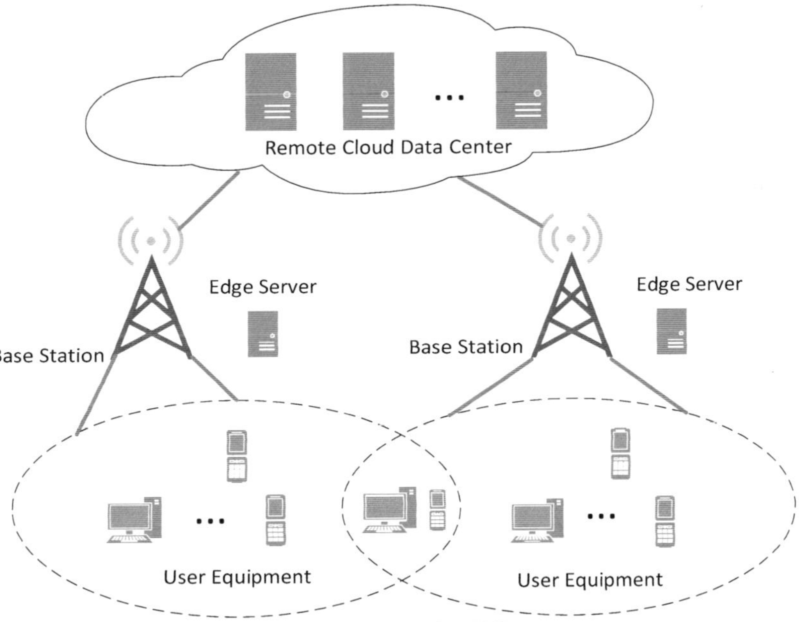

Fig.1 Network model of MEC system

3. Energy Consumption Optimization for Mobile Devices

Mobile devices in the MEC system can offload tasks to edge servers for processing, which effectively reduces the energy consumption of mobile terminals. This chapter studies the energy consumption optimization of mobile terminals in MEC based on the scenarios of multi-user-single-MEC server and multi-user-multi-MEC server respectively.

3.1. Energy Consumption Optimization Method Based on Multi-User-Single-MEC Server System

In the multi-user-single-MEC server scenario, there is only one MEC server, so the computing resources of the system and communication resources of the wireless channel are limited. Multiple users connected to the same MEC server compete for resources. In addition, since all users have self-interest, more resources will be wasted in the process of preempting computing resources. Therefore, how to optimize the energy consumption of mobile terminals in MEC is worth further study.

Literature [14] studied the joint allocation of computing and communication resources for the purpose of optimizing energy consumption in a multi-user-single- MEC server scenario in a cell. A greedy strategy and SMSEF algorithm are proposed to solve the problem, and it is compared with Priority-based algorithm and N-SMSEF algorithm. It is found that as the number of channels or the number of users increases, the SMSEF algorithm saves more energy than the other two algorithms.

In reference to the energy management problem of the MEC system, the literature [15] carried out research on the system energy consumption optimization problem in the scenario of a multi-user-single-MEC server equipped with energy harvesting equipment, and proposed an optimization algorithm based on the Lyapunov method and an alternative direction multiplication algorithm. The distributed optimization algorithm of ADMM achieves the optimization goal of minimizing system energy consumption under the condition of satisfying certain computing service quality and system queue stability.

Although the above two documents can optimize the energy consumption of the system, they do not consider the time delay problem. Most users in MEC also have relatively high requirements for time delay. In response to this problem, the following documents will consider time delay while optimizing energy consumption.

Literature [16] studied the centralized resource allocation by the MEC server in the multi-user-single-MEC server scenario. The goal is to minimize the total energy consumption of the system while meeting the task delay constraint. The optimal subcarrier allocation problem under the given task description and wireless channel conditions was mapped to a regression problem in machine learning, and an exhaustive search algorithm was used to construct a labeled data set, and then three sets of different regression models were trained, respectively It is a random forest, Xgboost, neural network, and predicts the allocation of sub-carriers on the test set. After obtaining the subcarrier allocation, calculate the resource allocation plan and the corresponding total energy consumption of the system according to the formula deduced in the preliminary conclusion [17]. The simulation experiment data shows that the distribution scheme obtained by the regression algorithm based on historical experience can obtain the system energy consumption close to the optimal solution.

Literature [18] proposed a system model for calculating migration energy optimization, including system model, migration model and migration time constraint model. The system model is a time division multiple access-based MEC migration system built in a cellular network environment. The system uses time slots as a mark to identify different user signals in the system. In any time slot, the base station is based on the time division multiple access to one of the users [19]. A subset of tasks are scheduled for complete or partial migration. Its core idea is to optimize the energy consumption in the multi-user computing migration system, and to balance the computing power of the system with the dynamic division of computing time as the constraint, to avoid the blind migration of a user in the multi-user migration system in order to save resources to ensure the system the total calculation time is the lowest. In addition, this document also proposes a Lagrangian-based energy consumption optimization model. According to the specific situation of the migration, the energy consumption of the terminal is divided into two parts: the local computing energy consumption and the computing migration energy consumption. The convexity proof of the energy consumption model is provided, the Lagrangian method is used to solve the model, and the optimal resource partitioning algorithm based on threshold is proposed [20].

In the multi-user-single-MEC server system proposed in this chapter, only one edge server is considered. If the MEC system wants to accommodate more users and provide more powerful MEC services, the resources of a single edge server are limited. Therefore, the following research can consider a multi-user-multi-MEC server system.

3.2. Energy Consumption Optimization Method Based on Multi-User-Multi-MEC Server System

There are multiple MEC servers in a multi-user-multi-MEC server system. The computing resources and communication resources of the wireless channel of the system are much more than that of a multi-user-single-MEC server system. This solves the problem of insufficient resources in a multi-user-single-MEC server system. Problems caused by use. Next, this chapter will study how to better optimize the energy consumption of the system in a multi-user-multi-MEC server system.

Aiming at the problem of how to make the time delay used in task execution shorter in the task offloading of MEC and how to further improve the energy efficiency of the MEC system, literature [21] proposed a computational offloading based on the population diversity-binary particle swarm optimization algorithm Strategy. This strategy needs to first establish the energy consumption model used by mobile terminal devices under multi-objective constraints when performing tasks, and then use the particle swarm optimization algorithm to turn the task unloading process into a particle optimization process, and finally Get the optimal unloading decision under the condition of seeking the lowest energy consumption [22]. Aiming at the problems of low energy efficiency of mobile terminal equipment and inflexible services due to the excessive concentration of computing tasks to be offloaded in the MEC system, the literature proposed a system energy efficiency optimization scheme based on energy harvesting. Under the condition that the unloaded transmission power is limited, it is necessary to analyze the state of energy collection and the power allocation of users, and then establish a joint optimization model for maximizing system energy efficiency [23]. In addition, the generalized fractional programming theory is used to transform the problem of unloading energy efficiency into a standard convex optimization problem, and then the objective function is iteratively optimized through the Lagrangian function to obtain the optimal energy index variables and power allocation.

In the multi-user-multi-MEC server system, there is still such a problem that the number of MEC servers cannot meet the needs of the mobile terminal, and the mobile terminal will compete for the MEC server, and the users in the multi-user system will interact with each other. Interference will increase the energy consumption of the system. Literature [24] proposed an energy efficiency optimization algorithm for mobile edge computing based on multi-dimensional games. An upper limit is set on the computing resources of the MEC server, and user queuing delay is added to solve the problem of insufficient energy efficiency of the mobile terminal in a resource-constrained multi-user system. The player of the game model in the algorithm is a mobile user, the decision space is a two-dimensional space composed of unloading decision and power control, and the revenue function is a function of energy efficiency and time delay. It is theoretically proved that the Nash equilibrium of the multidimensional game model exists and is unique.

Aiming at the problem of how to optimize the energy consumption in the process of system data offloading in MEC, literature [25] proposed an offloading strategy for minimizing the average energy consumption per unit data based on the optimal stopping theory. This strategy builds n task models for offloading data from mobile terminal equipment to m base stations. Combine local energy consumption, offload energy consumption and base station energy consumption for joint optimization. Then, the unconstrained problem without slack variables is obtained by defining the augmented Lagrangian function. Finally, the unconstrained problem is solved by an iterative formula and the solution of the original problem is further obtained to achieve the goal of minimizing the average energy consumption per unit data.

4. Energy Consumption Optimization Method for Base Station

The aforementioned research on energy consumption optimization in MEC mainly focuses on the research on energy consumption optimization of mobile terminals, while there are very few studies on energy consumption optimization of base stations. Only the joint optimization model in the literature

[25] mentioned Base station energy consumption optimization. This chapter mainly considers how to optimize the energy consumption of the base station in the MEC system.

At present, there is not much research on task offloading between base stations in the MEC field. The network model mainly used in existing research is a fully connected base station model with a control center similar to cloud computing. In the MEC system through wireless communication, the establishment of a control entity or virtual control center requires a large amount of wireless data communication between base stations, which not only takes up a large amount of channel resources, but also increases the energy consumption of the base station.

Based on the above problems, the literature [26] combined with the dynamic voltage adjustment technology, proposed and solved the problem of minimizing the energy consumption of the base station group cooperative processing tasks, and effectively reduced the energy consumption of the base station group processing tasks. In view of the low load of the base station task queue, dynamic voltage adjustment technology is introduced, and the energy consumption model of the base station group cooperative processing tasks is established. The task offloading strategy and computing resource allocation plan are respectively given, and the threshold of urgent tasks is determined to minimize the energy consumption of the processing tasks of the base station group is the optimization goal, the task offloading strategy is designed using the most optimized method, and the computing resource allocation scheme is designed using the idea of dynamic programming. In view of the high load of the base station task queue, the dynamic voltage adjustment technology is also introduced, and the task offloading strategy and the computing resource allocation plan are jointly considered, and the corresponding energy consumption model of the base station group cooperative processing tasks is established, giving a wider range of tasks Unloading judgment conditions: Through the analysis of the task fragmentation processing algorithm, the energy consumption models of the base station group cooperative processing tasks are divided into two types, and the optimization goal is also to minimize the energy consumption of the base station group processing tasks, and two nonlinearities are established Problem, a load balancing algorithm and a breakpoint search algorithm are designed to solve the problem.

The following paper is devoted to discussing the energy consumption optimization problem of the edge server deployed. In the study of energy consumption optimization based on multiple edge servers, the problem of energy waste caused by unbalanced load of edge servers is considered. Literature [27] proposed an edge server collaboration-sleep mechanism, which combined the sleep mechanism with the edge server collaboration mechanism. In the edge server, an edge server is selected as the leader server and needs to process tasks for all mobile terminals in the range. However, when the computing resources of the leader server are insufficient, the task can be offloaded to the remaining edge servers as collaborator servers for processing. The collaborator server in the system has two states: dormant and active. In the dormant state, it consumes less energy. When in the active state, the collaborator server receives tasks and processes them. Under this mechanism, the state of the collaborator is determined by the state of the leader. To control. This document optimizes the energy consumption of multiple edge servers.

Literature [28] proposed an algorithm based on semi-definite relaxation (SDR) to solve the problem of new computing tasks and frequent switching between multiple tasks. In the fixed CPU scenario, an algorithm based on linear programming relaxation is used as an alternative to SDR. In the flexible CPU scenario, the algorithm based on exhaustive search is used as the benchmark, and then the algorithm based on SDR is applied to find the near optimal solution, thereby reducing the energy consumption and task delay of the mobile device to reach the minimum.

In a MEC system with multiple mobile users, multiple smart mobile devices need to offload computing tasks to one MEC server. In order to reduce the energy consumption of smart mobile devices, it is necessary to coordinately optimize offload selection, radio resource allocation, and computing resource allocation. Literature [29] describes the energy minimization problem as a mixed interleaved nonlinear programming (MINLP) problem, which is subject to specific application delay constraints. The branch and boundary (RLTBB) method based on the reconstruction linearization technique is proposed to solve this problem. The method can get the optimal or sub-optimal result by setting the solution accuracy. Literature [28] and Literature [29] mainly study the optimization of resource allocation to achieve the purpose of reducing energy consumption, but they do not consider the selection of the optimal number of servers to be turned on.

Literature [30] proposed a mobile caching network model called Cachin Mobile. In the CachinMobile network, mobile users first send request content to edge nodes through device-to-device (D2D) communication. If the requested content cannot be found in the neighboring edge nodes, it will request such content from the eNodeB. According to this idea, the optimal content placement strategy between user edge nodes is proposed to minimize transmission energy consumption. This document considers the joint optimization of CPU frequency, but does not consider the off-sleep mode strategy of edge servers. For ultra-dense networks, no matter how the network load fluctuates, all edge server nodes are always in working state, which makes the system energy consumption excessive.

Literature [31] proposed a framework for a sequence solution, based on the communication and computing energy consumption perception selection law to select the corresponding number of servers to work when the energy consumption is optimal. The disadvantage is that it does not consider the switching energy consumption caused by server state switching. State switching is too frequent, and its consequences will not only produce a large amount of switching energy consumption but also shorten the service life of the device.

Based on these problems, literature [32] proposed a server switch dynamic switching algorithm without delay constraints, in which the user connection and service configuration of adjacent base stations are considered in the algorithm. Since the number of users at different times in the network will continue to change, a linear regression prediction mechanism is used to predict the number of user requests for each service in the next time slice to calculate the utilization rate of the edge server in the next time slice. In addition, a system is also designed. This kind of user migration mechanism is used for the user's service migration. Through this strategy, the switching energy consumption of edge servers can be reduced while reducing network energy consumption. Literature [32] also proposed a server switch dynamic switching algorithm that includes delay constraints. The delay constraint is a key indicator in the algorithm design. When selecting the server shutdown link, each iteration must try to shut down a server, and make a decision on the premise that the delay constraints of all users can be guaranteed until there is no edge server. Closed, which effectively saves the energy consumption of the MEC system. In addition, when migrating the edge server to be shut down, the user preferentially selects the base station with the most remaining delay in the base station request candidate list for service request, so as to ensure that the user obtains the service in a short time.

5. Conclusion

With the rapid development of 5G technology and Internet of Things technology, the issue of energy consumption optimization in the MEC system has also been paid more and more attention. Consumption optimization methods. Although the methods listed above have achieved very good research results in many aspects, their application research still faces many deficiencies and challenges in many aspects:

(1) Although the multi-user-multi-MEC server system solves the problem of limited computing and communication resources in the multi-user-single-MEC server scenario, it still has some shortcomings. The research in the multi-user-multi-MEC server system mainly focuses on the problems related to MEC in the static environment. There is no consideration for user devices moving back and forth between multiple servers.

(2) During the research on energy consumption minimization under MEC environment, when the amount of data generated is large, a large amount of data cannot be unloaded to MEC server for processing because of the limitation of upper limit of link bandwidth under existing wireless network conditions. It is bound to affect the development of 5G network and user experience. Our future research focus can be on improving the link bandwidth of hardware and combining hardware and software to better promote the development of network technology.

(3) The MEC system will generate more energy consumption with the improvement of the capacity of mobile terminals, edge servers and other devices. In the future, combined with the development of

green energy such as solar energy in recent years, these devices can be designed to consume energy and obtain energy at the same time.

In this paper, energy consumption optimization methods in MEC system are summarized, the shortcomings of these methods are summarized and the future prospects are given. It is hoped that scholars can conduct more in-depth research in the future.

Acknowledgments
This work was financially supported by the National Innovation and Entrepreneurship Training program for College students in Henan Province (202110479020), the Key Science and Technology Research Project of Anyang City (2021C01GX017 or ZK2021C01GX017), and the Research and Cultivation Fund Project of Anyang Normal University (AYNUKPY-2019-24). Jialei Liu is the corresponding author.

References

[1] Y. Fu et al. (2021) Energy-efficient offloading and resource allocation for mobile edge computing enabled mission-critical internet-of-things systems. EURASIP Journal on Wireless Communications and Networking, vol. 2021, no. 1, pp. 1-16.
[2] W. Shi and S. Dustdar. (2016) The Promise of Edge Computing, Computer, vol. 49, no. 5. pp. 78-81.
[3] X. S. Wang. (2019) Progress and Trends in Mobile Cloud Computing Research,Journal of Physics: Conference Series, vol. 1187. pp. 052004-.
[4] W. Shi, C. Jie, Z. Quan, Y. Li, and L. Xu. (2016) Edge Computing: Vision and Challenges,Internet of Things Journal, IEEE, vol. 3, no. 5. pp. 637-646.
[5] S. Singh, Y. C. Chiu, Y. H. Tsai, and J. S. Yang. (2017) Mobile Edge Fog Computing in 5G Era: Architecture and Implementation.in Computer Symposium.
[6] N. Ab Ba S, Z. Yan, A. Taherkordi, and T. (2017) Skeie.Mobile Edge Computing: A Survey.IEEE Internet of Things Journal, vol. PP, no. 99. pp. 1-1.
[7] H. Wu and J. Yan. (2021) "The mechanism of digitized landscape architecture design under edge computing," PLOS ONE, vol. 16.
[8] G. Jia, X. Gong, L. Jie, W. Wang, and X. Que. (2018) An Optimized Hybrid Unicast/Multicast Adaptive Video Streaming Scheme Over MBMS-Enabled Wireless Networks. IEEE Transactions on Broadcasting ,vol. PP, no. 4. pp. 1-12.
[9] R. Munk, M. Rost, H. Racke, and S. Schmid. (2021) It's Good to Relax: Fast Profit Approximation for Virtual Networks with Latency Constraints. in 2021 IFIP Networking Conference (IFIP Networking).
[10] W. Lee et al. (2018) BUQS: battery- and user-aware QoS scaling for interactive mobile devices.
[11] H. Luo, I. A. Hiskens, and Z. Hu. (2020) Stability Analysis of Load Frequency Control Systems with Sampling and Transmission Delay. IEEE Transactions on Power Systems, vol. PP, no. 99. pp. 1-1.
[12] Y. Sabucusandal, A. E. Pusane, G. K. Kurt, and F. Benedetto. (2020) Reputation Based Attacker Identification Policy for Multi-Access Edge Computing in Internet of Things. IEEE Transactions on Vehicular Technology, vol. PP, no. 99. pp. 1-1, 2020.
[13] S. Li. (2020) A task offloading optimization strategy in MEC - based smart cities. Internet Technology Letters.
[14] Z. Meng, H. Xu, L. Huang, X. Peng, and Y. Shuang. (2018) Achieving Energy Efficiency Through Dynamic Computing Offloading in Mobile Edge-Clouds. in 2018 IEEE 15th International Conference on Mobile Ad Hoc and Sensor Systems (MASS).
[15] H. Zhang, Z. Chen, W. Jia, Y. Deng, and Y. Xiao. (2018) Energy-Efficient Online Resource Management and Allocation Optimization in Multi-User Multi-Task Mobile-Edge Computing Systems with Hybrid Energy Harvesting. Sensors, vol. 18, no. 9, p. 3140.
[16] M. Hossain, M. A. Layek, T. Sultana, M. A. Hossain, and E. N. Huh. (2020) Orchestration-Based

Task Offloading for Mobile Edge Computing in Small-Cell Networks. in 2020 International Conference on Information Networking (ICOIN).

[17] D. Li. (2016) COMPOUND, 1, 3-DIOLEIN-2-LINOLEIN, AND PHARMACEUTICAL PREPARATION AND USE THEREOF.

[18] Z. X. Shi. (2019) Research on Computing Migration Strategy Based on Multi-objective Optimization in Fog Computing Environment. Journal of Xi'an University(Natural Science Edition).

[19] R. V. Bhat, R. Vaze, and M. Motani. (2021) Minimization of Age of Information in Fading Multiple Access Channels. IEEE Journal on Selected Areas in Communications, vol. PP, no. 99. pp. 1-1.

[20] V. Porkodi, A. R. Singh, A. Sait, K. Shankar, and G. P. Joshi. (2020) Resource Provisioning for Cyber–Physical–Social System in Cloud-Fog-Edge Computing using Optimal Flower Pollination Algorithm. IEEE Access, vol. PP, no. 99. pp. 1-1.

[21] F. Wang, J. Xu, X. Wang, and S. Cui. (2017) Joint Offloading and Computing Optimization in Wireless Powered Mobile-Edge Computing Systems. in 2017 IEEE International Conference on Communications (ICC).

[22] X. H. Dong, G. S. Yang, J. F. Tian, T. L. Rong, H. L. Jia, and H. Liu. (2018) Characteristics of deformation properties of frozen sandstone under lateral unloading condition. Rock and Soil Mechanics.

[23] L. You, Y. Huang, D. Zhang, Z. Chang, and X. Gao. (2021) Energy Efficiency Optimization for Multi-cell Massive MIMO: Centralized and Distributed Power Allocation Algorithms. IEEE Transactions on Communications, vol. PP, no. 99. pp. 1-1.

[24] Y. Li, P. Lee, and J. Lui. (2014) Stochastic modeling and optimization of garbage collection algorithms in solid-state drive systems. Queueing Systems, vol. 77. no. 2. pp. 115-148.

[25] W. Liu, S. Xiong, D. U. Wei, and W. Wang. (2019) Research on Cloudlet selection strategy for data streaming applications in mobile cloud environment. Journal on Communications.

[26] Y. Li and C. Jiang (2020) Distributed task offloading strategy to low load base stations in mobile edge computing environment. Computer Communications, vol. 164. pp. 240-248.

[27] X. Duan, B. Li, and W. Zhao. (2020) Energy Consumption Minimization for Near-Far Server Cooperation in NOMA-assisted Mobile Edge Computing System. IEEE Access, vol. PP, no. 99. pp. 1-1.

[28] P. Balakrishnan and C. K. Tham. (2014) Energy-Efficient Mapping and Scheduling of Task Interaction Graphs for Code Offloading in Mobile Cloud Computing. in IEEE/ACM International Conference on Utility & Cloud Computing.

[29] P. Zhao, H. Tian, C. Qin, and G. Nie. (2017) Energy-Saving Offloading by Jointly Allocating Radio and Computational Resources for Mobile Edge Computing. IEEE Access. pp. 1-1.

[30] S. Wang, X. Huang, L. Yi, and Y. Rong(2016) CachinMobile: An energy-efficient users caching scheme for fog computing. in IEEE/CIC International Conference on Communications in China.

[31] J. Opadere, Q. Liu, N. Zhang, and T. Han. (2019) Joint Computation and Communication Resource Allocation for Energy-Efficient Mobile Edge Networks. in ICC 2019 - 2019 IEEE International Conference on Communications (ICC).

[32] M. Huang, Q. Zhai, Y. Chen, S. Feng, and F. Shu. (2021) Multi-Objective Whale Optimization Algorithm for Computation Offloading Optimization in Mobile Edge Computing. Sensors, vol. 21, no. 8, p. 2628.

Operation Anomaly Monitoring of Customer Service Data Analysis Platform Based on Improved FP-Growth Algorithm

Jing Yang [1*], Lihua Gong [1], Kunpeng Liu [1] and Letian Xiu [1]

[1] State Grid Customer Service Center, Tianjin, China

[*]Corresponding author's e-mail: jing-yang@sgcc.com.cn

Abstract. Aiming at the problems of long time-consuming monitoring and poor monitoring accuracy in traditional customer service data analysis platform operation abnormality monitoring methods, a customer service data analysis platform operation abnormality monitoring method based on the improved FP-Growth algorithm is designed. Obtain customer service data sets, classify data types, filter customer behavior, identify the operating status of the data analysis platform, improve the FP-Growth algorithm to build a rule configuration model, set the platform safety factor threshold, and keep the reconstruction error of customer service data to a minimum Within the scope, optimize the abnormal monitoring mode. The experimental results show that the average recovery time of the proposed customer service data analysis platform operation anomaly monitoring method is 5.239s, and the average platform operation anomaly accuracy rate is 97.3%, indicating that the customer service data analysis platform integrated with the improved FP-Growth algorithm operates abnormally The monitoring method performs better.

1. Introduction

The improved FP-Growth algorithm is an algorithm used to describe the mining of frequent patterns without any candidate set [1,2]. According to the principle of data set classification and processing, after data scanning is completed, the extracted frequent sets are compressed until a frequent pattern tree with more branches is formed. Then, mining is carried out one by one according to the associated information between the frequent sets of data. The purpose of applying the improved FP-Growth algorithm to the abnormal monitoring of customer service data analysis platform operations is to explore a more complete data processing model. In addition to the application of platform operation abnormality monitoring in the field of fault diagnosis, it can also be used in fields such as mail filtering. It is currently one of the more widely used technologies. The customer service data analysis platform is based on the corresponding data processing technology. At the same time, when the amount of data shows a large-scale and rapid growth trend, the various types of frameworks and components of the customer service data analysis platform open source processing Tasks also need to be constantly adjusted and upgraded. However, during operation, in addition to the explosive growth of data, there are also potential risks including network attacks, server paralysis, and hardware crashes. Therefore, it is very necessary to monitor the abnormal operation of the customer service data analysis platform. The security of the current customer service data analysis platform is mainly provided by some infrastructure, and the upper-layer mechanism of the platform is relatively safe. However, for the lower-level facilities of the platform, the coverage of network abnormality monitoring cannot meet the security requirements. At this time, the platform architecture needs to be decomposed and optimized.

Due to the particularity of the customer service data analysis platform itself, the technology for abnormal monitoring from the perspective of the server host log is relatively mature, but for the continuity and variability of data, it cannot complete data processing at one time. , It is necessary to incorporate more advanced technologies at the data classification level. Academia has also accumulated some research results on abnormal monitoring, mainly focusing on the safe processing of data. In reference to the problem of temporal data congestion in the Industrial Internet of Things, the literature [3] disassembled the time series into multiple segments, and obtained a standardized process for data analysis; based on the calculation results of the correlation matrix of the data analysis, the time series correlation was constructed Graph model, and experimentally verify the time series in the model to obtain monitoring results, but in this process, the problem of unclear operation status is ignored. For the network security threat problem, literature [4] adopts the method of data segmentation pattern matching and semantic analysis to optimize and upgrade, extract the attack type, and locate the malicious code existing in the network level. After experimental testing, it is obtained. The experimental results are presented, but the description of the problem of unclear operation status still lacks more dimensions. Therefore, this paper proposes a method for monitoring abnormal operation of customer service data analysis platform based on the improved FP-Growth algorithm, and perfects the related topics of abnormal monitoring of customer service data analysis platform operation.

2. Operation anomaly monitoring of customer service data analysis Platform based on improved FP-Growth algorithm

2.1. Getting the customer service data set

Customer service data is mainly a platform-centered data set, which serves as the operation carrier of the whole platform [5,6]. Oriented by customer service needs, the corresponding statistical mode of data and statements is formed, which mainly includes topics such as user complaints, business handling, outbound statistics and customer service, and also includes several statistical applications under the four topics. Combined with the overall call volume of customer service, employee output, and business inquiry and processing data, summarized and analyzed the corresponding customer service data characteristics. In the current era of constantly changing market environment, customer service data also needs to meet the needs of the development of The Times, constantly adjust and upgrade, to achieve data management support. Through the abstract management of customer service data, the abstract representation of data set is obtained. In addition to reflecting the customer service data itself, the existing customer service data can be summarized by means of computing [7,8,9]. And according to the data and information relationship between customer service data, divide the data type, get more accurate data structure. The paradigm expression between customer service data is a kind of characteristic data that directly reflects the data relationship. The main idea of the standardized processing of customer service data is to eliminate the non-conforming data in the data set according to the corresponding processing rules in order to ensure that the relationship between data can be kept within a certain distance. In the platform, the attribute values of the data set are not unique in terms of the use of customer service data and have different meanings in different scenarios. At the same time, the attribute types of customer service data are also divided into major attribute and non-major attribute. As for the dependency relationship between data, any independent non-primary attribute data is completely subordinate to the primary key structure of the whole platform, but it also retains certain non-primary attributes. Taking into account the overall scalability requirements of customer data, the corresponding investment costs and management factors are taken into account. At the data management mode level of the platform, the delivery dependency between data is truncated. Combined with the representative data conversion method, using the tool software with integration function, based on the concept of database, establish a data system in line with the operation concept of the platform. Based on the service demand of customer service data, rationalize the internal scheduling of the platform. The data acquisition channel is used to improve the application and processing mode of customer service data. For customer service data analysis platform, standard and unified channel information, according to data application requirements and information retrieval requirements, the best solution.

2.2. Identifying operation status of data analysis platform

Generally, in some platforms with large data scale, the measurement method of the relationship between data will affect the overall operation status of the platform to a certain extent. In customer service data analysis platform, there is a linear relationship between some data sets and schedulable resources. Some external characteristics of customer service data, including performance indicators and availability indicators, can be used as a standard to measure the operating status of the platform. Given the combination of the number of platform components and space footprint, choosing to monitor the health of all customer service data can have a speed impact. In order to avoid the situation of monitoring index redundancy, the correlation between data is combined with centralized processing. When the platform operation is under normal operation conditions, the safety factor of the platform operation is set as 1, and the relationship between any two adjacent variables is positive correlation. In this case, there is a certain linear relationship between customer service data. When the operation status of the platform is abnormal, the value of the safety factor of the platform operation will change from 1 to -1, and the relationship between any two adjacent variables in the platform is completely negative correlation. According to this judgment standard, when the variable relationship is closely correlated, it indicates that the operation status of the platform is relatively normal, while when the variable relationship is alienated, it is necessary to carry out detailed comparison of various data of the platform operation. In order to reflect the impact of customers' recent behaviors on the operating status of the platform, the weight of customers' behaviors is set after filtering customers' behaviors based on their behavior characteristics. The specific expression formula is as follows:

$$L=\frac{1}{1+\delta\left|r_m-r_n\right|} \quad (1)$$

In formula (1), δ represents the sequence of customer service sessions, r represents the attenuation coefficient of the platform, m represents the early data of customer service, and n represents the current data of customer service. According to the attributes of sample variables in the data, the orthogonal matrix expression formula of platform operation data is obtained:

$$D=\frac{\sum d^{-1}}{\left[d_1,d_2\ldots d_z\right]_{z\times z}} \quad (2)$$

In formula (2), d represents the new random variable and z represents the dimension. The operating state of the platform is not constant across different uptime and application scenarios. In different state changes, Nengou captures some laws of customer service data analysis platform. The improved FP-growth algorithm is used to mine the historical operational metrics of the platform and use the aggregated data as the underlying data set. Based on the historical data of platform operation, reliable conclusions are drawn. The threshold setting of safety factor is also very important in the process of platform operation status identification. If the threshold range is higher than the actual value, the coefficients that should be relevant are more likely to be misjudged as irrelevant. Therefore, it is necessary to judge whether human intervention is needed according to the real-time operation of customer service data analysis platform. Based on this, the steps of identifying the operation status of the data analysis platform are completed.

2.3. Improve FP-growth algorithm to build rule configuration model

The improved FP-growth algorithm is an algorithm that upgrades the search strategy on the basis of the FP-growth algorithm and has the advantage of divide and conquer according to different types of data [10,11]. In the process of mining frequent itemsets, scan data, customer service will be a frequent itemsets, and then according to the setting of threshold platform security coefficient, will not satisfy the operation of the platform delete data sets, then get the new data set, and according to the new standard data set of frequent items, support count of descending order. In the actual application process, in order to improve the defect of scanning customer service database times, the improved FP-growth algorithm is used to adjust the arrangement of customer service data [12,13,14]. At the same time, according to the algorithm only need to traverse the characteristics of the database, separate from the original database for secondary mining, do not need to mining platform database again. In the process of building the rule configuration model, it is necessary to implement the rule from top to bottom in the order of conceptual model design, logical model design and physical model design step by step. Random vectors in customer service data are collected as the matrix of data samples, and the specific expression formula is as follows:

$$Q = \begin{bmatrix} q_{11} & q_{12} & \cdots & q_{1p} \\ q_{21} & q_{22} & \cdots & q_{2p} \\ & \cdots & \cdots & \\ q_{l1} & q_{l2} & \cdots & q_{lp} \end{bmatrix} \quad (3)$$

In formula (3), q represents the standard deviation of the random variable, p represents the coefficient vector, and l represents the load. On the basis of formula (3), model elements are transformed and expressed as follows:

$$k_{ij} = \begin{cases} x_{ij}, 0 \\ -x_{ij}, 1 \end{cases} \quad (4)$$

In formula (4), x represents the opposite indicator, $-x$ represents the opposite indicator, and ij represents the weight of the indicator. Based on the calculation results, you can determine the service coverage of the rule configuration after analyzing the platform operation requirements. On the basis of high generalization, the customer service data objects involved in the relevant business are classified. After dividing the subject domain, the subject of the data set is refined, and the rule attributes corresponding to each data set are clarified. Based on this, the steps of improving FP-growth algorithm to build the rule configuration model are completed.

2.4. *Optimizing the anomaly monitoring mode*

The anomaly monitoring mode of the customer service data analysis platform can discover the potential risks of the platform operation in a short period of time and take effective measures in time. In the process of platform operation, once the actual situation does not match the expected behavior, the problems are defined and the abnormal situation is described [15]. If an abnormal data is prominent in the sample set to be monitored, its observed value will also exceed the standard threshold range. Using the principle of the improved FP-growth algorithm, suspicious data can be found and marked for early warning. If the current statistical data are far from the expected outlier, then relevant effective data are defined, and the expression formula for the definition of mean value of abnormal data is as follows:

$$\lambda = \sum_{c=1}^{e} \frac{h_e}{c} \quad (5)$$

In formula (5), h represents the data set to be monitored, c represents the expected value of data outlier, and e represents the outlier. The purpose of defining the mean value of abnormal data is to simplify the original data of the platform and keep the reconstruction error of customer service data within the minimum range. In addition to the orthonormal vectors between data, noise and redundancy between data are the factors that affect the accuracy of monitoring. In the heterogeneous space of platform anomaly monitoring, the larger the volume of corresponding diagonal elements is, the greater the noise of customer service data is in the layer network, and the commonality of data redundancy is relatively high to a certain extent. Conversely, the smaller the element, the less likely it is to prove that there is noise and data redundancy. At the same time, a group of obvious linear feature vectors is found through linear transformation, and the extracted classification information is compressed to express the dimension of feature space in the form of the largest category distance. On the basis of high probability distribution, the expression formula of criterion function is obtained:

$$G(\eta)=\frac{\eta^{R}Ws}{\eta^{R}s} \quad (6)$$

In formula (6), η represents the total number of samples, W represents the maximum inter-class hash degree, s represents the signal separation speed of the data analysis platform, and R represents the minimum intra-class dispersion. According to the calculation results of Formula (6), the expression formula of projection coefficient of independent basis vector is obtained:

$$\varepsilon=\frac{F_{T-1}}{V_{T}} \quad (7)$$

In formula (7), F represents the linear combination coefficient, V represents the number of test data, and T represents the separated signal source of the data analysis platform. According to the calculation results of formula (5) to formula (7), anomaly monitoring steps in line with the platform operation and development law are formulated for data acquisition, modeling and monitoring stages respectively. Combined with the fault index and abnormal probability of anomaly monitoring, the corresponding parameter updating mechanism is integrated. In addition to detailed monitoring of the platform's internal operations, the operating environment outside the platform should also be monitored to ensure that the customer service operation platform is free from external interference. Based on the above description, complete the steps of optimizing the anomaly monitoring mode.

3. Experimental test

3.1. Building experimental environment

In order to verify the effectiveness of the operation anomaly monitoring method of the customer service data analysis platform designed this time, the experimental test was carried out and the corresponding experimental environment was built. Operating platform Select Linux platform, and take Map operator and Join operator as data set of operation anomaly monitoring. The experimental computing framework chooses Apache Flink as the computing engine with good performance and uses Nmon as the open source data tool. Nmon can automatically integrate the information of the platform server and the usage of data resources, and summarize them into a specific file according to a fixed format. It is a common analysis and use tool for monitoring platform operation anomalies in experimental environments. In addition, during the experiment, Nmon system has a low occupancy rate and does not cause system congestion. Under normal operating conditions, Nmon's CPU usage is between 1% and 2%. The software and hardware configuration of the experimental environment is shown in Table 1:

Table 1 Experimental software and hardware configuration

Serial number	Hardware	configure	Software
#1	disk	600G，7200RPM	Flink
#2	network	100MB/s	Zookeeper
#3	Memory	16GB	My Sql
#4	OS	Ubuntu	Nmon
#5	CPU	Intel(R)Core(TM) 64.bit i5-6200 3GHz	Intelli J IDEA

On the basis of Table 1, select the data set required for this experiment, mainly the real index data collected through nmon. Among them, there are 22300 normal data, accounting for 82% of the data set, and 7433 abnormal data, accounting for 24% of the data set. In addition, multiple indicators are set for the processor, disk, network memory and other resources of the platform, mainly including CPU, disk, I / O and memory. The relevant indicators and descriptions are shown in Table 2:

Table 2 Related indicators and descriptions

Indicator type	Indicator name	Index description
CPU	QL-memtotal	Physical memory occupies CPU space
	QL-memfree	Free memory occupies the total CPU space
	QL-membuff	The memory space of the kernel cache
	QL-memcach	Total cache memory occupies CPU space
	QL-memutil	Total physical memory space used by the platform
	QL-swpin	The amount of virtual memory space read from disk per second
	QL-swptotal	Total space of the platform exchange area
Disk and I / O	ci-tps	The number of requests sent per second by a data port
	ci-svctm	Port service time
	ci-qusize	Port queue Length
	ci-read	The number of bytes read per second by a platform port
	ci-util	An overview of platform operations
	ci-rqsize	Data volume size
Memory	qr-user	User space Space occupied by users
	qr-sys	Kernel space takes up space
	qr-hirq	Hardware interrupt request time takes up space
	qr-sirq	Software interrupt request time occupies space
network	Pz-packets	The number of packets sent by the platform
	tr-bytes	The number of bytes sent by the output port
	pz -dropped	Sending Port Number of packets lost when sending data
	tp -packets	The number of packets received by the platform

Combined with the data set information in Table 2, the monitoring indicators required by the experiment were measured, and experimental tests were conducted on the basis of reducing platform overhead.

3.2. *Experimental result*

The experimental scenario is as follows: the platform is injected with a single fault type, the corresponding monitoring period is set, and the CPU utilization in the monitoring period is collected. The collection time is set from 20:15:05 to: 23:15:05, and the fault injection time is 15 minutes and 10 minutes respectively. The process deadlock is selected as the fault exception injection method for the three operation anomaly monitoring methods of customer service data analysis platform. During the experiment, the input rate of the tuple was adjusted to 500-1500tuples /s, and the work task of the platform branch node was terminated. The operation anomaly monitoring method of customer service data analysis platform based on dynamic rule base and association rule are selected and compared with the operation anomaly monitoring method of customer service data analysis platform in this paper. Test the recovery time of three abnormal operation monitoring methods of customer service data analysis platform under different input rates. The experimental results are shown in Figure 1-3:

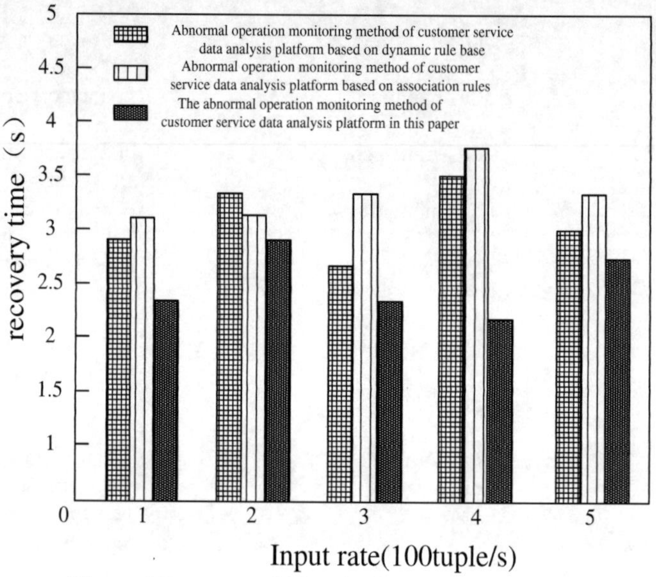

Figure 1 Input rate 500tuples/s recovery time (s)

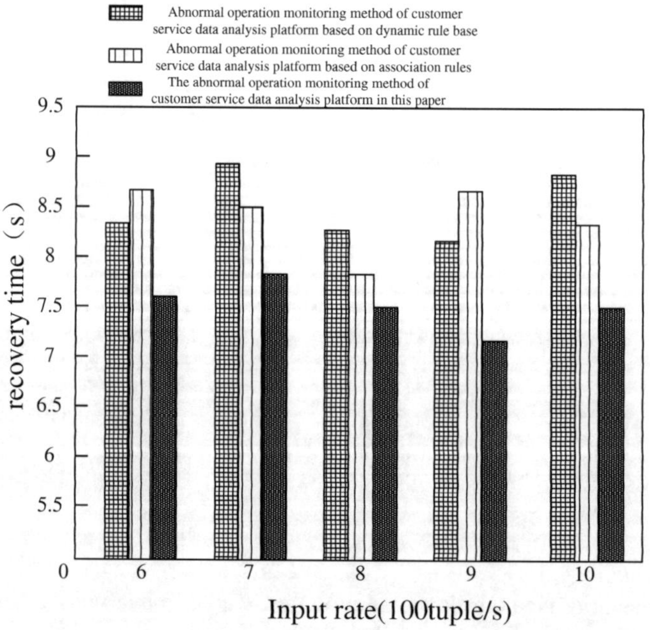

Figure 2 Input rate 1000tuples/s recovery time (s)

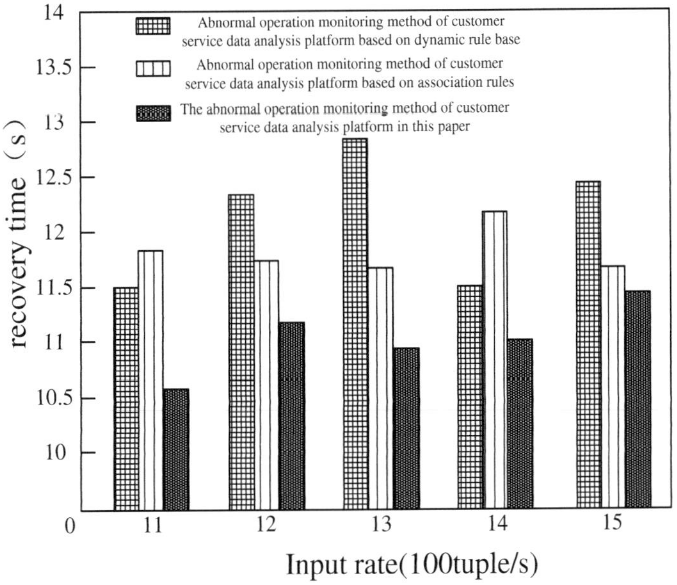

Figure 3 Input rate 1500tuples/s recovery time(s)

According to Figure 1-3, it can be concluded that when the input rate becomes 500~1000tuples/s, the recovery time difference of the three customer service data analysis platform operation abnormal monitoring methods is relatively small; when the input rate becomes 1500tuples/s , The recovery time gap of the three kinds of customer service data analysis platform operation anomaly monitoring methods is gradually increasing. From this, the average recovery time of the designed customer service data analysis platform operation abnormality monitoring method and the other two customer service data analysis platform operation abnormality monitoring methods are obtained, as shown in Table 3:

Table 3 Average recovery time of abnormal operation monitoring methods of three customer service data analysis platforms(s)

Input rate(tuples/s)	The designed customer service data analysis platform operation abnormal monitoring method	Operation anomaly monitoring method of customer service data analysis platform based on dynamic rule base	Operation anomaly Monitoring method of customer service data analysis platform based on association rules
100	1.321	1.983	2.036
200	1.416	2.372	2.408
500	2.561	3.316	3.322
700	4.321	5.069	5.834
900	5.644	6.358	6.889
1000	7.403	8.032	8.512
1200	8.226	9.784	10.205
1500	11.021	12.204	11.819

It can be seen from Table 3 that the average recovery time of the designed customer service data analysis platform operation abnormal monitoring method and the other two customer service data analysis platform operation abnormal monitoring methods are: 5.239s, 6.140s, 6.378s, indicating the design The customer service data analysis platform operates an abnormal monitoring method, and it takes less time to recover from an abnormal state.

The accuracy of the abnormal monitoring of the operation of the three customer service data analysis platforms was tested under the above test environment and conditions, and the results of the abnormal monitoring of the operation of the customer service data analysis platform were obtained as shown in Table 4.

Table 4 Anomaly monitoring accuracy rates of three customer service data analysis platform operations anomaly monitoring methods

Number of experiments	The designed customer service data analysis platform operation abnormal monitoring method	Operation anomaly monitoring method of customer service data analysis platform based on dynamic rule base	Operation anomaly Monitoring method of customer service data analysis platform based on association rules
1	97.5%	83.6%	79.6%
2	97.6%	82.5%	78.5%
3	97.4%	81.7%	77.9%
4	96.5%	83.5%	78.0%
5	96.8%	84.6%	80.5%
6	97.2%	80.5%	79.1%
7	97.8%	84.1%	78.6%
8	97.5%	85.2%	77.8%

It can be seen from Table 4 that the designed customer service data analysis platform operation abnormality monitoring method and the other two customer service data analysis platform operation abnormality monitoring methods have average abnormal monitoring accuracy rates of 97.3%, 83.2%, 78.6%, respectively. The designed customer service data analysis platform operation abnormality monitoring method has better abnormality monitoring accuracy.

4. Conclusion

The customer service data analysis platform designed in this paper is built on the basis of customer service data, using the improved FP-Growth algorithm to carry out the work of frequent itemsets mining in more detail. Experimental tests have proved that this method has improved the traditional platform operation abnormal monitoring method that takes too long to recover from abnormal conditions, and the monitoring accuracy of abnormal operation conditions is higher, which lays a foundation for the academic community to carry out research on related topics. The theoretical basis and practical basis of the platform, and promote the development process of the field of platform operations. Due to the limited research conditions, the article has not divided the scenes of abnormal monitoring of platform operations in detail. The relevant details will be continuously improved in the future.

References

[1] Ls A , Fv C , Rya B , et al. (2020) A guided FP-Growth algorithm for mining multitude-targeted item-sets and class association rules in imbalanced data[J]. Information Sciences, doi: 10.1016/j.ins. 2020.10.020

[2] Wu Y , Zhang J . (2020) Building the electronic evidence analysis model based on association rule mining and FP-growth algorithm[J]. Soft Computing, 24(2), doi : 10.1007/s00500-019-04032-0

[3] Ding Xiaoou,Yu Shengjian,Wang Muxian, et al. (2020) Anomaly Detection on Industrial Time Series Based on Correlation Analysis[J]. Journal of Software, 31(03): 726-747.

[4] Li Yuan, Wang Guoyin, Li Zhixing, et al. (2019) A Sequential Attention Based Convolutional Neural Network for Anomaly Detection[J]. Journal of Zhengzhou University(Natural

Science Edition), 51(2): 17-22.

[5] Misra A , Bhuiyan M , J Mahmud, et al. (2019) Using Structured Representation and Data: A Hybrid Model for Negation and Sentiment in Customer Service Conversations[J]. Tenth Workshop on Computational Approaches to Subjectivity.

[6] Li Z , Guo H , Wang W M , et al. (2019) A Blockchain and AutoML Approach for Open and Automated Customer Service[J]. IEEE Transactions on Industrial Informatics, 2019:1-1.

[7] Black A . (2019) 'Bringing People Together' - How Microsoft Teams and Other Collaboration Tools are Transforming Customer Service[J]. Database and network journal, 49(3):7-7.

[8] Zhang C , Ren M . (2021) Customer service robot model based on e-commerce dual-channel channel supply coordination and compensation strategy in the perspective of big data[J]. International Journal of System Assurance Engineering and Management, 2021:1-11.

[9] Thiriku J M , Kegoro H O , Mandal P C . (2019) Strategic leadership on customer service delivery in the ministry of industry, investment and trade in kenya[J]. Noble International Journal of Business and Management Research, 2019, 3.

[10] Bagui S , Devulapalli K , Coffey J . (2020) A Heuristic Approach for Load Balancing the FP-Growth Algorithm On MapReduce[J]. Array, 2020:100035. doi : 10.1016/j.array.2020.100035

[11] Jia Y , Liu L , Chen H , et al. (2020) A Chinese unknown word recognition method for micro-blog short text based on improved FP-growth[J]. Pattern Analysis and Applications, 2020, 23(2). doi: 10.1007/s10044-019-00833-z

[12] Liu S , Yun S U , Zhang Y . (2019) Open-line Fault Diagnosis and Positioning Method for 10kV Power Distribution Network Branch Line Based on FP-growth Algorithm[J]. Power System Technology, 43(12), 4575-4892.

[13] Jiang X , Xiao X , Wen W , et al. Operation Optimization of Thermal Power Plant Based on Parallel FP-Growth Algorithm[J]. Power System Technology, 2019.

[14] Zhang J , Sun H , Sun Z , et al. (2020) Reliability Assessment of Wind Power Converter Considering SCADA Multistate Parameters Prediction Using FP-Growth, WPT, K-means and LSTM Network[J]. IEEE Access, 2020, PP(99):1-1, doi: 10.1109/access.2020.2992089.

[15] Liu Qinglian, Wang Xueping. (2019) Real Time Monitoring Simulation of Abnormal State of Large Data In Smart Grid[J]. Computer Simulation, 36(3): 364-367.

The Improved Quantum Secret Sharing Protocol Based on Grover Algorithm

Zhuo Yu[1a*]

University of Warwick, Physics Department, Coventry, England, CV4 7AL

[a*]E-mail:u2008160@live.warwick.ac.uk

Abstract—Quantum secret sharing(QSS) is a cryptographic protocol based on the uncertainty principle and quantum no-cloning theorem and is an essential aspect in quantum computing. This paper describes the Grover search algorithm and how to improve QSS protocol by Grover Algorithm. The QSS based on the Grover algorithm is more secure and easier to operate compared to old QSS protocols. Some scientists improved the security of the first QSS protocol based on QSA by changing the check mode. After that, a protocol with quite different procedures was proposed. The improved one enhances the feasibility of the protocol and it could also make sure that the communication is secure from the attacks.

1. Introduction

Secret sharing is a consequential topic in quantum information science. Nowadays, people pay more and more attention on the privacy and at the same time, there is a crisis of confidence in the society. Many people try their best to protect their own secrets, but want to gain others' secrets. So the multi-party protocols are required. One of the most critical protocols is the secret sharing. In the secret sharing protocol, a boss splits the secret into several parts and sends them to the agents. As a result, any agent can not read the secret alone, and they must work together to get the secret. Quantum secret sharing was proposed in 1999 [1]. Compared to the Classical secret sharing, quantum secret sharing is more secure because the QSS is based on the uncertainty principle and quantum no-cloning theorem.

Quantum search algorithm is another crucial part of quantum computing. Due to the quantum superposition, the state of a qubit could be superposed. Then the system could complete several assignment simultaneously. This makes the efficiency of the quantum searching task much higher than that of the classical search, especially for the large space. For an unsorted database of size N, the QSA needs $\sqrt{N}$ operation, while N steps are required to the target classically. The Grover algorithm is one of the most significant quantum searching algorithm and is discovered by Grover in 1996 [2]. The Grover algorithm can not ensure that the answer is true and it can only offer the correct outcome with a high probability. However, some scientists proposed an upgraded Grover algorithm that can achieve a 100% success rate [3].

Many scientists focus on QSS protocols based on the QSA since this kind of the protocols play an important role in quantum computation. Hsu tried to come up with a QSS protocol based on the QSA in 2003 [4]. But in 2010, Hao et al. published a eavesdropping and the solution in the Hsu's protocol [5]. For these two protocols, the boss needs to prepare the initial state $|S_i>$ and divide them into two parts, S_1 and S_2. Then the boss sends S_1 to one agent and sends S_2 to the other. However, there are some defects. For example, agents need to store photons, which is extremely challenging especially when the time duration is long. Besides, face-to-face cooperation or the third party's help is necessary [6]. This makes the process more complicated.

Some scientists introduced another QSS protocol based on the Grover algorithm. Two problems above are solved and the scientists proved that this protocol is secure [6]. The research on the improvement of this kind of protocols is necessary because it need to be optimized in order to make sure that it can be more practical and must be secure when there is eavesdropping.

2. Grover Algorithm

2.1. Process of the Grover Algorithm

The Grover's quantum search algorithm is designed for the search task in an unsorted database with 2^n elements. This is followed by the detailed process.

(1) The first step is to use a quantum oracle O operation. An oracle can be considered as a "black box" operation. A target state should be chosen and the target state reverses in this step. But other states do not change.

(2) Secondly, the Hadamard transformation is required. In this step, Hadamard gate is operated on every qubit. Hadamard gate could be represented as $\frac{1}{\sqrt{2}}\begin{pmatrix} 1 & 1 \\ 1 & -1 \end{pmatrix}$.

(3) In the third step, |0> remains invariant and other states add -1 phase. This means |0> → |0>, |x> → -|x> if x does not equal 0. This step is equal to 2|0><0| - I .

(4) The fourth step is repeating the second step.

In conclusion, a single grover iteration equals G = (2|ψ><ψ| - I)O (1) where ψ is the superposed state with equal weight.

2.2. Defects and Improvement of the Grover Algorithm

The Grover Algorithm has some drawbacks. For instance, the probability of finding the target can not be 1 and can only approach 1 for the standard Grover algorithm. But there are also some solutions. In 2001, G. L. Long came up with an improved Grover algorithm in which we could find the target state with certainty. In this improved version, the phase inversion is replaced by phase rotation [7].

3. Quantum Secret Sharing

In recent years, many scientists are working on quantum secret sharing and trying to combine the quantum search algorithm and quantum secret sharing. The following section will focus on some types of QSS based on the QSA, especially the advanced quantum secret sharing protocol based on the Grover algorithm. And there will also be some detailed comparison of these different protocols.

3.1. QSA-based Protocols

The first QSS protocol based on the QSA is for one boss and two agents and this is based on the two-qubit Grover algorithm. Firstly, the definition "threshold scheme" needs to be introduced. For a (k, n) threshold theme, there are n shares and it is enough to reconstruct the data by any k shares. In this protocol, a (2, 2) threshold scheme is used [4]. The procedure of this protocol will be discussed.

(1) The initial states(or the carrier states) $|S_i>$ (i = 1, 2, 3,...,16) should be prepared by the boss(Alice). Sixteen initial states are required and they are produced by |+>, |->, |+i>, |-i>, where |+> = $\frac{1}{\sqrt{2}}$(|0>+|1>) (2) , |-> = $\frac{1}{\sqrt{2}}$(|0>-|1>) (3), |+i> = $\frac{1}{\sqrt{2}}$(|0>+i|1>) (4)and |-i> = $\frac{1}{\sqrt{2}}$(|0>-i|1>) (5). Then Alice need to to use the unitary operation U_w, which is equal to (1-2|w><w|). So the code state is $|S_i>_w = U_w |S_i>$ (6).

(2) Two qubits are sent by Alice to Bob and Eve. Bob and Eve should announce that they receive the qubits.

(3) Alice need to confirm that the qubits are received through classical communication.

(4) Alice publishes her carrier states $|S_i>$.

(5) Bob and Eve have to combine their qubits and perform a decoding operation in order to get the key state |w>. According to the Alice's announcement, Bob and Eve know the initial state, so they

could determine which operation should use. The decoding operation is $2|S_i><S_i| - 1$ denoted as U_{Si}. Then the key state could be obtained with reference to the grover algorithm. The key states in this protocol are |00>, |01>, |10>, |11> . The states |01> and |10> are the message states, which are used to encode, while the states |00> and |11> are used to detect the eavesdropping.

(6) Bob and Eve need to make the measurement on their qubits respectively. Then they have to discuss whether or not their outcomes are perfectly correlated.

(7) If the results of Bob and Eve are perfectly correlated, they will both inform Alice through the classical channel. And then Alice needs to analysis that whether there is an eavesdropper.

However, Hsu's protocol is not secure theoretically. In 2010, Hao et al. proposed a dense-coding attack on Hsu's protocol and they also gave an improved protocol [5]. The attack and solution will be discussed briefly.

Eve is assumed to be the dishonest agent in this attack here. Besides, the definition "Bell states" (or "EPR pairs") should be introduced. The Bell states are two qubits with four entangled quantum states. And it is possible to create the Bell states via quantum circuits.

In this attack, Eve needs to block all the qubits from Alice. At the same time, Eve has to prepare a EPR pair in $\frac{1}{\sqrt{2}}(|01> + |10>)$ and sends one of them to Bob, leaving himself another qubits. Since Bob actually receives the qubit (although it is the incorrect one), both Bob and Eve will make the announcement and Alice would confirm the reception. Then Alice announces the initial states in public. As a result, Eve obtains $|S_i>$ and can decode through the operation U_{Si} . Eve could get the key states. Besides, Eve needs to transform the prepared qubit (the one of the EPR pair) into $|S_i>_w$ by operations. In the following steps, all the results are correct and both Alice and Bob will not discover the eavesdropping.

Hao et al. also gave the solution of this attack. The check mode is changed. Instead of announcing the decoding operation, Alice asks Bob and Eve to measure their qubit in random basis which might be the plus-basis or the cross basis, or the circular basis. And both Bob and Eve should publish their measuring results. Alice will analysis whether their results conform to her code state and if the results accord with Alice's states, then Alice could make sure that there is no cheating and the communication is secure.

3.2. *An Enhanced QSS Protocol based on the Grover Algorithm*

Although the protocol by Hao et al. improves the security, this version also has some problems during the procedure. For example, agents need to store the qubits from the boss, which is not convenient or even impossible. So Tseng et al. proposed an improved protocol (still based on the Grover algorithm) which could make the process more realistic [6].

In this enhanced version, Bob and Eve do not need to store photons. Instead, they need to prepare a random N-bits string, respectively. The two random strings are denoted as K_B and K_E . And they have to prepare a sequence of photons respectively, represented as S_B and S_E . The "0" in the string corresponds to |+> and the "1" corresponds to |-> . Apart from that, Bob and Eve need to prepare abundant decoy photons. The decoy photons should be randomly chosen from one of the four states which are |+>, |->, |0> and |1> . These prepared decoy photons are inserted into the sequence of photons, S_B and S_E in order to obtain a new sequence, denoted as S_B' and S_E' . Then these two new sequences are sent to Alice.

Secondly, Alice is confirmed to receive the sequences. After that, there should be a security check. Bob and Eve need to publish the positions, bases and values of the decoy photons and Alice makes the measurement according to the agents' announcement. If the error rate outstrips a threshold value, this means there exists an eavesdropper. Alice will stop the transmission and restart the communication. If the error rate is tolerable, the communication could continue.

In the third step, photons of S_B and S_C on the same position in the sequence could be combined by Alice to obtain two-qubit initial states, represented as |S> . Alice has the secret, so she knows the |w> . Alice is able to perform the correct operation U_w on the initial state . The secret "0" corresponds to the unitary operation $U_{00} = I - 2|00><00|$ (7)or $U_{11} = I - 2|11><11|$ (8). Otherwise, the corresponding operation is $U_{01} = I - 2|01><01|$ (9) or $U_{10} = I - 2|10><10|$ (10) . After that, operation U_S, which equals $2|++><++| - I$, need to be performed. And then Alice measures in the Z basis and obtains the results. Measurement |00> or |11> leads to the result "0", and otherwise, the measurement would lead to measurement "1". After these procedures, Alice gets a N-bit string, denoted as MR_A and sends to Bob and Eve. Then Bob and Eve could decode the secret by computing $K_B \oplus K_C \oplus MRA$.

4. Security of the improved QSS protocol

Tseng et al. also proposed some attacks that might exist. The attacks could come from both the agents and eavesdroppers who do not take part in the activity. However, the agents could gain more information and they could cheat in the public discussion. So it is essential to focus on the insiders' cheating. We could make sure that the protocol is secure if the attacks from the agents could be avoided [6]. In this section, the existing attacks by agents and solutions will be discussed and Eve will be assumed as the eavesdropper.

4.1. Existing Attacks and the solutions of the improved QSS

One of the existing attacks is the Intercept-and-Resend Attack [8]. Eve intercepts S_B' and resends the false photons to Alice. But it is extremely possible to be detected during the check procedure since Eve does not know the positions, bases and values of the decoy photons and there may be some errors. If there are copious errors, the eavesdropping will be discovered and the communication will stop.

The Entangle-and Measure attack is another kind of existing attacks on this protocol [9], but it also have faults. In this attack, auxiliary photons are necessary. Eve should prepare some auxiliary photons, represented by $E = \{|E_1>, |E_2>, \ldots , |E_n>\}$. And Eve also needs to perform a unitary operation on the decoy photons in order to make the auxiliary photons be entangled with S_C'. The consequence of the operation is below [6].

$U|0>|E_n> = a|0>|e_{00}> + b|1>|e_{01}>$ (11),

$U|1>|E_n> = c|0>|e_{10}> + d|1>|e_{11}>$ (12),

$U|+>|E_n> = \frac{1}{2}[|+> (a|e_{00}> + b|e_{01}> + c|e_{10}> + d|e_{11}>) + |-> (a|e_{00}> - b|e_{01}> + c|e_{10}> - d|e_{11}>)]$ (13),

$U|->|E_n> = \frac{1}{2}[|+> (a|e_{00}> + b|e_{01}> - c|e_{10}> - d|e_{11}> + |-> (a|e_{00}> - b|e_{01}> - c|e_{10}> + d|e_{11}>))]$ (14),

where $|a|^2 + |b|^2 + |c|^2 + |d|^2 = 1$.

In order to eavesdrop without the detection, there are some conditions which are needed:

1. b should equal 0 ;
2. c should equal 0 ;
3. $a|e_{00}> - b|e_{01}> + c|e_{10}> - d|e_{11}> = 0$ (15);
4. $a|e_{00}> + b|e_{01}> - c|e_{10}> - d|e_{11}> = 0$ (16).

While if the first two conditions are satisfied, $a|e_{00}>$ will be equal to $d|e_{11}>$ according to the third condition. And then these two states can not be distinguished [6].

5. Conclusion

This paper concludes some quantum secret sharing protocols based on the Grover algorithm and how this kind of protocol is improved. Hsu starts the quantum secret sharing based on the quantum searching algorithm and this protocol was improved by Hao et al. Then the protocol proposed by Tseng et al. is not only secure from the existing attacks, but it also have some other advantages during the communication: the agents do not need to store photons and the secret of the boss is easy to decode for the agents. The QSS protocols based on QSA are vital for the quantum computation and can avoid some practical problems during the communication. This kind of protocols is useful and suitable for many situations. However, maybe there will be some attacks for this protocol, which should be discovered in the future.

Acknowledgments
The author would like to thank Dr. Ming-Deh A. Huang and my thesis advisor for the help.

References

[1] Hillery, M., Buek, V., Berthiaume, A., Quantum secret sharing [J] Phys. Rev. A, vol.59, no.3, 1829, 1999.

[2] Lov K. Grover, A fast quantum mechanical algorithm for database search [C] Proc. of the 28th annual ACM Symp on Theory of Computing. New York, USA: ACM Press, July 1996, pp.212-219.

[3] G. L. Long, Grover algorithm with zero theoretical failure rate [J] Phys. Rev. A, vol.64, 022307, 2001.

[4] Li-Yi Hsu, Quantum secret-sharing protocol based on Grover's algorithm [J] Phys. Rev. A, vol.68, 022306, 2003.

[5] Hao Liang, Li Junlin and Long Guilu, Eavesdropping in a quantum secret sharing protocol based on Grover algorithm and its solution [J] Science China Physics, Mechanics and Astronomy, vol.53, 2010, pp.491-495.

[6] Hsin-Yi Tseng, Chia-Wei Tsai, Tzonelih Hwang and Chuan-Ming Li, Quantum Secret Sharing Based on Quantum Search Algorithm [J] International Journal of Theoretical Physics, vol.51, 2012.

[7] Long Guilu, Li Yansong, Xiao Li, Tu Changcun and Sun Yang, Phase matching in quantum searching and the improved Grover algorithm [J] Nuclear Physics Review, vol.21, 2004, pp.114-116.

[8] Chun-Wei Yang, Chia-Wei Tsai and Tzonelih Hwang, Thwarting intercept-and-resend attack on Zhang's quantum secret sharing using collective rotation noises [J] Quantum Information Processing, vol.11, 2012, pp.113-122.

[9] Chia-Wei Tsai, Shih-Hsueh Wang and Tzonelih Hwang, Comment on "N quantum channel are sufficient for multi-user quantum key distribution protocol between n users" [J] Optics Communications, vol.283, no.24, 2010, pp.5285-5286.

Research On FPGA-based High-speed Data Optical Fiber Transmission

Xianbo Wang

Mechanical and Electrical Engineering,Chengdu University of Technology,Chengdu 610000,China

2019050631@stu.cdut.edu.cn

Abstract—This article briefly introduces the principles and advantages of optical fiber transmission and the characteristics of the integrated IP core developed by Xilinx. Aiming at the advantages of optical fiber communication, Xilinx ZYNQ7000 series FPGA chips are used to design a high-speed data optical fiber transmission scheme based on FPGA. This solution uses the method of creating Block Design for design and development. It uses Verilog hardware description language and Vivado development software for experimental verification. In the FPGA simulation environment, the code is converted into a binary file and written to the chip through the JTAG interface for testing, and the experimental data is analyzed. As a result, it meets the design requirements of high-speed data transmission, which is of great significance to the research of high-speed data optical fiber transmission.

1. Introduction

With the rapid development of modern science and technology, people have made more in-depth research on communication technology. The high speed of communication and the high quality of communication are more important aspects. In future communication, whether it is wired communication or wireless communication, there are higher requirements for data exchange and processing capabilities. Optical fiber communication has the advantages of fast transmission speed, low loss, low capacity, low occupied bandwidth, and anti-electromagnetic interference. It is very suitable for high-speed data transmission. This article mainly uses the Aurora 64/66B IP core developed by Xilinx company to realize the fiber loopback communication of different interfaces on the board under the environment of the FPGA tool Vivado2018.3 software, the rate can reach 10Gbps, and the bit error rate test is carried out on this.

2. Optical fiber transmissione

Optical transmission is a communication method that uses optical fibers as the propagation medium and light waves as carrier waves to transmit information, so it can also be called optical fiber transmission. Its transmission principle is to first convert the modulated electrical signal into an optical signal through a photoelectric conversion module, and then use light waves as the carrier in the entire optical fiber transmission system for information transmission. In optical fiber communication, the entire optical fiber system not only works It is only a single optical fiber, but many optical fibers gather to transmit information in the form of optical cables [1][2]. Optical fiber transmission has been developed for nearly a hundred years. Optical fiber transmission mainly takes large-capacity, high-speed, and long-distance transmission as the development advantages, which can be the only way out for data communication systems after the 21st century.

3. Aurora 64 / 66B IP core

There are multiple versions of Aurora. According to the differences in encoding methods, the most popular ones are 8B10B and 64B66B[3][4]. Different encoding methods have different natural throughput rates. Among them, the 64/66B version has a different throughput rate. Highest.

3.1 Interface signal

Aurora IP is an IP packaged by Xilinx and available to users, as shown in Figure 1; there are many clock interfaces, data interfaces, and status signals. Among them, the physical layer reference clock refclk1_in and init_clk need to be based on the clock provided by the development board. The frequency is determined, and user_clk and sync_clk are generated by the shared clock logic module; GT_AERIAL_RX and GT_AERIAL_TX are the data transceiver interface of the module; in addition, CORE_STATUS is the status signal of the module.

Aurora64B/66B is a communication protocol with multiple links. At the same time, the serial communication technology using the Aurora64/66B protocol can solve the problems of parallel communication technology such as multiple lines, complex wiring, and low anti-interference ability. The Aurora protocol can be used to transmit data between devices with GTX transceivers. The Aurora IP core released by Xilinx can use 16 effective continuous GTX, GTH or GTY transceivers at any supported line rate. Run to provide a low-cost, universal data channel. The integration of FPGAs is getting higher and higher thanks to the rapid development of chip manufacturing processes. Each high-speed serial transceiver can have one channel, and in units of blocks, each block contains a four-channel phase-locked loop and four serial ports. High-speed transceiver, four transceivers share the same reference clock[5][6].

3.2 Interface signal

Xilinx's aurora IP provides two different usage modes, one is the frame interface mode, and the other is the data stream interface mode. The frame interface mode is that all data is valid only when it is framed. At the same time, the Aurora interface can indicate the transmission and end of frame data through the tready, tvalid and tlast signals; the data stream interface mode is a streaming data transmission, as long as the tready and tvalid signals At the same time, the Aurora core is ready for data transmission, and in order to avoid data loss, when the data reaches rx, it must be read immediately. Here, the Aurora interface adopts the frame interface mode.

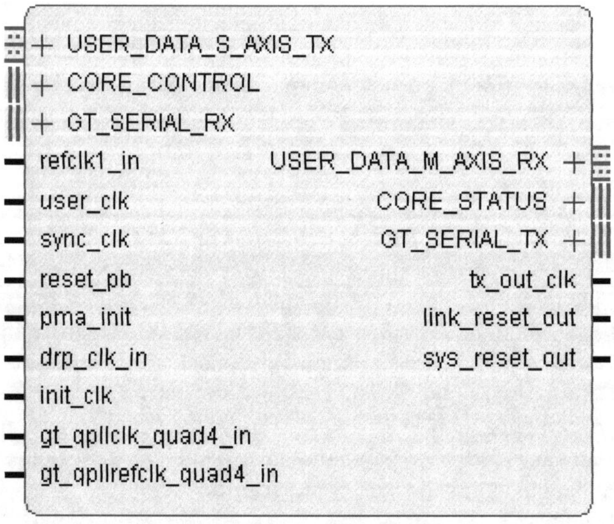

Fig.1Aurora 64B/66B IP

4. Design

The overall solution includes a ZYNQ module, two cache modules, and two Aurora modules. Among them, the ZYNQ module has a data movement module, which can move data from the source address to the destination address. The logical relationship between them is shown in Figure 2.

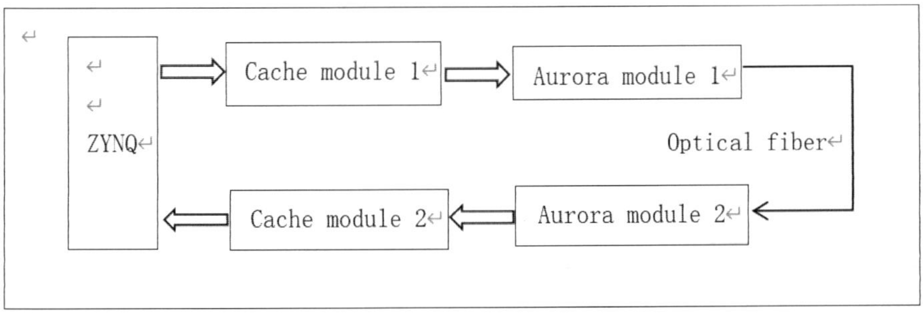

Fig.2 Scheme design drawing

It can be seen from Figure 2-1 that the ZYNQ module acts as a data processing center. Data is generated in the ZYNQ module. Cache module 1 and Aurora module 1 together form a data sending module, and cache module 2 and Aurora module 2 together form a data receiving module. The data received by ZYNQ can be compared with the data sent.

In the environment of FPGA tool Vivado2018.3 software development, create a Block Design, add ZYNQ7 Processing System and DMA module to it. It should be noted that there are multiple DMA modules packaged by Xilinx, and AXI Direct Memory Access IP should be selected here. The AXI interface can make development more efficient and more convenient for later development. Also add the FIFO module as a buffer module, and for the FIFO module, you can choose AXI4-Stream Data FIFO IP, which also uses the AXI interface protocol. In addition, Aurora 64B66B IP must be added. The IP setting parameters here should be selected according to the clock provided by the specific development board, and the interface mode should be frame mode. Finally, create a top-level file, loop back the high-speed interface, and instantiate the Block Design into the top-level file.

5. Bit error rate test

ChipScope Pro integrated bit error rate test core controller IP core is designed for different FPGA series and is customizable to evaluate and monitor the working status of GTX/GTH/GTP transceivers[7]. The bit error rate test is the IBERT test. IBERT is a tool provided by Xilinx to test high-speed serial transceivers. We can use the IBERT IP in Vivado 18.3 to generate IBERT test projects and perform IBERT tests on selected solutions.

6. Test results and analysis

From the IBERT test result Figure 3, it can be seen that there are two channels in the Link state, corresponding to each Aurora module having one receiving and one sending, so it is two Link states; in addition, you can see that the rates of the two Link states are both It is 10.000Gbps, and the bit error rate is $10^{(-13)}$, which meets the design requirements.

Name	TX	RX	Status	Bits	Errors	BER	BERT Reset	TX Pattern	RX Pattern
Ungrouped Links (0)									
Link Group 0 (4)							Reset	PRBS 7-bit	PRBS 7-bit
Link 0	MGT_X0Y28/TX	MGT_X0Y28/RX	10.000 Gbps	2.591E12	0E0	3.86E-13	Reset	PRBS 7-bit	PRBS 7-bit
Link 1	MGT_X0Y29/TX	MGT_X0Y29/RX	10.000 Gbps	2.591E12	0E0	3.86E-13	Reset	PRBS 7-bit	PRBS 7-bit

Fig.3 IBERT test chart

7. Conclusion
This article is based on the FPGA environment and uses the IP construction mode to create a Block design in Vivado 18.3 for development. The development efficiency is greatly improved, and the Aurora interface protocol is briefly introduced, the high-speed data fiber transmission scheme is designed, and the design verification is carried out at the same time to realize the difference in the board Optical fiber transmission of high-speed data between interfaces is of great practical value to optical fiber high-speed data transmission.

References
[1] Cai, X. (2015) Optical fiber transmission technology. Communication University of China Press, The Beijing.
[2] Li,L.(2017)Research and Implementation of High Speed Optical Fiber Communication Data Transmission Technology Based on FPGA.Jilin University.
[3] Wang, Y.P., Wang, X.G., Wang, X.J. (2019) T Research on high-speed data transmission based on Aurora. Technological wind, 30: 94-94.
[4] Liu,Y.F.(2019)Research on Key Technology of 10 Gigabit Network Protocol Conversion System.Hebei University of Science and Technology.
[5] Xilinx.(2018)PG074:Aurora 64B/66B v11.2. https://www.xilinx.com/support/documentation/ip_documentation/aurora_64b66b/v11_2/pg074-aurora-64b66b.pdf
[6] Ke, Y. (2020) Design and development of high-speed data transmission board based on FPGA. Huazhong Normal University.
[7] Liu,D.H.(2013) Debug and verify IP core.In:HongFeng,T.(Eds.), Xilinx series FPGA chip IP core detailed explanation. Electronic Industry Press,Beijing.510-534.

Power Grid Field Violation Recognition Algorithm Based on Enhanced YOLOv5

Wenwen Zhu[1a], Yuanzhong Shu[1b*], Shuo Liu[2a]

[1]College of Information Engineering, Nanchang Hangkong University, Nanchang, Jiangxi, 330063, China

[2]College of Electrical Engineering, Anhui Polytechnic University, Wuhu, Anhui, 241000, China

[1b*]36012@nchu.edu.cn, [1a]1904081200104@stu.nchu.edu.cn, [2a]lindsayshuo@foxmail.com

Abstract—Aiming at the problem that Yolov5 is difficult to accurately detect backlighting and small samples in national grid power transmission and distribution operations, studies the illegal behavior technology of "two wear and one wear" in power grid operation, and proposes a violation recognition algorithm based on the enhanced YOLOv5. First, add a new detection layer and use the BiFPN (bi-directional feature pyramid network) layer for feature fusion, so that feature layers of different scales can better learn the weight distribution and enhance the fusion ability. Secondly, add CBAM (Convolutional Block Attention Module) module before the output detection layer feature map to make full use of channel and spatial information to achieve better model accuracy and recall. The experimental results on the six self-made data sets show that the mAP of this method is 91.0%, which is 5.6% higher than the original algorithm, indicating that the model has stronger predictive ability and robustness for the identification of personnel and safety tools in transmission and distribution scenarios.

1. Introduction

The power grid operation site has large space, high risk, and difficult management and control. In recent years, the convolutional neural network model has developed rapidly and has been widely applied[1,2]. Many scholars have published a large number of excellent models based on deep learning[3]. It focuses on the intelligent control of on-site operation safety, and timely reminds the operation risks and automatically alarms based on the recognition of violations in the grid operation site. It is divided into Two-Stage algorithm and one-stage detection algorithm mainly according to whether there is candidate box generation. The former has high detection accuracy, large computation and slow speed, which is difficult to meet the needs of real-time target detection. The latter has the advantages of less computation, faster speed, and stronger real-time performance and practicability. For example, the R-CNN series methods proposed by Girshick[4,5,6] et al., such two-stage method firstly generates candidate regions based on feature extraction, and then determines the target category and position by regression of candidate regions. The detection rate can reach 5fps.

The SSD (Single Shot multibox Detector) method proposed by Liu et al. [7], the target detection algorithm based on YOLO (You Only look once) proposed by Redmon et al. [8,9,10], V2 uses darknet-19 network model to improve mAP value from 63.4% to 78.6%, while V3 uses the new Darknet-53 as feature extractor, which has the same accuracy as SSD while 3 times faster, and also significantly improves detection of small objects. Alexey et al. [11] proposed the YOLOv4 algorithm, in which CSPDarknet53 draws on the network structure of cross-stage partial network (CSPnet) and SPP, as well as data enhancement using the Mosaic method. The aforementioned One-Stage algorithm directly predicts the coordinates and category of the bounding box based on the features extracted by the network, and the detection speed reaches 45fps. Literature[12] reduces the dimensionality of the convolutional layer of the YOLOv5 network and improves the post-processing method, which improves the detection performance of the quality of inkjet characters. Literature[13] improves YOLOv5 as a lightweight backbone network to meet faster and more accurate vehicles detection.

This paper proposes a recognition algorithm of illegal behavior based on improved YOLOv5, aiming at the problems of difficult supervision of power grid construction site, high personnel cost and difficult judgment of illegal behavior. The specific improvement scheme is as follows: add multi-scale detection layer for small targets in YOLOv5 network under pytorch framework and use BiFPN[14] connection to make four detection layers learn important features and use CBAM[15] attention. The performance of small target detection is improved, and the sensitivity of the model to background noise is reduced, so that the model for identifying violation behaviors achieves a higher degree of fit. Finally, the practicability and robustness of the proposed method are verified by experiments.

2. Violation Recognition Based on Improved YOLOv5 Algorithm

2.1. YOLOv5 algorithm principle

YOLOv5 is similar to the channel and layer control factors of EfficientNet [16], which can flexibly configure models of different complexity. According to the model depth multiple and layer channel multiple, it can be divided into four types: YOLOv5s, YOLOv5m, YOLOv5l, YOLOv5x. The YOLOv5 network structure is shown in Fig.1.

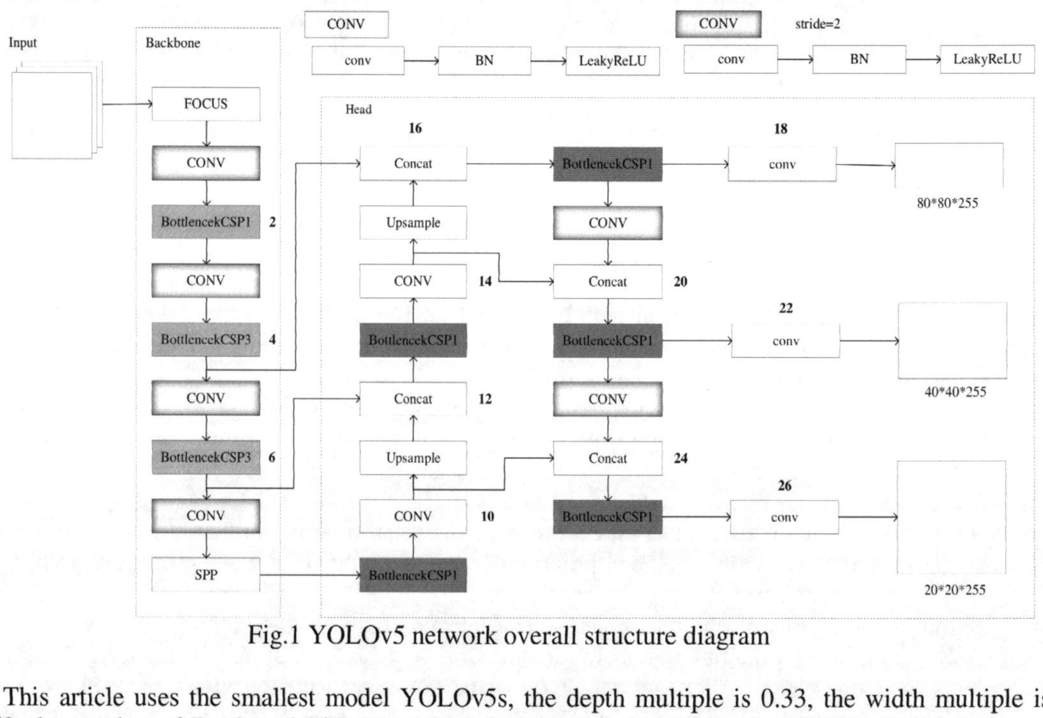

Fig.1 YOLOv5 network overall structure diagram

This article uses the smallest model YOLOv5s, the depth multiple is 0.33, the width multiple is 0.50, the number of BottleneckCSP (True) is 1, 3, 3, the number of BottleneckCSP (False) is 1, and the number of convolution kernels is 32, 64, 128, 256, 512, the fastest and easy to deploy quickly.

Input uses Mosaic data enhancement, randomly using 4 pictures, random zooming, random cropping and random arrangement for splicing, enriching the data set, and significantly improving the recognition ability of small objects. The data of 4 pictures can be calculated at the same time, without a large Mini-batch, which reduces the consumption of memory. The use of k-means clustering to calculate the optimal anchor value of the data set and the use of adaptive image scaling reduce the black edges and improve the detection speed. Backbone mainly adopts FOCUS, CSPnet and SPP structure, as shown in Fig.2. Focus performs 4 slices, and then passes through a convolutional layer with a convolution kernel number of 32, turning the initial 3x640x640 image into a 32x320x320 feature map. It is used to reduce the amount of calculation and the number of layers to increase the speed, and at the same time minimize the loss of information for down-sampling. CSPnet first divides the feature mapping of the base layer into two parts, and then splices the two parts through a cross-stage hierarchical structure, which reduces the amount of calculation while ensuring accuracy. SPP uses 4 convolution kernels of different sizes for maximum pooling, and the number of hierarchically fixed windows, each window corresponds to a layer of the pyramid, and the window size is adjusted to achieve multi-level pooling and fixed output.

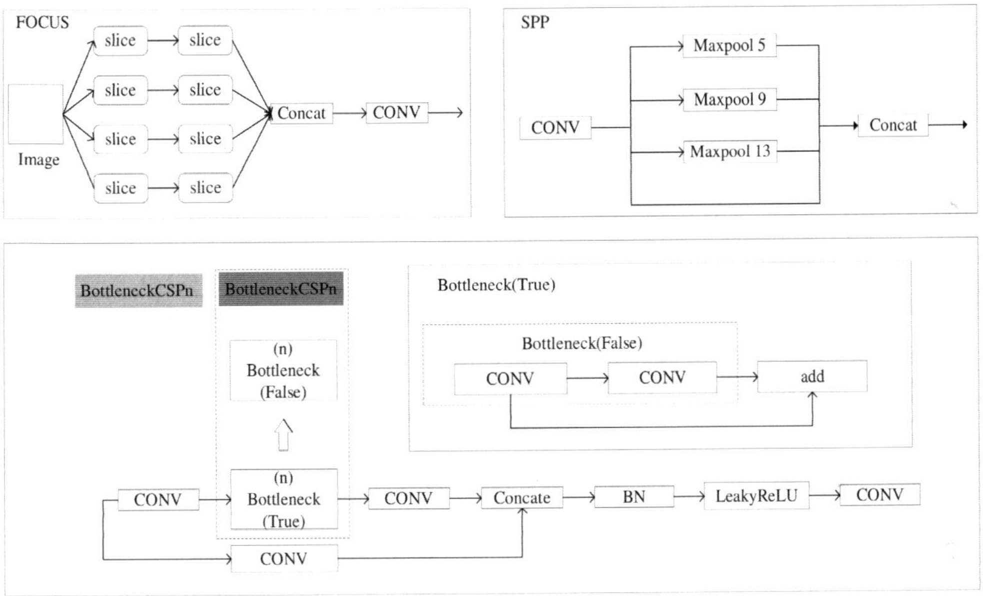

Fig.2 Backbone component structure diagram

Head uses feature pyramid networks (FPN) and Path Aggregation Network (PAN). FPN transfers features from top to bottom to solve multi-scale problems. PAN aggregates shallow features from the bottom-up path, and different feature layers are fully integrated, which speeds up the flow of information. Output uses the binary cross-entropy loss function to calculate the category probability and target confidence score loss. When calculating the bounding box loss, GIOU_ Loss [17] is used, as shown in Equation 1:

$$L_{GIoU} = 1 - IoU + \frac{|A_c - U|}{|A_c|} \tag{1}$$

Among them, IoU (Intersection over Union) is the intersection ratio of the real box and the predicted box, A_c is the minimum closure area of the real box and the predicted box, and U is the intersection of the real box and the predicted box.

Through real-time streaming protocol (RTSP) access to outdoor job site cameras and indoor webcam video stream, combined with YOLOv5 algorithm for real-time analysis, to identify wearing helmets, wearing work clothes and safety belts. YOLOv5 has better detection results and real-time performance. However, in the outdoor high-altitude operation scene in this article, the helmet and safety belt targets under the long-range lens are often small. The original down-sampling 80*80, 40*40, 20*20 three This feature layer is difficult to detect targets smaller than 8*8, that is, there will be a certain degree of missed detection and false detection. In the original Head branch, features of different scales are equally fused, and the importance of feature layers of different scales is not considered.

2.2. *Improved multi-scale feature layer*

It can be seen from Fig.1 that YOLOv5 performs 3, 4, and 5 down-sampling of pictures with an input of 640*640 to obtain three detection layers of 80*80, 40*40, and 20*20, which are used to detect small , medium and large targets of different scales. Considering that the detected helmets and safety belt cameras are mostly small in the long view, which is easy to cause missed detection and misdetection, a feature layer for detecting smaller targets is added. As shown in Fig.3, the first step is to add BottleneckCSP and CONV to extract features on the basis of the original 16 layers, and then use Upsample to upsampling to obtain a feature map of 160*160. In the 20th layer and the second layer of the backbone network, 160* 160 feature maps are fused. In the second step, we continue to use BottleneckCSP, firstly obtain the feature map of the 22nd layer for predicting the smaller target through convolution, and then down-sampling by the CONV module with a step size of 2 to obtain the 80*80 feature map. In the 24th layer and The 80*80 feature map of the 18th layer is fused. After completing the improvement of the multi-scale feature layer through the above two steps, it can be found that the model now uses four-scale feature maps for prediction, making full use of the resolution of the bottom layer and the semantic features of the high layer.

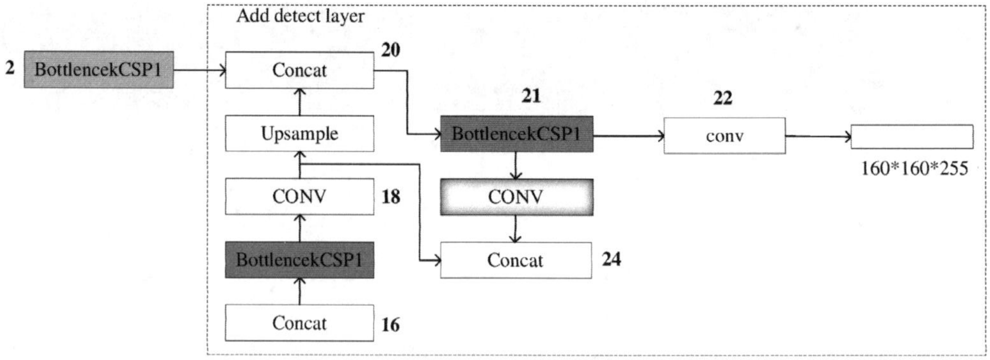

Fig.3 Added multi-scale feature layer structure map

2.3. *Add feature fusion of BiFPN and CBAM*

In this paper, the BiFPN layer is used to replace the original tensor splicing Concat layer, which corresponds to the 12th, 16th, and 20th layers in the upsampling and the 24th, 28th, and 32th layers in the downsampling, all replaced with BiFPN. Combining the down-sampling features of the backbone network with the up-sampling features in the Head, and then fusing the down-sampling features in the Head, so that the network has stronger semantic information, thereby solving the missed detection caused by small targets and repeated feature extraction in backlight scenes As well as the problem of misdetection, the model has a better feature extraction effect on targets with fuzzy features. The improved feature fusion and BiFPN structure are shown in Fig.4.

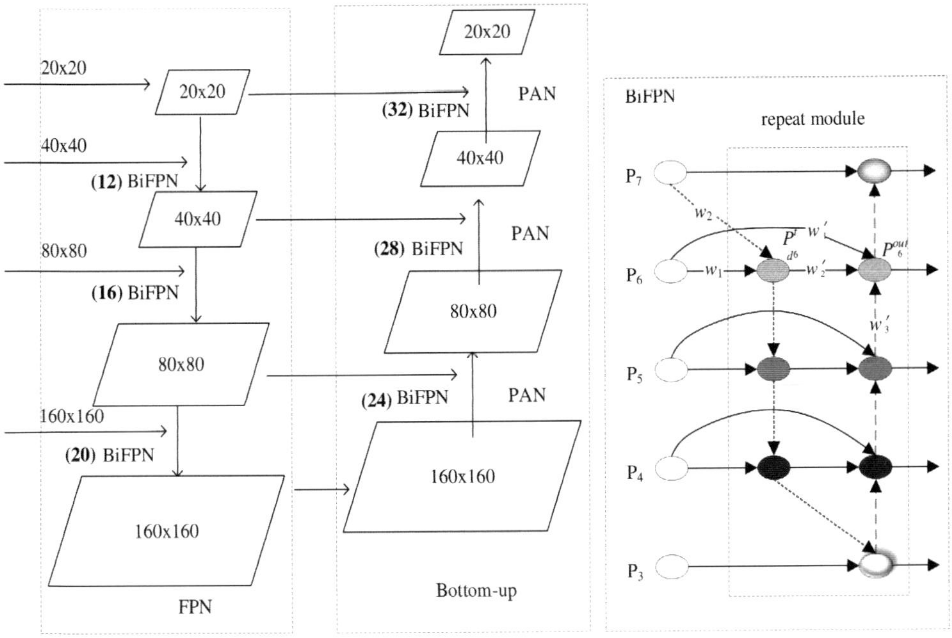

Fig.4 Head feature fusion structure diagram

Mainly adopt two-way cross-scale connection and fast normalization fusion. BiFPN removes nodes with only one input edge. This node has very little contribution to the feature network that integrates different features. Add additional edges at the same level of input to output nodes to fuse more features without adding too much cost. , And repeat each bidirectional path as a feature network layer multiple times to achieve a higher level of feature fusion. The formulas for fast normalized fusion and BiFPN are shown in equations 2 and 3.

$$O = \sum_i \frac{w_i}{\varepsilon + \sum_j w_j} \cdot I_i, w_i \geq 0 \tag{2}$$

$$P_6^{td} = Conv\left(\frac{w_1 \cdot P_6^{in} + w_2 \cdot Resize(P_7^{in})}{w_1 + w_2 + \varepsilon}\right)$$

$$P_6^{out} = Conv\left(\frac{w_1' \cdot P_6^{in} + w_2' \cdot P_6^{td} + w_3' Resize(P_5^{out})}{w_1' + w_2' + w_3' + \varepsilon}\right) \tag{3}$$

In formula (2), w_i is a learnable weight, and the Relu activation function is used after each w_i. The normalized weight is between 0 and 1, and the ε value is 0.0001, which is used to avoid numerical instability. Equation (3) takes the 6th level as an example, P_6^{td} is the intermediate feature of the 6th level on the top-down path, P_5^{out}, P_6^{out} is the output feature of the 5th and 6th levels in the bottom-up path, P_6^{in}, P_7^{in} is the input of the 6th and 7th levels feature. Resize is usually up-sampling or down-sampling, and Conv is usually a convolution operation of feature processing.

The convolutional attention module CBAM makes the model focus on key areas through channels and spaces, enhances important features in the feature map, and suppresses useless features. It can be divided into a channel attention module and a spatial attention module. The channel attention module is shown in Fig.5(a), which performs maximum pooling and average pooling on the input feature map. Through sharing MLP, the activation function is used to output the channel attention after addition, and then multiplying to get the final channel attention. The spatial attention module is shown in Fig.5(b). After the feature map of the channel attention module is obtained, the maximum pooling and average pooling are performed respectively, and the spatial attention is obtained through convolution and activation after splicing, and then the channel attention The force feature map is multiplied to obtain the final output feature map.

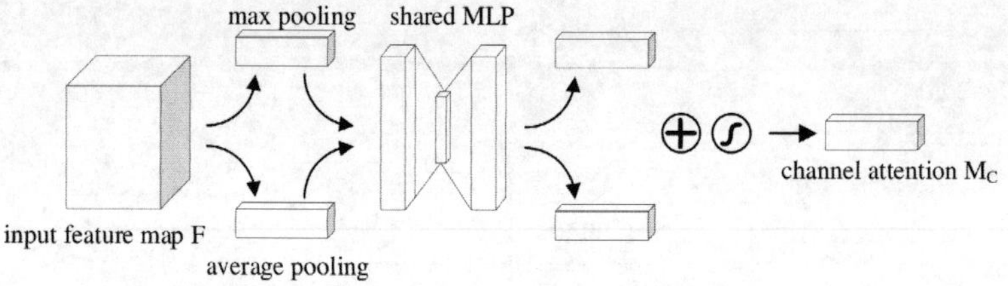

(a) Channel attention module diagram

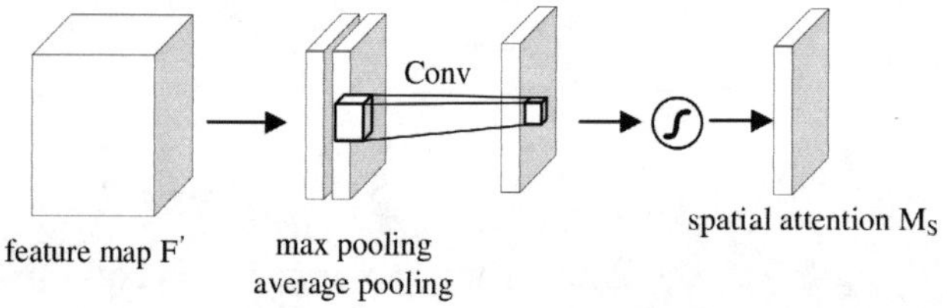

(b) Spatial attention module diagram

Fig.5 Channel attention and spatial attention module diagram

Consider the following ways to add CBAM modules: The first is to add the last layer at the end of the backbone network extraction of features, the second is to add after the BottleneckCSP in the head sub-sampling stage, and the third is to integrate the first two methods. . The experiment adopts the second scheme, which is mainly to add CBAM fusion before convolution and output 4 detection feature layers in the Head branch, namely the 22nd, 27th, 32nd, and 37th layers.

Through the above-mentioned feature fusion operation, the expression ability of the target feature is better, the target has better robustness, and the recognition accuracy of the network is improved. After labeling training on 12088 samples and testing on 2418 test sets, it is found that the recall rate, accuracy rate and mAP of the improved network are significantly increased.

3. Experiment

3.1. Data set and experimental method

According to Pascal VOC[18]2007 data set format, video data were obtained from cameras in various complex environments such as power transmission and distribution lines and substations, and labelimg tool was used to mark categories and locations. Compliance behavior refers to the correct wearing of safety helmet and standard wearing of overalls and safety belts with the corresponding labels of helmet, workcloth and belt. Violation behavior mainly refers to the improper wearing of safety helmet, overalls and safety belts with the corresponding labels of head, person and nobelt. A data set containing 12088 pictures has been established. And use a script to convert it into a txt file required to train the yolo model.

Add a multi-scale detection layer to the YOLOv5s network, use BiFPN and CBAM modules for feature fusion, improve the traditional YOLOv5s, and use the same data set for ablation comparison experiments. The experiment uses GeForce RTX3080 graphics card, Ubuntu 18.04 host with 10G video memory, based on pytorch 1.7.1 framework, configuration cuda11.0, using python language to improve the training of the YOLOv5s network model by migration learning on the self-made data set, and set the batch size to 8. The learning rate is 0.01 and the epoch is 300.

In order to measure the real-time performance of the algorithm, FPS (Frame Per Second) is used to test the detection speed, that is, the number of pictures detected per second. In addition, the problem studied in this paper belongs to multiple classifications, so mAP is introduced to evaluate the improved model. mAP is the sum of the average accuracy of all categories divided by all categories. The higher the mAP, the better the classifier. As a relatively good evaluation index, mAP is calculated as shown in formulas (4) - (7):

$$AP = \frac{1}{101} \sum_{r \in \{0,0.01,\ldots,1\}} p_{\text{interp}}(r) \tag{4}$$

$$p_{\text{interp}}(r) = \max_{\tilde{r}:\tilde{r} \geq r} p(\tilde{r}) \tag{5}$$

$$\mathrm{p} = \frac{TP}{TP+FP} \tag{6}$$

$$\mathrm{r} = \frac{TP}{TP+FN} \tag{7}$$

Among them: TP is a positive sample for correct prediction, TN is a negative sample for correct prediction, FP is a negative sample for incorrect prediction, and FN is a positive sample for incorrect prediction. p (presicion) is the accuracy rate, r (recall) is the recall rate. After the training, compare the recognition effect of the model before and after the improvement, and measure the performance index of the algorithm.

3.2. Experimental results

During the experiment, the changes of location loss, classification loss and confidence loss are shown in Fig.6. The abscissa is the training batch, the ordinate is the loss value, GIoU Loss predicts the error between the bounding box and the real box, the Objectness Loss predicts the target of the box, and the Classification Loss corresponds to the loss of the classification. It can be seen that the curve of the loss function drops rapidly in the first 50 epochs, and tends to balance after 250 epochs, and the model converges.

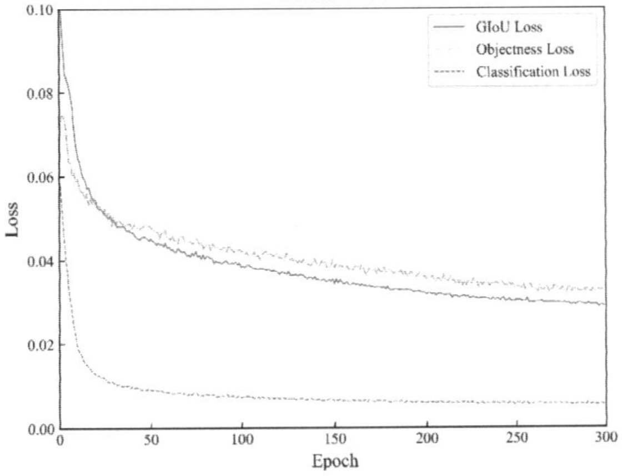

Fig.6 Loss function graph

The IoU threshold set in the experiment is 0.5. When IoU>0.5, the sample is considered to be a positive sample, otherwise the sample is considered to be a negative sample. The improved YOLOv5 algorithm performs two wears (tooling, safety belt) and one wears a (helmet) to detect compliance and violations, and the recognition results of the improved algorithm and the initial algorithm are shown in Fig.7.

YOLOv5 improved YOLOv5

Fig.7 Network recognition results before and after improvements

For dense scenes and pictures in the long view of the camera, YOLOv5 has missed and repeated detections for small targets wearing helmets. As shown on the left side of Fig.7, small targets wearing helmets and violations without helmets all have leaks. In addition, the occluded tooling is not recognized. After the addition of multi-scale feature layers, the detection of small targets is more accurate. The improved network of feature fusion detects both compliance and violations of the helmet wearing. Tooling is also accurately identified. In addition, in the backlit scene under high-altitude operations in the second column, the confidence in the detection of helmets and safety belts is also higher.

3.3. Comparative experiment analysis

Use 2418 test set images for testing. Method 1 adds a multi-scale feature layer to YOLOv5, Method 2 uses BiFPN layer for feature fusion based on Method 1, and Method 3 adds CBAM on the basis of 2, to get mAP and other performance before and after improvement The indicators are shown in Table 1:

Table 1. Test results of different network models.

Method	Multi-scale feature layer	BiFPN	CBAM	mAP(%)	recall(%)	FPS
YOLOv5				85.4	93.0	90.08
Method 1	√			86.1	93.6	80.80
Method 2	√	√		86.5	94.1	78.07
Method 3	√	√	√	**91.0**	**95.0**	77.39

As can be seen from Table 1, the mAP of feature layer used for detecting small targets in Method 1 is slightly improved. Compared with the original YOLOv5, the mAP and recall of Method 3 are improved by 5.6% and 2.0% respectively, that is, the network is more excellent and the generalization ability is also improved, but the value of FPS is decreased to some extent. Considering that the algorithm has met the real-time demand, the calculation cost is sacrificed for higher accuracy. For Method 3, add attention module to Head branch, method 4 adds attention module to backbone network during feature extraction, and method 5 adds attention module to backbone network and Head branch at the same time. The test results are shown in Table 2.

Table 2. Attention Ablation Experiment.

Method	Backbone	Head	mAP(%)
Method 3		√	**91.0**
Method 4	√		90.6
Method 5	√	√	89.8

The mAP value of the attention module used in Method 3 in Table 2 is slightly higher than that in Method 4, which is less than the attention module added in Method 5, and achieves a higher accuracy rate. In summary, finally adopt method 3 as the final improvement network, and train on the VOC07+12 data set to test the performance of the algorithm before and after the improvement. As shown in Table 3, the VOC data set is of higher quality than the data set obtained from the camera in this article, Method 3 increases the mAP value by 9.3% and 6% compared with the original method, and the network performance is better.

Table 3. VOC07+12 test results comparison.

Method	mAP(%)	recall(%)
YOLOv5	68.8	90.0
Method 3	**78.1**	**96.0**

4. Conclusion

It is difficult for the power grid safety supervisor to judge whether the construction personnel are wearing safety hats and safety belts and whether the number of construction personnel meets the demand through the camera at the operation site. In this paper, multi-scale feature layer is added to YOLOv5s network, and BiFPN and attention CBAM modules are improved for feature fusion, and a violation recognition algorithm based on enhanced YOLOv5 is proposed. The above problems are solved, and it is also convenient for the management of safety supervision personnel, and provides a guarantee for the identification of violations in power transmission and distribution scenarios.

Acknowledgments

This work was financially supported by the National Natural Science Foundation of China (71761028, 61572032); the Science and Technology Research Project of Jiangxi Provincial Department of Eduction (209924); the University Synergy Innovation Program of Anhui Province (GXXT-2020-070).

References

[1] YAN C M， WANG C. (2021) Development and application of convolutional neural network model. J. Journal of Frontiers of Computer Science and Technology, 15(01): 27-46.
[2] LIU S, GE Y, LI Y Y, et al. (2021) Recognition Algorithm of Unmanned Ship at Sea Based on Improved SSD. J. Transducer and Microsystem Technologies, 40(02): 120-123.
[3] GAO Y, XIAO D. (2020) Small object detection algorithm based on multi-feature fusion.J. Computer Engineering and Design, 41(07): 1905-1909.
[4] Girshick R, Donahue J, Darrell T, et al. (2014) Rich feature hierarchies for accurate object detection and semantic segmentation. In: IEEE Computer Society Conference on Computer Vision and Pattern Recognition. Columbus. pp. 580–587.
[5] Girshick R. (2015) Fast r-cnn. In: IEEE International Conference on Computer Vision. Santiago. pp. 1440–1448.
[6] REN S Q, HE K M, Girshick R, et al. (2015) Faster r-cnn: Towards real-time object detection with region proposal networks. J. IEEE Transactions on Pattern Analysis and Machine Intelligence, 39(6): 1137–1149.
[7] LIU W, Anguelov D, Erhan D, et al. (2016) SSD: Single shot multibox detector. In: European Conference on Computer Vision, Amsterdam. pp. 21–37.
[8] Redmon J, Divvala S, Girshick R, et al. (2016) You only look once: Unified, real-time object detection. In: IEEE Conference on Computer Vision and Pattern Recognition. Las Vegas. pp. 779–788.
[9] Redmon J, Farhadi A. (2017) YOLO9000: Better, faster, stronger. In: IEEE Conference on Computer Vision and Pattern Recognition. Honolulu. pp. 6517–6525.
[10] Redmon J, Farhadi A. (2018) YOLOv3: An incremental improvement. J. arXiv: 1804. 02767, 2018.
[11] Bochkovskiy A, Wang C Y, Liao H M. (2020) YOLOv4: Optimal speed and accuracy of object detection. J. arXiv preprint arXiv: 2004. 10934, 2020.
[12] GE Q, LIANG Q K, ZOU K L, et al. (2021) Detection Method of Inkjet Code Quality Based on Lightweight Network and Embedded. J/OL. Control Engineering of China, 1-9[2021-11-04]. https://doi.org/10.14107/j.cnki.kzgc.20200903.
[13] WANG C P, WANG H Q, YU F J, et al. (2021) A high-precision fast smoky vehicle detection method based on improved Yolov5 network. In: IEEE International Conference on Artificial Intelligence and Industrial Design (AIID). Guangzhou. pp. 255-259.
[14] TAN M X, PANG R M, LE Q. (2020) EfficientDet: scalable and efficient object detection. In: IEEE/CVF Conference on Computer Vision and Pattern Recognition (CVPR). Seattle. pp. 10778-10787.
[15] Woo S, Park J, Lee J-Y, et al. (2018) Cbam: convolutional block attention module. In: European conference on computer vision(ECCV). Munich. pp. 3-19
[16] TAN M, LE Q. (2019) Efficientnet: Rethinking model scaling for convolutional neural networks. In: International Conference on Machine Learning. Long Beach. pp. 6105-6114.
[17] Rezatofighi H, Tsoi N, Gwak J, et al. (2019) Generalized intersection over union: a metric and a loss for bounding box regression. In: IEEE/CVF Conference on Computer Vision and Pattern Recognition. Long Beach. pp. 658–666.
[18] Everingham M, Gool L, Williams C K, et al. (2010) The Pascal visual object classes (VOC) challenge. J. International Journal of Computer Vision, 88(2): 303–338

AUTHOR INDEX